Mathematical Sciences Research Institute
Publications 13

W.-M. Ni L. A. Peletier J. Serrin
Editors

Nonlinear Diffusion Equations and Their Equilibrium States II

Proceedings of a Microprogram held August 25-September 12, 1986

With 34 Illustrations

Springer-Verlag
New York Berlin Heidelberg London Paris Tokyo

W.-M. Ni
James Serrin
School of Mathematics
University of Minnesota
Minneapolis, MN 55455

L.A. Peletier
Department of Mathematics and
Computer Science
University of Leiden
The Netherlands

Mathematical Sciences Research Institute
1000 Centennial Drive
Berkeley, CA 94720
USA

The Mathematical Sciences Research Institute wishes to acknowledge support by the National Science Foundation.

Mathematics Subject Classification (1980): 35JXX, 35KXX

Library of Congress Cataloging-in-Publication Data
Nonlinear diffusion equations and their equilibrium states : proceedings from a conference held August 25–September 12, 1986 / W.-M. Ni, L. Peletier, J. Serrin, editors.
p. cm. — (Mathematical Sciences Research Institute publications ; 12–13)
Includes bibliographies.
ISBN-13: 978-1-4613-9610-9 e-ISBN-13: 978-1-4613-9608-6
DOI: 10.1007/978-1-4613-9608-6
1. Differential equations, Partial—Congresses. 2. Differential equations, Nonlinear—Congresses. 3. Diffusion—Mathematical models—Congresses. I. Nim W.-M. (Wei-Ming) II. Peletier, L. (Lambertus) III. Serrin, J. (James), 1926– . IV. Series.
QA377.N645 1988
515.3'53—dc19 86–12370

Softcover reprint of the hardcover 1st edition 1988

Camera-ready text prepared by the Mathematical Sciences Research Institute using PC TEX.

9 8 7 6 5 4 3 2 1

Preface

In recent years considerable interest has been focused on nonlinear diffusion problems, the archetypical equation for these being

$$u_t = \Delta u + f(u).$$

Here Δ denotes the n-dimensional Laplacian, the solution $u = u(x,t)$ is defined over some space-time domain of the form $\Omega \times [0,T]$, and $f(u)$ is a given real function whose form is determined by various physical and mathematical applications. These applications have become more varied and widespread as problem after problem has been shown to lead to an equation of this type or to its time-independent counterpart, the elliptic equation of equilibrium

$$\Delta u + f(u) = 0.$$

Particular cases arise, for example, in population genetics, the physics of nuclear stability, phase transitions between liquids and gases, flows in porous media, the Lend-Emden equation of astrophysics, various simplified combustion models, and in determining metrics which realize given scalar or Gaussian curvatures. In the latter direction, for example, the problem of finding conformal metrics with prescribed curvature leads to a ground state problem involving critical exponents. Thus not only analysts, but geometers as well, can find common ground in the present work.

The corresponding mathematical problem is to determine how the structure of the nonlinear function $f(u)$ influences the behavior of the solution. Since the function $f(u)$ in almost all applications seriously departs from linearity it might be thought that the basic problem would be nearly intractable, but with increasing experience, gained from recent research, certain important patterns have emerged. At the same time, new directions of investigation, which might have appeared impossibly difficult only a few years ago, are now open and inviting.

The papers which comprise the two volumes of this work attest to this growing level of interest in nonlinear diffusion equations. Their content and subject matter is broadly directed, but nevertheless can be roughly classified into several individual categories of particular interest:

Ground state problems, critical exponent problems and critical dimension problems associated with the elliptic equation $\Delta u + f(u) = 0$.

Singularity theory for solutions in both the elliptic and parabolic cases.

Asymptotic behavior of solutions in both space and time.

Blow-up phenomena associated with the non-continuability of solutions as time increases, as well as compact support questions when the operator arises in the theory of porous media.

Studies of similarity solutions for elliptic and parabolic problems, leading to a number of fascinating questions in the theory of ordinary differential equations.

Finally, the underlying Laplace operator in the problems discussed above can be replaced by more generalized variational operators of quasi-linear type, or by degenerate elliptic operators, leading to still further ramifications of the basic theory.

The above list is by no means comprehensive, including neither uniqueness nor convexity questions, nor related work in critical point theory or the use of specialized phase plane analysis.

In summary, the present volumes bring together for the intended reader a wealth of stimulating ideas in an important and rapidly expanding branch of differential equations and applied analysis.

The organizers of the Microprogram are pleased to thank the Mathematical Sciences Research Institute for the opportunity to present this work, both as lectures at the original Microprogram at the Institute in 1986 as well as in the form of the papers included here. We also wish to thank the Office of Naval Research, Department of Defense, for their generous financial support of the Microprogram itself.

Our special and particular thanks also go to Professor Irving Kaplansky, who worked capably and imaginatively and with ever-present patience and good humor to ensure the success of the project from the moment of its inception, to Arlene Baxter who coordinated the editorial effort for the many papers involved, and to David Mostardi who carried out the immense effort of transcribing the original manuscripts into this final published version.

To all these, and to the authors of the papers included here, the organizing committee conveys its thanks.

James Serrin, Chairman
Wei-Ming Ni
Lambertus A. Peletier

Nonlinear Diffusion Equations and their Equilibrium States

TABLE OF CONTENTS – VOLUME 2

Nonlinear Diffusion Equations and their Equilibrium States

Table of Contents – Volume 1

M. Bertsch, R. Dal Passo and M. Ughi

NONUNIQUENESS AND IRREGULARITY RESULTS FOR A NONLINEAR DEGENERATE PARABOLIC EQUATION

M. Biroli

EXISTENCE AND MEYERS ESTIMATES FOR SOLUTIONS OF A NONLINEAR PARABOLIC VARIATIONAL INEQUALITY

M. Bramson

CONVERGENCE TO TRAVELING WAVES FOR SYSTEMS OF KOLMOGOROV-LIKE PARABOLIC EQUATIONS

C. Budd and J. Norbury

SYMMETRY BREAKING IN SEMILINEAR ELLIPTIC EQUATIONS WITH CRITICAL EXPONENTS

K.-C. Chang

REMARKS ON SADDLE POINTS IN THE CALCULUS OF VARIATIONS

M. Chipot and F. B. Weissler

ON THE ELLIPTIC PROBLEM $\Delta u - |\nabla u|^q + \lambda u^p = 0$

C. V. Coffman

NONLINEAR ELLIPTIC BOUNDARY VALUE PROBLEMS: LYUSTERNIK-SCHNIRELMAN THEORY, NODAL PROPERTIES AND MORSE INDEX

E. Di Benedetto

HARNACK-TYPE INEQUALITITES FOR SOME DEGENERATE PARABOLIC EQUATIONS

A. Eydeland and J. Spruck

THE INVERSE POWER METHOD FOR SEMILINEAR ELLIPTIC EQUATIONS

Uniqueness of non-negative solutions of a class of semi-linear elliptic equations

by HANS G. KAPER AND MAN KAM KWONG

Abstract. This article is concerned with boundary value problems of the type

(BVP) $u'' + g(r)u' + f(u) = 0, \quad r > 0; \quad u'(0) = 0, \quad \lim_{r \to \infty} u(r) = 0;$

where $f(0) = 0$. Such problems arise in the study of semi-linear elliptic differential equations in $\mathbb{R}^n$. It is shown that (BVP) has at most one non-negative non-trivial solution under appropriate conditions on f and g. The conditions are weaker than those given by Peletier and Serrin [6], who considered the special case $g(r) = (n-1)/r$, $n = 2, 3, \cdots$.

§ 1. Introduction.

In this article we are concerned with boundary value problems of the type

(1.1) $u'' + g(r)u' + f(u) = 0, \quad r > 0; \quad u'(0) = 0; \quad \lim_{r \to \infty} u(r) = 0.$

Such problems arise in the study of semi-linear elliptic differential equations in $\mathbb{R}^n$. For example, radial solutions of the boundary value problem

$$\Delta u + f(u) = 0, \quad x \text{ member } \mathbb{R}^n; \qquad \lim_{x \to \infty} u(x) = 0; \tag{1.2}$$

satisfy (1.1), where $r = |x|$ and $g(r) = (n-1)/r$. Radial solutions of the more general problem

$$\Delta u + a(|x|)\, f(u) = 0, \quad x \text{ member } \mathbb{R}^n; \qquad \lim_{x \to \infty} u(x) = 0; \tag{1.3}$$

satisfy (1.1), where r is a weighted radial variable, $dr = \sqrt{a(|x|)}\, d|x|$, and $g(r) = (d/dr) \ln \left(|x|^{n-1}\sqrt{a(|x|)}\right)$.

If $f(0) = 0$ -- as we assume throughout this investigation --, then (1.1) has at least one solution, viz., the trivial solution $u \equiv 0$. In addition, there may be other solutions. Our interest focuses on the uniqueness of non-negative non-trivial solutions of (1.1). We use the following notation:

$$F(u) = \int_0^u f(s)\, ds, \tag{1.4}$$

$$G(r) = \exp\left(\int_a^r g(s)\, ds\right), \tag{1.5}$$

where a is some fixed positive number. Notice that $G'(r) = g(r)G(r)$ for all r. Furthermore,

$$\alpha = \inf\{u > 0 : f(u) > 0\}, \tag{1.6}$$

$$\beta = \inf\{u > 0 : F(u) > 0\}. \tag{1.7}$$

These constants satisfy the ordering relation $0 \le \alpha \le \beta$.

The boundary value problem (1.1) with $g(r) = (n-1)/r$ and the corresponding problem (1.2) have been the subject of numerous investigations; cf. Gidas, Ni, and Nirenberg [**1**] (radial nature of the solutions); Berestycki and Lions [**2**], and Berestycki, Lions and Peletier [**3**] (existence of positive solutions); Peletier and Serrin [**4**], and McLeod and Serrin [**5**] (uniqueness of positive solutions); Peletier and Serrin [**6**] (uniqueness of non-negative non-trivial solutions); and the references cited therein. Our investigation was, in fact, motivated by the results of [**4**] and [**6**].

In [**4**], Peletier and Serrin showed that (1.2) has at most one positive radial solution if f satisfies the following conditions (we follow the notation of [**4**]):

(H1) f is locally Lipschitz continuous on $(0, \infty)$;

(H2) $\lim_{u \to 0} f(u)/u = -m$ for some $m > 0$;

(H3) $F(u) > 0$ for some $u > 0$; and

(S) $f(u)/(u - \beta)$ is a monotone non-increasing function of u on the set $\{u > \beta : f(u) > 0\}$.

The conditions (H2) and (H3) concern the behavior of f for small values of the argument, (S) its behavior for large values of the argument; (S) is equivalent with the condition that the region $D = \{(u, v)$ *member* $\mathbb{R}^2 : u > \beta, 0 < v < f(u)\}$ is star-shaped with respect to the point $(\beta, 0)$.

In [**6**], the same authors showed that the condition (H2) is stronger than necessary. They proved that (1.2) has at most one non-negative non-trivial radial solution if (H1) through (H3) are replaced by the following conditions (we follow the notation of [**6**]):

(A1) f is locally Lipschitz continuous on $(0, \infty)$;

(A2) $F(u) > 0$ for some $u > 0$;

(A3) $\lim_{u \to 0} f(u) = 0$ and $f(0) = 0$; and

(H*) $\beta > 0$ if $n = 2$, $\alpha > 0$ if $n = 3, 4, \cdots$.

If n is considered as a continuous variable, then the condition $\beta > 0$ in (H*) is sufficient for all n in the range $3/2 \leq n \leq 2$, while the condition $\alpha > 0$ applies to all $n > 2$.

We note that the original condition (H2) implies that $f(u) \to 0$ as $u \to 0$, and also that $f(0) = 0$. As these properties do not necessarily follow from the condition (H*), the condition (A3) must be added for consistency.

The new condition (H*) is clearly weaker than the original condition (H2). However, it will follow from the present investigation that it is still stronger than necessary and that the single inequality $\beta > 0$ is sufficient for all values $n = 2, 3, \cdots$ (or $n \geq 3/2$, if n is continuous).

In this article we shall prove that (1.1) has at most one non-negative non-trivial solution if f and g satisfy the following conditions:

(F1) f is locally Lipschitz continuous on $(0, \infty)$, $\lim_{u \to 0} f(u) = 0$, and $f(0) = 0$;

(F2) $\beta > 0$;

(F3) $f(u)/(u - \beta)$ is a monotone non-increasing function of u on the set $\{u > \beta : f(u) > 0\}$;

(G1) g is (strictly) positive and monotone, with $g' \leq 0$, on $(0, \infty)$;

(G2) GG' is monotone, with $(GG')' \geq 0$, on $(0, \infty)$;

(G3) $\int^{\infty} \frac{ds}{G(s)} = \infty$, or else

$$\left(\int_r^{\infty} \frac{ds}{G(s)G'(s)}\right)^{1/2} = o\left(\int_r^{\infty} \frac{ds}{G(s)}\right) \quad \text{as } r \to \infty. \tag{1.8}$$

In the case of the boundary value problem (1.2), where $g(r) = (n - 1)/r$, the conditions (G1), (G2), and (G3) are satisfied for all $n = 2, 3, \cdots$ (or $n \geq 3/2$, if n is continuous).

The plan of this article is as follows. In Section 2 we establish several general properties of solutions of the boundary value problem (1.1). Some of these are straight-forward generalizations of the corresponding properties derived in [6], but in particular those given in Lemma 4 are new. In Section 3 we give a new proof of the fact that the graphs of two distinct solutions of the boundary value problem (1.1) must intersect (Theorem 1). The proof is based on the introduction of a new dependent variable, which is the solution of a nonlinear initial value problem. A uniqueness result for this type of initial value problem was shown by the authors in [7]; it makes a direct comparison of the two solution graphs possible. The Monotone Separation Lemma [6, Lemma 9] is an immediate consequence of Theorem 1 of this section. In Section 4 we show that the graphs of two non-negative non-trivial solutions of the boundary value problem (1.1) cannot intersect, unless the solutions are identical (Theorems 2 and 3). Thus we arrive at the desired uniqueness result, which is stated in Section 5 (Theorem 4).

§ 2. General Properties of Solutions.

Let $[r_1, r_2]$ be any compact subinterval of $[0, \infty)$ and let u satisfy the differential equation

$$u'' + g(r)u' + f(u) = 0 \tag{2.1}$$

on $[r_1, r_2]$. Multiplying (2.1) by u' and G^2u', respectively, and integrating the resulting equations over $[r_1, r_2]$, we obtain the following identities:

$$\Big[\frac{1}{2}u'(r)^2 + F(u(r))\Big]_{r=r_1}^{r=r_2} = -\int_{r_1}^{r_2} g(s)u'(s)^2\, ds, \tag{2.2}$$

$$\Big[G(r)^2\big(\frac{1}{2}u'(r)^2 + F(u(r))\big)\Big]_{r=r_1}^{r=r_2} = 2\int_{r_1}^{r_2} G(s)G'(s)F(u(s))\, ds. \tag{2.3}$$

We use these identities in the proofs of the following lemmas, which give some general properties of solutions of the boundary value problem

$$u'' + g(r)u' + f(u) = 0, \quad r > 0; \quad u'(0) = 0; \quad \lim_{r\to\infty} u(r) = 0. \tag{2.4}$$

In the proofs of Lemmas 1 and 2 we assume that f and g satisfy the conditions (F1) and (G1). In the proof of Lemma 3 we require, in addition, that f satisfies the condition (F2). Lemma 4 holds if f satisfies the conditions (F1) and (F2), and g satisfies the conditions (G1) and (G3).

LEMMA 1. *Let u be a non-negative non-trivial solution of (2.4). Then (i) $\lim_{r\to\infty} u'(r) = 0$; and (ii) u satisfies the identity*

$$\frac{1}{2}u'(r)^2 + F(u(r)) = \int_r^\infty g(s)u'(s)^2\, ds, \qquad r \geq 0. \tag{2.5}$$

PROOF. Let $r \geq 0$ be fixed. We take any R *member* (r, ∞) and apply (2.2) to u on the compact interval $[r, R]$,

$$\Big[\frac{1}{2}u'(R)^2 + F(u(R))\Big] - \Big[\frac{1}{2}u'(r)^2 + F(u(r))\Big] = -\int_r^R g(s)u'(s)^2\, ds. \tag{2.6}$$

Since g is positive, the expression in the right member tends to some negative constant, possibly $-\infty$, as $R \to \infty$. The same must then be true for the expression in the left member. Since r is kept fixed, it must therefore be the case that the quantity $\frac{1}{2}u'(R)^2 + F(u(R))$ tends to some constant as $R \to \infty$. This constant may be $-\infty$, but it is certainly less than ∞. Because $\lim_{R \to \infty} u(R) = 0$ and $\lim_{u \to 0} F(u) = 0$, the quantity $F(u(r))$ tends to 0 as $R \to \infty$, so it must then be the case that $u'(R)^2$ tends to some constant as $R \to \infty$. Clearly, this constant cannot be negative, so $u'(R)^2$, and therefore $u'(R)$, tends to some non-negative finite constant as $R \to \infty$. In fact, the only value compatible with the limiting condition $\lim_{R \to \infty} u(R) = 0$ is zero, so $\lim_{R \to \infty} u'(R) = 0$. This proves (i). The identity (2.5) follows from (2.6) as $R \to \infty$. □

Evaluating the identity (2.5) at $r = 0$, we find that $F(u(0)) > 0$. Hence, taking into account the definition (1.5) of the constant β, we conclude that any non-negative non-trivial solution u of (2.4) must necessarily satisfy the inequality $u(0) > \beta$.

LEMMA 2. *If u is a non-negative non-trivial solution of the boundary value problem (2.4), then either*

$$(2.7) \qquad u > 0 \quad on \ [0, \infty), \qquad u' < 0 \quad on \ (0, \infty);$$

or there exists an R member $(0, \infty)$ such that

$$(2.8) \quad u > 0 \quad on \ [0, R), \qquad u' < 0 \quad on \ (0, R), \qquad u = 0 \quad on \ [R, \infty).$$

PROOF. Let u be a non-negative non-trivial solution of (2.4). Let R be the right-most point of the support of u:

$$(2.9) \quad R = \inf\{r \ member [0, \infty) : u(s) = 0 \ \text{ for all } s \ member (r, \infty)\}.$$

Clearly, $R > 0$, because u is non-trivial; R may be finite or infinite.

Let a be defined by

$$(2.10) \ a = \inf\{r \ member [0, R) : u'(s) < 0 \ \text{ for all } \ s \ member (r, R)\}.$$

Because u is non-negative, we have $u(a) > 0$; furthermore, $u'(a) = 0$.

Suppose that $a > 0$ and u has a local maximum at a. Then there exists a point b *member* $[0, a)$, such that $u'(b) = 0$ and $u' > 0$ on (b, a). Because $u(R) = 0$, there must be a point c *member* $(a, R]$ where $u(c) = u(b)$. Evaluating (2.5) at b and at c and subtracting the resulting equations, we obtain the identity

$$(2.11)\qquad -\frac{1}{2}u'(c)^2 = \int_b^c g(s)u'(s)^2\,ds.$$

But here we have a contradiction: the expression on the left-hand side is obviously negative, while the expression on the right-hand side is positive. We must therefore conclude that either $a = 0$ or, if $a > 0$, then $u(r) \geq u(a)$ for all r *member* $[0, a]$. We claim that the latter configuration is impossible.

Suppose that $a > 0$ and $u(r) \geq u(a)$ for all r *member* $[0, a]$. Then $u' \leq 0$ on $(a - \eta, a)$ for some $\eta > 0$. Also, $u'(a) = 0$, as we have seen above. If $u' = 0$ on $(a - \eta, a)$, i.e., u is constant there, then the C^2-continuity of u at a implies that $u''(a) = 0$. If $u' < 0$ on $(a - \eta, a)$, then u has an inflection point at a, so $u''(a) = 0$ as well. In either case, $u'(a) = u''(a) = 0$. Since u satisfies the differential equation and g is continuous at a, it must be the case that $f(u(a)) = 0$. We shall show that these properties together imply that $u(a) = 0$.

Let $u_a = u(a)$ and let the constant function $\tilde{u}$ be defined on $[a, \infty)$ by the expression

$$(2.12)\qquad \tilde{u}(r) = u_a, \quad r \text{ member} [a, \infty).$$

Both u and $\tilde{u}$ are solutions of the initial value problem

$$(2.13)\quad y'' + g(r)y' + f(y) = 0, \quad r > a; \qquad y(a) = u_a, \quad y'(a) = 0.$$

Because we have assumed $a > 0$, it follows from the condition (F1) that f is Lipschitz continuous on $[a, \infty)$; hence, the solution of (2.13) is unique, and we conclude that $u(r) = \tilde{u}(r) = u_a$ for all $r \geq a$. But $\lim_{r \to \infty} u(r) = 0$, so it must be the case that $u_a = 0$, i.e., $u(a) = 0$.

This conclusion contradicts the earlier statement that $u(a) > 0$, which followed from the definition (2.10). Hence, the configuration $a > 0$ and $u(r) \geq u(a)$ for all r *member* $[0, a]$ is impossible, as claimed. It must therefore be the case that $a = 0$. □

LEMMA 3. *Let u be a non-negative non-trivial solution of the boundary value problem (2.4). Then there exists a number $K \geq 0$, such that (i) $\lim_{r \to \infty} G(r)^2[\frac{1}{2}u'(r)^2 + F(u(r))] = K$; (ii) u satisfies the identity*

$$(2.14)\quad G(r)^2\left[\frac{1}{2}u'(r)^2 + F(u(r))\right] = K - 2\int_r^\infty G(s)G'(s)F(u(s))\,ds, \quad r \geq 0.$$

PROOF. Let $r \geq 0$ be fixed. We take any R *member* (r, ∞) and apply (2.3) to u on the compact interval $[r, R]$,

$$(2.15)\quad G(R)^2\Big[\frac{1}{2}u'(R)^2 + F(u(R))\Big] - G(r)^2\Big[\frac{1}{2}u'(r)^2 + F(u(r))\Big] = 2\int_r^R G(s)G'(s)F(u(s))\,ds.$$

The first term in the left member is always positive; cf. (2.5). Consequently, (2.15) implies that

$$(2.16)\quad 2\int_r^R G(s)G'(s)(-F(u(s)))\,ds \leq G(r)^2\Big[\frac{1}{2}u'(r)^2 + F(u(r))\Big].$$

Since $u(s) \to 0$ as $s \to \infty$, $u(s)$ is less than or equal to β for all sufficiently large s, so $F(u(s))$ is zero or negative beyond a certain point. (Here, we use the condition (F2).) >From that point on, the integrand in (2.16) is zero or positive. These arguments show that the integral in the left member of (2.16) is a monotone non-decreasing function of R for R sufficiently large. Because the expression in the right member is positive and fixed, the integral is also bounded above by some (positive) constant. It must therefore be the case that the integral converges to some finite limit as $R \to \infty$. Going back to (2.15), where we now know that the expression in the right member converges as $R \to \infty$, we conclude that the same must be true for the expression in the left member. Since the second term is fixed, it must be the case that the first term converges as $R \to \infty$. Property (i) states that K is its limiting value. The identity (2.14) follows from (2.15) as $R \to \infty$. □

LEMMA 4. *Let u be a non-negative non-trivial solution of (2.4). Let K be the number defined in Lemma 3(i). If $\int^\infty \frac{ds}{G(s)} = \infty$, then $K = 0$; otherwise,*

$$(2.17)\quad \lim_{r\to\infty}\Big[u(r) \Big/ \int_r^\infty \frac{ds}{G(s)}\Big] = \sqrt{2K}.$$

PROOF. First, we consider the case where $\int^\infty \frac{ds}{G(s)} = \infty$. Suppose that $K \neq 0$. Then necessarily $K > 0$, so taking any K' *member* $(0, K)$ we have

$$\frac{1}{2}u'(r)^2 > \frac{K'}{G(r)^2} - F(u(r)) \geq \frac{K'}{G(r)^2}$$

for all sufficiently large r. Hence,

$$(2.18) \qquad - u'(r) > \frac{\sqrt{2K'}}{G(r)}.$$

Since the right-hand side is not integrable at infinity, whereas the left-hand side is, we have a contradiction and conclude that $K = 0$, as claimed.

Next, we consider the case where the integral $\int^{\infty} \frac{ds}{G(s)}$ converges. Let ε be arbitrarily small positive. It follows from Lemma 3(i) that

$$G(r)^2\Big[\frac{1}{2}u'(r)^2 + F(u(r))\Big] > K - \varepsilon$$

for all sufficiently large r. Hence, $- u'(r) > \sqrt{2(K - \varepsilon)}/G(r)$; so, upon integration, we find the inequality

$$(2.19) \qquad u(r) > \sqrt{2(K - \varepsilon)} \int_r^{\infty} \frac{ds}{G(s)}$$

for all sufficiently large r. On the other hand, we also have the inequality

$$G(r)^2\Big[\frac{1}{2}u'(r)^2 + F(u(r))\Big] < K + \varepsilon$$

for all sufficiently large r, so $- u'(r) < \big(2(K + \varepsilon)/G(r)^2 - 2F(u(r))\big)^{1/2}$. Since $u(r) \to 0$ as $r \to \infty$, $u(r)$ is less than or equal to β beyond a certain value of r, and from that point on $F(u(r))$ is zero or negative. Using the elementary inequality $(a^2 + b^2)^{1/2} \le a + b$ for $a, b \ge 0$, we find that $- u'(r) < \sqrt{2(K + \varepsilon)}/G(r) + (- 2F(u(r)))^{1/2}$ for all sufficiently large r. Hence, upon integration,

$$(2.20) \qquad u(r) < \sqrt{2(K + \varepsilon)} \int_r^{\infty} \frac{ds}{G(s)} + \int_r^{\infty} (- 2F(u(s)))^{1/2}\, ds.$$

We estimate the last integral by means of Holder's inequality:

$$(2.21) \qquad \int_r^{\infty} (- 2F(u(s)))^{1/2}\, ds$$
$$<= s+6\Big(\int_r^{\infty} G(s)G'(s)(- 2F(u(s)))\, ds\Big)^{1/2} \Big(\int_r^{\infty} \frac{ds}{G(s)G'(s)}\Big)^{1/2}.$$

Lemma 3 implies that the first integral in the right member becomes arbitrarily small as r tends to infinity, so by taking r sufficiently large we can certainly achieve the inequality

$$(2.22) \qquad \int_r^{\infty} (- 2F(u(s)))^{1/2}\, ds \le \varepsilon\Big(\int_r^{\infty} \frac{ds}{G(s)G'(s)}\Big)^{1/2}.$$

We recall from the condition (G3), that the asymptotic relation (1.8) is satisfied in the present case. Hence, the expression in the right member of

(2.22) is of the same order of magnitude as ε times the integral $\int_r^\infty \frac{ds}{G(s)}$ as $r \to \infty$. By increasing r further if necessary, we can therefore certainly achieve the inequality

$$\int_r^\infty (-2F(u(s)))^{1/2}\, ds \le \varepsilon \int_r^\infty \frac{ds}{G(s)}. \tag{2.23}$$

Combining (2.20) and (2.23), we find that

$$u(r) < \left(\sqrt{2(K+\varepsilon)} + \varepsilon\right)\int_r^\infty \frac{ds}{G(s)} \tag{2.24}$$

for all sufficiently large r. The relation (2.17) follows from (2.19) and (2.24) as ε tends to zero. □

§ 3. Distinct Solutions Must Intersect.

Throughout this section we assume that u and v are two distinct non-negative non-trivial solutions of (1.1). Our purpose here is to show that the graphs of u and v have at least one point in common.

According to Lemma 2, u and v are (strictly) decreasing on their respective supports, so they can be inverted there. We denote the inverse functions by r and s, respectively. Thus, r is defined on the interval $[0, u(0)]$, s is defined on the interval $[0, v(0)]$, and

$$u(r(t)) = t, \quad t \ member [0, u(0)]; \quad v(s(t)) = t, \quad t \ member [0, v(0)]. \tag{3.1}$$

Let the functions R and S be defined by

$$R(t) = u'(r(t))^2, \quad t \ member [0, u(0)]; \tag{3.2}$$

$$S(t) = v'(s(t))^2, \quad t \ member [0, v(0)].$$

Clearly, R and S are non-negative on their respective domains of definitions. In fact, $R(0) = S(0) = 0$, because of Lemma 1(i), and $R(u(0)) = S(v(0)) = 0$, because of the boundary condition that is satisfied by u and v at 0. Furthermore, R and S are (strictly) positive in the open interval $(0, u(0))$ and $(0, v(0))$, respectively, because of Lemma 2. We also observe that $u'(r) = -\sqrt{R(u(r))}$ and $v'(r) = -\sqrt{S(v(r))}$ for all $r \ge 0$.

One readily verifies that $R'(t) = 2u''(r(t))$ and $S'(t) = 2v''(s(t))$, so R and S satisfy the initial value problems

$$R'(t) = 2\{g(r(t))\sqrt{R(t)} - f(t)\}, \quad t > 0; \qquad R(0) = 0; \tag{3.3}$$

and

$$S'(t) = 2\{g(s(t))\sqrt{S(t)} - f(t)\}, \quad t > 0; \qquad S(0) = 0; \tag{3.4}$$

respectively. We prove the following theorem.

THEOREM 1. *Let f and g satisfy the conditions (F1), (F2), (G1), and (G2). If u and v are two distinct non-negative non-trivial solutions of the boundary value problem (1.1), then $u(r) = v(r)$ for at least one value r member $(0, \infty)$.*

PROOF. Suppose that $u(r) > v(r)$, say, for all r *member* $(0, \infty)$. Then $r(t) > s(t)$ for all t *member* $(0, v(0))$. Because g is non-increasing, it follows that $g(r(t)) \le g(s(t))$ for all t *member* $(0, v(0))$. Hence, R satisfies the same initial value problem as S, but with a differential inequality:

$$R'(t) \le 2\{g(s(t))\sqrt{R(t)} - f(t)\}, \quad t > 0; \qquad R(0) = 0. \tag{3.5}$$

The theory of differential inequalities enables us to compare the *maximal* solutions of the initial value problems (3.3) and (3.4). For our purpose, such a comparison is not meaningful, unless there are no other solutions besides the maximal solutions. The conditions of the theorem indeed imply that (3.3) and (3.4) have at most one non-negative solution.

The initial value problems (3.3) and (3.4) are of the type

$$x' = p(t)x^{1/2} + q(t), \quad t > 0; \qquad x(0) = 0; \tag{3.6}$$

where $p(t) = 2g(r(t))$ and $p(t) = 2g(s(t))$, respectively, and $q(t) = -2f(t)$. Uniqueness of non-negative non-trivial solutions of this type of initial value problem has been established by the authors in [7] under the following conditions: (i) the coefficients p and q are integrable near 0; (ii) p and the integral q_1, defined by $q_1(t) = \int_0^t q(s)\,ds$, $t \ge 0$, are non-negative near 0; and (iii) for every $t > 0$, there is a point τ *member* $(0, t)$, where $q_1(\tau) > 0$.

One readily verifies that the conditions (i) and (ii) are satisfied by the coefficients p and q above. The condition (iii) is certainly satisfied if $\beta > 0$ and f does not vanish identically near 0. If f vanishes identically near 0, $f = 0$ on $[0, \gamma]$ for some $\gamma > 0$, say, then the differential equation (3.6) reduces to $x' = p(t)x^{1/2}$ on $[0, \gamma]$. The solution that satisfies the initial condition $x(0) = 0$ is either trivial or strictly positive on $(0, \gamma]$. In the former case, we consider the same initial value problem for $t \ge \gamma$. Since $\gamma < \beta$, the condition (iii) is satisfied to the right of γ, so there is at most one non-negative non-trivial solution of (3.6) for $t \geq \gamma$. When combined with the trivial solution on $[0, \gamma]$, this solution becomes the unique non-negative non-trivial solution for $t \geq 0$. In the latter case, when the solution is already strictly positive at γ, we can use the Lipschitz continuity of f at γ to show that the solution has a unique continuation beyond γ. Thus, we conclude that the initial value problem (3.6) has at most one non-negative non-trivial solution.

A simple comparison argument then shows that

$$R(t) \leq S(t), \qquad t \textit{ member} [0, v(0)]. \tag{3.7}$$

If $u(0) > v(0)$, then $R(v(0)) > 0$, while $S(v(0)) = 0$. This would clearly contradict (3.7), so at this point we must conclude that $u(0) = v(0)$.

The inequality (3.7) implies that $|u'(r(t))| \leq |v'(s(t))|$. Furthermore, $0 < g(r(t)) \leq g(s(t))$, so

$$g(r(t))\, |u'(r(t))| \leq g(s(t))\, |v'(s(t))|, \qquad t \textit{ member} [0, u(0)]. \tag{3.8}$$

Next, we apply (2.5) to u and v at $r = 0$ and subtract the resulting equations. We find

$$\int_0^\infty g(r) u'(r)^2\, dr = \int_0^\infty g(s) v'(s)^2\, ds, \tag{3.9}$$

or, after a transformation of variables,

$$\int_0^{u(0)} g(r(t))\, |u'(r(t))|\, dt = \int_0^{v(0)} g(s(t))\, |v'(s(t))|\, dt. \tag{3.10}$$

We recall that $u(0) = v(0)$ and conclude that the inequality (3.8) is compatible with the identity (3.10) if and only if $u'(r(t)) = v'(s(t))$ for all $t \textit{ member} [0, u(0)]$. This equality, in turn, implies that u and v coincide everywhere. But here we have arrived at a contradiction, since we had assumed that u and v were distinct. Hence, if u and v are distinct, their graphs must intersect at some point $r \textit{ member} (0, \infty)$. □

The Monotone Separation Lemma of Peletier and Serrin [**6**, Lemma 9] is an immediate consequence of Theorem 1. We formulate it as a corollary.

COROLLARY. *Let the conditions of Theorem 1 be satisfied. If u and v are two distinct non-negative non-trivial solutions of the boundary value problem*

(1.1) and $u(r) = v(r) = \tau$ *for some* r *member* $(0, \infty)$, *then* $r(t) - s(t)$ *is a monotone non-increasing function of* t *on* $[0, \tau]$.

PROOF. Suppose $u(r) > v(r)$ beyond the point of intersection of the two graphs. According to (3.7), we have $R(t) \le S(t)$ for all t *member* $[0, \tau]$. That is, $-u'(r(t)) \le -v'(s(t))$ for all t up to the ordinate of the point of intersection. Since $u'(r(t))r'(t) = 1$ and $v'(s(t))s'(t) = 1$, the last inequality is equivalent with $r'(t) \le s'(t)$ for all t *member* $[0,\tau]$, which shows that the difference $r - s$ is monotone on $[0, \tau]$, as claimed. □

§ 4. Intersecting Solutions are Identical.

In the previous section we showed that the graphs of two non-negative non-trivial solutions of (1.1) either coincide everywhere or intersect in at least one point r *member* $(0, \infty)$. In this section we prove that the latter alternative is excluded by the conditions (F1) through (F3) and (G1) through (G3).

THEOREM 2. *Let* f *and* g *satisfy the conditions (F1), (F3), and (G1). Let* u *and* v *be two non-negative non-trivial solutions of the boundary value problem (1.1). If* $u(r) = v(r) > \beta$ *for some* $r \ge 0$, *then* $u(r) = v(r)$ *for all* $r \ge 0$.

PROOF. The theorem follows from condition (F3) and is shown in the same way as the Sturm comparison theorem.

Suppose that $u(a) = v(a) = \tau$ for some $a \ge 0$ and that $u(r) > v(r)$ for $0 \le r < a$. If $\tau > \beta$, we obtain the following inequality from the condition (F3):

$$(v(r) - \beta) f(u) - (u(r) - \beta) f(v) \le 0, \qquad 0 \le r \le a.$$

Since u and v satisfy (1.1), this inequality can be rewritten,

$$\big(G(r)(-(v(r) - \beta)u'(r) + (u(r) - \beta)v'(r))\big)' \le 0.$$

Hence, upon integration over $[0, a]$,

$$(-u'(a) + v'(a))(\tau - \beta) \le 0.$$

Our assumptions imply that $-u'(a) > -v'(a)$, so the expression in the left member is positive. Clearly, we have a contradiction, unless $u'(a) = v'(a)$. But if both $u(r) = v(r)$ and $u'(r) = v'(r)$ at $r = a$, then $u(r) = v(r)$ for all r; cf. the last part of the proof of Lemma 2. □

THEOREM 3. *Let f and g satisfy the conditions (F1), (G1), (G2), and (G3). Let u and v be two non-negative non-trivial solutions of the boundary value problem (1.1). If $u(r) = v(r) \leq \beta$ for some $r \geq 0$, then $u(r) = v(r)$ for all $r \geq 0$.*

PROOF. We prove the theorem in two steps. In the first step we rule out the possibility that the graphs of two distinct solutions u and v have more than one point in common, once they are at or below the horizontal line $t = \beta$. In the second step we show that they cannot even have a single point in common. The symbols r, s, R, and S below have the same meaning as in Section 3; cf. (3.1) and (3.2).

Step 1. Suppose that there are two points a and b, where $u(a) = v(a) \leq \beta$ and $u(b) = v(b) \leq \beta$. Without loss of generality, we may assume that $a < b$ and $u(r) > v(r)$ on (a, b). Then $R(u(a)) < S(v(a))$ and $R(u(b)) > S(v(b))$.

By continuity, there exists a pair of points c, d, with $a < c < d < b$, such that $u(d) = v(c) = \tau$, with $u(b) < \tau < u(a)$, and $R(\tau) = S(\tau)$. If more than one such pair exists, we take the one for which τ is largest. Thus, $R(t) < S(t)$ for all t *member* $(\tau, u(a))$.

We apply the identity (2.3) to u on $[a, d]$ and to v on $[a, c]$ and subtract the resulting equations,

$$\begin{aligned} &\left[G(d)^2 - G(c)^2\right]\left[\frac{1}{2}u'(d)^2 + F(u(d))\right] \\ &\quad - \frac{1}{2}G(a)^2\left[R(u(a)) - S(v(a))\right] \\ &= 2\int_{\tau}^{u(a)} \left(\frac{G(r(t))G'(r(t))}{R(t)^{1/2}} - \frac{G(s(t))G'(s(t))}{S(t)^{1/2}}\right)F(t)\, dt. \end{aligned} \tag{4.1}$$

The first term in the left member is the product of two positive terms, so it is positive. The second term in the left member is negative, by assumption. Hence, the expression in the left member of (4.1) is certainly positive. Since $r(t) > s(t)$ on (a, b) and the product GG' is non-decreasing, we have $G(r(t))G'(r(t)) \geq G(s(t))G'(s(t))$ for all t *member* $(\tau, u(a))$. Furthermore, $R(t) < S(t)$ there, so the expression inside the parentheses under the integral sign is positive. On the other hand, $F(t)$ is zero or negative on the entire range of integration, so the integral is negative. But now we have a contradiction. The possibility of two points of intersection is thus ruled out.

Step 2. Suppose that $u(a) = v(a) \leq \beta$ at some point $r = a$. Without loss of generality we may assume that $u(r) > v(r)$ for all $r > a$.

According to Lemma 3(i), the following limits exist:

$$K = \lim_{r \to \infty} G(r)^2\Big[\frac{1}{2}u'(r)^2 + F(u(r))\Big], \tag{4.2}$$

$$L = \lim_{r \to \infty} G(r)^2\Big[\frac{1}{2}v'(r)^2 + F(v(r))\Big]. \tag{4.3}$$

Applying Lemma 3(ii) to u and v on $[a, \infty)$ and subtracting the resulting equations, we obtain

$$\begin{aligned} &\frac{1}{2}G(a)^2\big[R(u(a)) - S(v(a))\big] \\ &\quad = K - L - 2\int_0^{u(a)} \Big(\frac{G(r(t))G'(r(t))}{R(t)^{1/2}} - \frac{G(s(t))G'(s(t))}{S(t)^{1/2}}\Big)F(t)\,dt. \end{aligned} \tag{4.4}$$

The expression in the left member is negative. Under the integral sign, the expression inside the parentheses is positive, while $F(t)$ is zero or negative, so the integral is certainly negative. Hence, if $K \geq L$, the expression in the right member of (4.4) is positive and we have a contradiction.

It remains to investigate the case $K < L$. If $K < L$, we take $\varepsilon < (L - K)/8$ and choose r sufficiently large that

$$G(r)^2\Big[\frac{1}{2}u'(r)^2 + F(u(r))\Big] < K + \varepsilon, \tag{4.5}$$

and

$$G(r)^2\Big[\frac{1}{2}v'(r)^2 + F(v(r))\Big] > L - \varepsilon. \tag{4.6}$$

It follows from (4.5) that

$$u(r) < \big(\sqrt{2(K + \varepsilon)} + \varepsilon\big) \int_r^\infty \frac{ds}{G(s)}; \tag{4.7}$$

cf. the derivation of (2.24) in the proof of Lemma 4. By reducing ε if necessary, we can certainly achieve that $\sqrt{2(K + \varepsilon)} + \varepsilon < \sqrt{2(L - \varepsilon)}$. Thus,

$$u(r) < \sqrt{2(L - \varepsilon)}\int_r^\infty \frac{ds}{G(s)}. \tag{4.8}$$

On the other hand, it follows from (4.6) that

$$v(r) > \sqrt{2(L - \varepsilon)}\int_r^\infty \frac{ds}{G(s)}; \tag{4.9}$$

cf. the derivation of (2.19) in the proof of Lemma 4. The inequalities (4.8) and (4.9) together imply that $u(r) < v(r)$ for r sufficiently large. But this conclusion contradicts the earlier assumption that $u(r) > v(r)$ for all $r > a$. Thus, the possibility that the graphs of u and v intersect is ruled out. □

§ 5. Conclusion and Discussion.

It is clear that the conclusions of Sections 3 and 4 are compatible only if the solutions u and v of the boundary value problem (1.1) coincide. We have therefore shown the following uniqueness result.

THEOREM 4. *If f and g satisfy the conditions (F1) through (F3) and (G1) through (G3), then the boundary value problem (1.1) has at most one non-negative non-trivial solution.*

This uniqueness result generalizes and strengthens the earlier result of Peletier and Serrin [**6**]. It generalizes it, because the boundary value problem (1.1) is a (non-trivial) generalization of the problem considered in [**4**]. It strengthens it, because the condition (F2) on the nonlinear term f is weaker than the corresponding condition (H*) in [**6**].

The crucial step in the proof presented here is the transformation of variables (3.2). The new variables R and S satisfy the nonlinear initial value problems (3.3) and (3.4), and if (F2) holds, then R and S are uniquely determined. Thus, a direct comparison of their graphs is possible.

The results obtained here can be generalized to the more general boundary value problem

$$(5.1)\qquad \begin{gathered} (A(|u'|)u')' + g(r)A(|u'|)u' + f(u) = 0, \quad r > 0; \\ u'(0) = 0; \\ \lim_{r\to\infty} u(r) = 0; \end{gathered}$$

where the real-valued continuous function A is such that the product $pA(p)$ is strictly increasing on $[0, \infty)$. Problems of this type arise, for example, in the study of the semi-linear elliptic equation

$$(5.2)\quad \nabla\cdot A(|u'|)\nabla u + f(u) = 0, \quad x \text{ member } \mathbb{R}^n; \qquad \lim_{x\to\infty} u(x) = 0;$$

cf. the article by Franchi, Lanconelli, and Serrin [**8**] in these proceedings. Details will be presented elsewhere.

Acknowledgment.

The authors wish to thank the organizers, Profs. Wei-Ming Ni, L. A. Peletier, and J. Serrin, for their kind invitation to participate in the Microprogram on Nonlinear Diffusion Equations and Their Equilibrium States at the MSRI. They wish to express their particular appreciation to Profs. Peletier and Serrin for their interest in this work and for several stimulating discussions during the Microprogram.

This work was supported by the Applied Mathematical Sciences subprogram of the Office of Energy Research, U.S. Department of Energy, under contract W-31-109-Eng-38.

Permanent address: Department of Mathematical Sciences, Northern Illinois University, DeKalb, IL 60115.

Mathematics and Computer Science Division, Argonne National Laboratory, Argonne, IL 60439-4844, U.S.A.

REFERENCES

1. B. Gidas, W.-M. Ni and L. Nirenberg, *Symmetry and related properties via the maximum principle.* Commun. Math. Phys. **68** (1979), pp. 209-243.
2. H. Berestycki and P.-L. Lions, *Existence d'ondes solitaires dans des problèmes non-linéaires du type Klein-Gordon.* C. R. Acad. Sci. Paris, Série A **287** (1978), pp. 503-506.
3. H. Berestycki, P. L. Lions and L. Peletier, *An ODE approach to the existence of positive solutions for semilinear problems in* $\mathbb{R}^N$. Indiana Univ. Math. J. **30** (1981), pp. 141-157.
4. K. McLeod and J. Serrin, *Uniqueness of the ground state solution for* $\Delta u + f(u) = 0$. Proc. Nat. Acad. Sci. USA **78** (1981), pp. 6592-6595.
5. L. A. Peletier and J. Serrin, *Uniqueness of positive solutions of semilinear equations in* $\mathbb{R}^n$. Arch. Rat. Mech. Anal. **81** (1983), pp. 181-197.
6. L. A. Peletier and J. Serrin, *Uniqueness of non-negative solutions of semilinear equations in* $\mathbb{R}^n$. J. Diff. Eq. **61** (1986), pp. 380-397.
7. H. G. Kaper and M.-K. Kwong, *Uniqueness for a class of non-linear initial value problems.* J. Math. Anal. Applic. (to appear).
8. B. Franchi, E. Lanconelli, and J. Serrin, *Existence and uniqueness of ground state solutions of quasilinear elliptic equations.* Preprint (1986).

Some qualitative properties of nonlinear partial differential equations

BY BERNHARD KAWOHL

The purpose of my lecture is to point out the usefulness of curvilinear coordinates in dealing with nonlinear partial differential equations. I shall present three pieces of evidence on this matter, and therefore one might say that I give three "microlectures" in this microprogram.

The first microlecture is on the evolution of surfaces of mean-curvature type

$$u_t - \operatorname{div}((1+|\nabla u|^2)^{-1/2}\nabla u) + f(u) = 0 \qquad \text{in } \Omega \times (0,\infty).$$

The main result presented here is a simple decay estimate for $|\nabla u|$ under suitable assumptions on f and under a curvature assumption on $\partial\Omega$. This result is part of joint work with Hans Engler and Stefan Luckhaus [**EK**].

The second microlecture is on a free boundary problem which occurs in the modelling of potential flow under Bernoulli's boundary condition. This problem is also addressed in the lecture of P. Korman, and I give a short proof of the convexity of the free boundary. The result is already known, [**T**,**H**], but the old proofs are of a function theoretic nature and the new proof seems to be more transparent (at least to me).

The third microlecture is on a question which can be traced back to Saint Venant's paper from 1856 [**SV**]. Suppose u solves an elliptic problem such as $\operatorname{div}((1+|\nabla u|^2)^{-1/2}\nabla u) = -1$ in Ω, $u = 0$ on $\partial\Omega$, and that Ω is bounded and convex and has symmetry properties. Consider the function $|\nabla u|$ on $\partial\Omega$. Can one predict those points in which $|\nabla u|$ attains its maximum? This problem will be treated with the moving plane method as developed by Alexandrov [**Al**], Serrin [**S0**,**S1**], Gidas, Ni and Nirenberg [**GNN**] and Caffarelli, Friedman and Pozzi [**CFP**].

As mentioned before, the proof of all three problems will have the use of curvilinear coordinates as a common ingredient. Other common features are the use of the maximum principle and the search for those points in which $|\nabla u|$ becomes maximal.

Throughout the paper I assume that $\Omega \subset \mathbb{R}^n$, $n \geq 2$, is a bounded domain with smooth boundary $\partial\Omega$. $u(x,t)$ or $u(x)$ denote real functions on $\Omega \times (0,\infty)$ or Ω. ∇u denotes the spatial gradient in $\mathbb{R}^n$ and div the spatial divergence operator in $\mathbb{R}^n$.

1. Gradient estimates for evolutionary surfaces of prescribed mean curvature.

Consider the quasilinear parabolic equation

$$u_t - \operatorname{div}((1+|\nabla u|^2)^{-1/2}\nabla u) + f(u) = 0 \qquad \text{in } \Omega \times (0,\infty) \tag{1.1}$$

under Dirichlet boundary conditions

$$u(x,t) = 0 \qquad \text{on } \partial\Omega \times (0,\infty) \tag{1.2}$$

and with initial data

$$u(x,0) = u_0(x) \qquad \text{in } \Omega. \tag{1.3}$$

Equation (1.1) is an example of a more general class of parabolic equations, namely

$$u_t - \operatorname{div}(g(|\nabla u|^2)\nabla u) + f(u) = 0 \qquad \text{in } \Omega \times (0,\infty), \tag{1.4}$$

with natural assumptions on g. The extensions of the result in this paragraph to such more general equations or even to systems is more or less a technical question which is treated in [**EK**]. The key ideas of the proof, however, can be demonstrated in the special case of equation (1.1) in a more condensed way.

In order to explain Theorem 1 below I have to introduce the notion of a pseudoconvex domain Ω. Let $n(x)$ denote the exterior normal to $\partial\Omega$ and let $-n(x)$ be the normal vector on the oriented surface $\partial\Omega$. Then Ω is called pseudoconvex if and only if the mean curvature H of $\partial\Omega$ is everywhere nonnegative. A convex domain, for example, is always pseudoconvex, but also a torus in $\mathbb{R}^3$ can be pseudoconvex.

THEOREM 1. *Let u be a classical solution to problem (1.1)(1.2)(1.3), and suppose that f is of class C^1, $f(0) = 0$, and there is a real constant M such that $f'(u) \geq -M$. Finally assume that Ω is pseudoconvex and that $u_0(x)$ is in $C^1(\bar{\Omega})$.*

Then for every $p \geq 1$ the following estimate holds:

$$\|\nabla u(.,t)\|_{L^p(\Omega)} \leq e^{Mt} \|\nabla u_0(.)\|_{L^p(\Omega)}. \tag{1.5}$$

PROOF: The existence and regularity of solutions to problem (1.1)(1.2)(1.3) was studied by Gerhardt [**G**] and Ecker [**E**]. It follows from their work that all the formal differentiations below can be justified, since u is sufficiently regular.

First I discuss the case $p = \infty$ which is necessary to establish the uniform ellipticity of the operator defined by $\tilde{u} \to \operatorname{div}((1 + |\nabla u|^2)^{-1/2} \nabla \tilde{u})$. For notational brevity I set $g(s) := (1+s)^{-1/2}$ and $v(x,t) := |\nabla u(x,t)|^2$. Then I proceed in the spirit of Payne and Philippin [**PP**] or Sperb [**Sp**] and calculate

$$v_t - Lv + \vec{b} \cdot \nabla v - 2\, M\, v < 0 \qquad \text{in } \Omega \times (0, \infty),$$

with

$$L\psi = g(|\nabla u|^2) \Delta \psi + 2\, g'(|\nabla u|^2) \nabla u \nabla^2 \psi \nabla u,$$

with $\nabla^2 \psi$ denoting the Hessean matrix of ψ and with

$$\vec{b}(x) = \left\{2\, g'(|\nabla u|^2)(\Delta u \nabla u + \nabla v) + g''(|\nabla u|^2)(\nabla u \nabla v) \nabla v\right\}.$$

L is a uniformly strongly elliptic operator on a finite space-time cylinder, say $Q_T := \bar{\Omega} \times [O,T]$, $0 < T < \infty$. So we may apply the parabolic maximum principle [**PW**] to $w = e^{2Mt} v$ and conclude that w restricted to Q_T attains its maximum for $t = 0$ or in a point x^0 on $\partial\Omega \times (O,T]$. In the latter case we have

$$\frac{\partial w}{\partial n}(x^0) > 0 \tag{1.6}$$

according to Hopf's Lemma. Now let us introduce tangential and normal coordinates in x^0 such that u_n denotes differentiation in direction n. Then we can use the identity

$$\begin{aligned} \operatorname{div}(g(|\nabla u|^2)\nabla u) &= g(|\nabla u|^2)\Delta u + 2g'(|\nabla u|^2) u_{nn} u_n^2 = \\ &= \left\{g(|\nabla u|^2) + 2g'(|\nabla u|^2)|\nabla u|^2\right\} u_{nn} + g(|\nabla u|^2)(n-1)H(x^0) u_n \end{aligned} \tag{1.7}$$

as follows. We observe that $u(x^0) = u_t(x^0) = f(u(x^0)) = 0$, so the expression in (1.7) vanishes, i.e. u_n and u_{nn} have opposite signs in x^0. therefore we reach the contradiction to (1.6)

$$w_n = 2\ \exp(2MT)u_n u_{nn} < 0 \qquad \text{on } \partial\Omega,$$

and thus w can only attain its maximum over Q_T at $t = 0$. Now by induction with respect to k the same can be shown for w restricted to Q_{kT}, $k \in \mathbb{N}$, which proves the theorem for $p = \infty$.

It remains to discuss the case $1 \leq p < \infty$. Using the differential equation (1.1) and integration by parts we calculate

$$\begin{aligned}
\frac{d}{dt}\|\nabla u\|_{L^p(\Omega)} &= \|\nabla u\|_{L^p(\Omega)}^{1-p} \int_\Omega |\nabla u|^{p-2}\nabla u \nabla u_t dx \\
&= \|\nabla u\|_{L^p(\Omega)}^{1-p} \Big\{ -\int_\Omega \operatorname{div}(|\nabla u|^{p-2}\nabla u) \operatorname{div}(g(|\nabla u|^2)\nabla u) dx \\
&\qquad - \int_\Omega |\nabla u|^{p-2}\nabla u f'(u)\nabla u dx \Big\}
\end{aligned}$$

Notice that the first integral in the curly brackets contains two prototypes of quasilinear elliptic operators. Suppose we can show

$$\int_\Omega \operatorname{div}(|\nabla u|^{p-2}\nabla u) \operatorname{div}(g(|\nabla u|^2)\nabla u) dx > 0, \tag{1.8}$$

then

$$\frac{d}{dt}\|\nabla u\|_{L^p(\Omega)} \leq M\|\nabla u\|_{L^p(\Omega)}$$

and the proof of Theorem 1 is complete. So it remains to show property (1.8). We use the following Green's identity for vector-valued functions ψ and $\psi \in C^1(\bar{\Omega}, \mathbb{R}^n)$

$$\begin{aligned}
&\int_\Omega \operatorname{div}\psi \operatorname{div}\psi dx \\
&= \int_\Omega \operatorname{tr}(\nabla\psi\nabla\psi^t)dx + \int_{\partial\Omega} \{(\psi\cdot n)\operatorname{div}\psi - n\cdot\nabla\psi\cdot\psi\}\, dx,
\end{aligned} \tag{1.9}$$

where tr denotes the trace and ^{t} the transpose of a matrix. Moreover we use

LEMMA 1. *Let $u \in C^2(\bar{\Omega})$, $u = 0$ on $\partial\Omega$, and let g and $h : \mathbb{R}_0^+ \to \mathbb{R}$ be of class C^1. Then, setting $s = |\nabla u|^2$,*

$$\begin{aligned}
&\int_\Omega \operatorname{div}(g(s)\nabla u)\operatorname{div}(h(s)\nabla u)dx = \\
&\int_\Omega \left\{h(s)\left[g(s)|\nabla^2 u^2| + 2\,g'(s)|\nabla^2 u\nabla u|^2\right] + \right.\\
&\left.+2\,h'(s)\left[g(s)|\nabla^2 u\nabla u|^2 + 2\,g'(s)(\nabla u\nabla^2 u\nabla u)^2\right]\right\}dx \\
&+\int_{\partial\Omega} g(s)h(s)s(n-1)H\,dx
\end{aligned} \tag{1.10}$$

To prove Lemma 1 we set $\psi = g(|\nabla u|^2)\nabla u$ and $\psi = h(|\nabla u|^2)\nabla u$ in (1.9) and note that on $\partial\Omega$ we have $\nabla u = u_n \cdot n$ and $\Delta u = u_{nn} + (n-1)Hu_n$, since $u \equiv 0$ on $\partial\Omega$.

To prove (1.8) we apply Lemma 1 to $h(s) = (s+\varepsilon)^{(p-2)/2}$ and to $g(s) = (1+s)^{-1/2}$ and note that $h'(s) \geq 0$ for $p \geq 2$. This and the ellipticity of the mean curvature operator implies that the first integral on the right-hand side of (1.10) is nonnegative. Also, the boundary integral in (1.10) is nonnegative because Ω is pseudoconvex by assumption. After passing to the limit ($\varepsilon \to 0$) in (1.10) one establishes (1,8) for $p \geq 2$. the case $1 \leq p \leq 2$ can be treated similarly, since then the expression in curly brackets in (1.10) can be bounded below by the nonnegative product

$$[g(s) + 2\,s\,g'(s)]\,[h(s) + 2\,s\,h'(s)]\,|\nabla^2 u\nabla u|^2.$$

□

REMARK 1: Theorem 1 can be improved. If Ω is no longer pseudoconvex, certain weaker *a priori* estimates hold, see [**EK**]. If $|\nabla u|$ decays to zero as t approaches infinity, then the mean curvature operator converges to the Laplace operator. In this case, a local decay estimate would be $\exp(-\lambda_1 + M)t$, with λ_1 being the lowest eigenvalue of the Laplacean on Ω under Dirichlet boundary conditions. Questions of this type were studied by Wiegner in [**W**].

One could also replace the Dirichlet condition (1.2) by the Neumann condition $u_n = 0$ on $\partial\Omega \times (0,\infty)$ and the pseudoconvexity assumption of Theorem 1 remains true. In fact, this was shown for the case $g(s) = |s|^{m-1}$, $m > 1$, in equation (1.4) by Alikakos and Rostamian [**AR**].

Among the technical difficulties which have to be overcome for such degenerate equations are that the solutions are not sufficiently smooth to do our integrations by part. Therefore, g in (1.4) and (1.8) has to be regularized and another passage to limits has to be studied.

Finally, one should note that if G denotes a primitive of g and if F denotes a primitive of f, and if u solves problem (1.4)(1.2)(1.3), then the Ljapunov functional

$$I(u) = \int_{\Omega} \{G(|\nabla u|^2) + F(u)\}\, dx$$

is nonincreasing in t independent of the geometry of Ω.

2. On free boundaries satisfying a Bernoulli condition.

Let $\Omega \subset \mathbb{R}^2$ be a bounded convex domain with smooth boundary $\partial\Omega$ and consider the problem of finding a function u and a smooth closed arc Γ which is the boundary of a domain $D \supset \Omega$, and such that

$$(2.1) \qquad \Delta u = 0 \qquad \text{in } G = D \setminus \bar{\Omega},$$

$$(2.2) \qquad u = 1 \qquad \text{on } \partial\Omega,$$

$$(2.3) \qquad u = 0, \quad \frac{\partial u}{\partial n} = \text{const.} \qquad \text{on } \Gamma.$$

This problem occurs in the modelling of potential flow. For its physical background, one can consult [**H**], among others. If Γ is sufficiently smooth, then one can infer the uniqueness of (u, Γ) and the convexity of D, see [**T**,H]. Here I want to prove the convexity of D without using conformal maps. The initial hope to generalize the convexity result to the analogous problem in higher dimensions (D and Ω are subsets of $\mathbb{R}^n$ with $n \geq 3$) is dampened by several obstacles. First of all, one has nonuniqueness and singular free boundaries, as was shown by Alt and Caffarelli [**AC**]. Secondly, there are no conformal mapping techniques available. My proof will explain why the two-dimensional case is indeed a special one. One of the reasons is that for $n = 2$ the word "pseudoconvex" can be replaced by "convex".

THEOREM 2. *Let u be a classical solution to the problem (2.1) (2.2)(2.3) with smooth free boundary. If Ω is convex, then D is convex.*

For the proof one has to proceed in two steps. We consider the function $v(x) := |\nabla u(x)|^2$ and derive the following lemmata.

LEMMA 2. *Under the assumptions of Theorem 2 the function $v(x)$ is positive on the set $\bar{G}$.*

PROOF: The shortest proof I know for this Lemma uses conformal maps (see [**T, H**]). The proof given here works also in higher dimensions.

We know from Hopf's Lemma that v is positive on ∂G, since $0 < u < 1$ in G. so it remains to study v in G.

As Alt and Caffarelli found out [**AC**], the solution of problem (2.1)(2.2)(2.3) is the (unique) global minimizer of the functional

$$J(v) := \int_{\{v>0\}} \{|\nabla u|^2 + 1\}\, dx \tag{2.4}$$

over the admissible set of functions

$$A := \{v \in W^{1,2}(\mathbb{R}^n) \mid v = 1 \text{ on } \bar{\Omega}\}. \tag{2.5}$$

In more than two dimensions the functional J can have other critical points which also solve (2.1)(2.2)(2.3) in a weak sense. However, the minimizing ones have to have nonvanishing gradient. If $y \in \Omega$, then the auxiliary function $w(x) := (y - x)\nabla u(x)$ is nonnegative due to the results in [**K**] on star shaped rearrangements. Moreover w is harmonic in G so that w must be positive in G. Consequently v is positive in G as well. This completes the proof of Lemma 2, but more can be said about v and u if $n = 2$.

The function v satisfies the differential inequality

$$\nabla v = 2\ \nabla u \nabla \Delta u + 2 \sum_{i,j=1}^{n} u_{ij} \geq 0 \qquad \text{in } G, \tag{2.6}$$

and, if $n = 2$, it satisfies even the differential equality

$$\Delta v - 2\left(\frac{\nabla v}{v}\right) \nabla v = 0 \qquad \text{in } G, \tag{2.7}$$

as can be seen, e.g., from [**Sp**, p. 69]. So for $n = 2$ the function v attains its (positive) minimum on ∂G. Finally, for $n = 2$, the free boundary is known to be sufficiently smooth [**AC**] to prove uniqueness for problem (2.1)(2.2)(2.3). This was done in [**T**].

LEMMA 3. *Let u satisfy the assumptions of Theorem 2. If $v(x)$ is positive on G and if $n = 2$, then D is convex provided Ω is convex.*

Notice that the combination of Lemma 2 and Lemma 3 implies Theorem 2. To prove Lemma 3 we observe that due to (2.7) the function v must attain its minimum on $\partial G(=\partial\Omega\cup\Gamma)$. But where on ∂G does this occur? The minimum cannot be attained on $\partial\Omega$, since we would have

$$(2.8)\qquad -2\, u_n u_{nn} < 0 \qquad \text{on } \partial\Omega.$$

This would contradict the fact that u is harmonic up to $\bar{G}$:

$$(2.9)\qquad \Delta u = u_{nn} + (n-1)H(x)u_n = 0 \qquad \text{on } \bar{G}.$$

For the definitions of H and u_n one may consult the paragraph above Theorem 1. Consequently, the minimum of v must be attained on the free boundary Γ. But on Γ the function v is constant. So v attains its minimum in any point on the free boundary. If this boundary is sufficiently smooth, which we know from [**AC**] in two dimensions, then we can again use (2.9) and Hopf's Lemma,

$$(2.10)\qquad 2\, u_n u_{nn} < 0 \qquad \text{on } \Gamma,$$

to conclude that $H(x)$, the mean curvature of the free boundary, is positive. For $n = 2$ this means that D is strictly convex.

REMARK 2: As mentioned above, if the space dimension exceeds 2, as proof of Theorem 1 seems to be hard. In this case (2.7) is no longer available — only (2.6) is, and this is not sufficient for my proof of Lemma 3.

The case that $n = 2$ and Ω is not convex has been thoroughly studied by Acker [**A**]. He derived more relations between the shapes of $\partial\Omega$ and Γ.

3. On points of maximal stress in the torsion of elastic bars.

Consider classical solutions, i.e. $u \in C^2(\Omega)\cap C^0(\bar{\Omega})$ of the following class of quasilinear elliptic Dirichlet problems

$$(3.1)\qquad a(u,|\nabla u|)\Delta u + h(u,|\nabla u|)u_i u_j u_{ij} = f(u,|\nabla u|) \qquad \text{in } \Omega,$$

$$(3.2)\qquad u > 0 \qquad \text{in } \Omega,$$

$$(3.3)\qquad u = 0 \qquad \text{on } \partial\Omega,$$

where $\Omega \subset \mathsf{R}^2$ is a bounded domain with boundary of class $C^{3,\alpha}$. The subscripts i and j in (3.1) denote derivatives with respect to x_i and x_j, and we use the summation convention that i and j run from 1 to 2. The functions a, h and f are of class C^1 with respect to their arguments and, in order to ensure ellipticity, we furthermore require the inequality

$$a(u,p) + h(u,p)p^2 > 0 \qquad \text{for every } u \geq 0,\, p \geq 0$$

to hold. The existence and regularity of solutions to problem (3.1)(3.3) was studied in the papers [**S1**] and [**JS**]. Among other things it was shown that, in general, classical solutions to problems of type (3.1)(3.3) can only exist if Ω is convex (see [**JS**, Theorem 5]). Of course, this does not mean that for a special equation there are solutions also for nonconvex domains. A special case of equation (3.1) is

$$\Delta u = -1 \tag{3.4}$$

and what I am about to show can in fact for this special equation be derived from work of Caffarelli, Friedman and Pozzi [**CFP**].

My work is motivated by an observation of Saint Venant. In his famous work "Memoire sur la torsion des prismes ... " [**SV**] he calculated the solutions of (3.3)(3.4) for various domains $\Omega \subset \mathbb{R}^2$, which are symmetric with respect to the cartesian x_1 and x_2-axes, for instance for ellipses, rectangles, astroids, etc. For all his examples he noted that the function $|\nabla u(x)|$ becomes maximal in those points of $\partial\Omega$, which have minimal distance from the origin. These points are of physical significance, because the vector $\nabla u(x)$ contains the nonvanishing components of the stress tensor of an elastic cylindrical bar with cross-section Ω. As the amount of torsion increases, the material loses its elastic behavior (since $|\nabla u|$ gets too large) and becomes plastic.

A lot of things have been discovered since 1856 about solutions of (3.3)(3.4), but for the location of the "fail points", i.e. points of maximal stress, there seem to be only rules of thumb. The following is quoted from [**Go**]: "In general [the maximal stress] occurs near one of the points of contact of the largest inscribed circle, and of these at that one where the boundary is least convex or most concave."

Under the assumptions (3.5) and (3.6) on the geometry of Ω or for regular polygons Ω, one can turn this rule of thumb into a theorem. Moreover, the theorem holds for the general class of problems (3.1)(3.2)(3.3) and not just for (3.3)(3.4).

$$\text{(3.5)}\begin{cases} \Omega \subset \mathbb{R}^2 \text{ is a convex bounded domain and symmetric} \\ \text{with respect to the } x_1 \text{ and } x_2\text{-axes.} \end{cases}$$

$$(3.6)\begin{cases}\text{The boundary } \partial\Omega \text{ is of class } C^{3,\alpha} \text{ and the curvature} \\ \text{of } \partial\Omega \cap \{x_1 > 0, x_2 > 0\} \text{ is monotone nondecreasing in } x_1.\end{cases}$$

THEOREM 3. *Let u be a classical solution of problem (3.1)(3.2) (3.3). If Ω satisfies the assumptions (3.5)(3.6) then the function $|\nabla u(x)|$ attains its maximum over $\partial\Omega$ only in those points of $\partial\Omega$ which have minimal distance to the origin. Furthermore, unless $\partial\Omega$ is a circle, there are precisely two such points, namely the points in which the x_2-axis intersects $\partial\Omega$.*

THEOREM 4. *Let u be a classical solution of problem (3.1)(3.2) (3.3). If Ω is a regular polygon in $\mathbb{R}^2$, then the function $|\nabla u(x)|$ attains its maximum over $\partial\Omega$ only in the midpoints of the sides of the polygon and is nondecreasing as one approaches a midpoint from a corner.*

Before I start with a proof of these Theorems I want to elaborate a little bit on the history of the result for problem (3.3)(3.4). One can learn from [**TP**, Part I, §160 and §162] that as early as 1859 Saint Venant was aware of a counterexample to his apparently general observation. This example of a doubly symmetric nonconvex domain is depicted in Figure 1. It was copied from Filon [**F**, p. 342] who is incorrectly credited as the first one to discover it by people like Love [**L**] or Timoshenko and Goodier [**TG**]. Nonetheless the counterexample is mentioned by Thomson and Tait [**TT**] or Todhunter and Pearson [**TP**] and it was published in [**N**]. The fail points are marked with "F" in Figure 1.

A heuristic idea for a proof of Theorem 1 in the case of problem (3.3)(3.4) is due to Boussinesq [**Bo**]. He studied laminar flow in pipes of cross-section Ω and ran across the same observation. To explain the location of the fail points he argued as follows. First of all, he took it for granted that the function $u(x)$ attains its maximum in the origin and center of symmetry of Ω. This assumption is not correct for the domain in Figure 1, but thanks to [**GNN**] it is justified for domains which satisfy (3.5). So if one follows the function $u(x)$ along a line segment from the origin to $\partial\Omega$, this function has to decay from its maximal value $u(0)$ to zero. The shorter the line segment is, the steeper is the decay of u, i.e. the larger is $|\nabla u|$. This heuristic reasoning does in fact suggest the validity of Theorems 3 and 4.

To prove Theorem 3 we have to introduce some notation. We parametrize the portion of $\partial\Omega$ which lies in $\{x_1 > 0, x_2 > 0\}$ with respect to arc length in such a way that x_1 increases when s decreases. For a fixed point x on

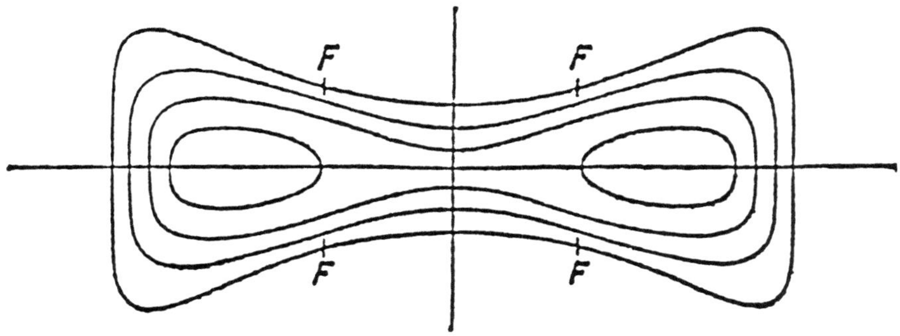

Figure 1

this portion of the boundary we denote the exterior normal to $\partial\Omega$ by $n(x)$ and the tangent vector (pointing in direction of increasing arc length) by t.

In a first step we want to show that

$$\left(\frac{d}{ds}\right)(|\nabla u|^2) = 2\left(\frac{\partial u}{\partial n}\right)\left(\frac{d}{ds}\right)\left(\frac{\partial u}{\partial n}\right) \geq 0 \tag{3.7}$$

along $\partial\Omega$. The factor $\frac{\partial u}{\partial n}$ is negative and the second factor can be rewritten as follows (see [**Sp**, p. 45f]):

$$\left(\frac{d}{ds}\right)\left(\frac{\partial u}{\partial n}\right) = \left(\frac{\partial}{\partial n}\right)\left(\frac{\partial u}{\partial t}\right) + \kappa\frac{\partial u}{\partial t}.$$

Here κ denotes the (nonnegative) curvature of the boundary, and

$$\left(\frac{\partial}{\partial n}\right)\left(\frac{\partial u}{\partial t}\right) := \lim_{\varepsilon\to 0}\frac{1}{\varepsilon}\left(\frac{\partial u}{\partial t}(x) - \frac{\partial u}{\partial t}(x-\varepsilon n)\right)$$

Since $\frac{\partial u}{\partial s} = 0 = \frac{\partial u}{\partial t}$ on $\partial\Omega$, it remains to study the sign of $\frac{\partial u}{\partial t}(x-\varepsilon n)$. This sign is positive due to the results of Gidas, Ni and Nirenberg [**GNN**] as long as a certain reflection property holds. The normal through x cuts off a cap (below the normal) from Ω. The reflection of this cap across the normal has to be contained in Ω. The situation is depicted in Figure 2.

Therefore, (3.7) holds once we can verify this reflection property. In their paper [**CFP**] Caffarelli, Friedman and Pozzi have shown that the reflection property holds for domains satisfying (3.5)(3.6) so that (3.7) holds and $|\nabla u|$ attains its maximum on the intersection of $\partial\Omega$ with the x_2-axis.

In a second step of the proof we have to show that there are no other locations of the "fail points" unless $\partial\Omega$ is a circle. If there are more than

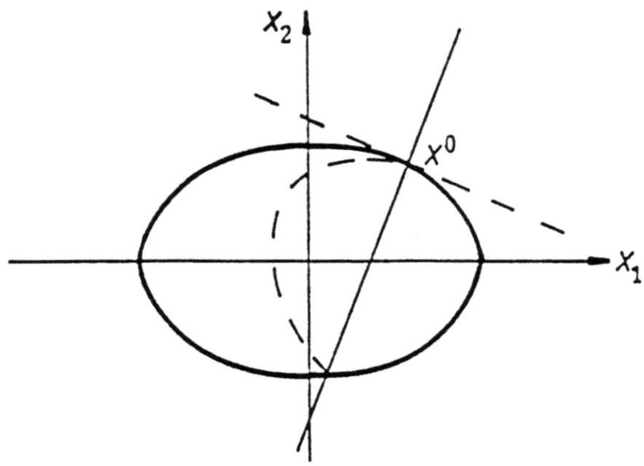

Figure 2

two such points, then due to (3.7) there is a portion of $\partial\Omega$ with positive arclength, such that $u = 0$ and $|\nabla u| = \text{const.}$ along this segment γ of $\partial\Omega$. Such a situation was investigated by Serrin in [**S2**]. He showed, using his corner lemma, that in this case Ω is symmetric with respect to reflections across the normal through x, where x is a point in the interior of the just-defined segment γ. Due to (3.6) we can conclude that γ is a circular arc, and after finitely many reflections, that $\partial\Omega$ is a circle. This concludes the proof of Theorem 3.

The proof of Theorem 4 goes along the same lines and is left as an exercise to the reader.

REMARK 3: It seems to be an open problem whether the first assertion of Theorem 3 remains valid if one merely assumes (3.5) and not (3.6).

ACKNOWLEDGEMENT: This work was financially supported in part by the Mathematical Sciences Research Institute, by my wife and, last but not least, by the Deutsche Forschungsgemeinschaft (DFG) under the auspices of the SFB 123. I want to express my gratitude to all of them.

SFB 123, Univ. Heidelberg, Im Neuenheimer Feld 294,
D 6900 Heidelberg, GERMANY

Research at MSRI supported in part by NSF Grant 812079-05.

REFERENCES

[**A**] A. Acker, *On the geometric form of free boundaries satisfying a Bernoulli boundary condition*, I and II, Math. Methods Appl. Sci. **6** (1984), 449–456;, **8** (1986), 387–404.

[**Al**] A.D. Alexandrov, *Uniqueness theorems for surfaces in the large V*, Vestnik Leningrad Univ. **13** (1958), 5–8; *Translated in Amer. Math. Soc. Translations (Series 2)*, **21**, 412–416.

[**AR**] N.D. Alikakos and R. Rostamian, *Gradient estimates for degenerate diffusion equations*, Math. Annalen **259** (1982), 53–70.

[**AC**] H.W. Alt and L.A. Caffarelli, *Existence and regularity for a minimum problem with free boundary*, J. Reine Angew. Math. **325** (1981), 105–144.

[**CFP**] L.A. Caffarelli, A. Friedman and G. Pozzi, *Reflection methods in the elastic plastic torsion problem*, Indiana Univ. Math. J. **29** (1980), 205–228.

[**E**] K. Ecker, *Estimates for evolutionary surfaces of prescribed mean curvature*, Math. Z. **180** (1982), 172–192.

[**EK**] H. Engler, B. Kawohl and S. Luckhaus, *Decay in Sobolev spaces for solutions of parabolic equations and systems*, Preprint 394, SFB 123, Heidelberg (1986).

[**G**] C. Gerhardt, *Evolutionary surfaces of prescribed mean curvature*, J. Differ. Equations **36** (1980), 139–172.

[**GNN**] B. Gidas, W.M. Ni and L. Nirenberg, *Symmetry and related properties via the maximum principle*, Comm. Math. Phys. **68** (1979), 209–243.

[**Go**] J.N. Goodier, *Torsion*, in "Handbook of Engineering Mechanics," W. Flügge, Ed., McGraw Hill, New York, 1962.

[**H**] R.S. Hamilton, *The inverse function theorem of Nash and Moser*, Bull. Amer. Math. Soc. **7** (1982), 65–222.

[**JS**] H. Jenkins and J. Serrin, *The Dirichlet problem for the minimal surface equation in higher dimensions*, J. Reine Angew. Math. **229** (1968), 170–187.

[**K**] B. Kawohl, *On star shaped rearrangements and applications*, Trans. Amer. Math. Soc. **296** (1986), 377–386.

[**K2**] B. Kawohl, *On the location of maxima of the gradient for solutions to quasilinear elliptic problems and a problem raised by Saint Venant*, J. of Elasticity (to appear).

[**PP**] L.E. Payne and G.A. Philippin, *Some maximum principles for nonlinear elliptic equations in divergence form with applications to capillary sufaces and to surfaces of constant mean curvature*, Nonlin. Anal. TMA **3** (1979), 193-211.

[**PW**] M. Protter and H. Weinberger, "Maximum principles in differential equations," Prentice Hall, Englewood Cliffs, 1967.

[**SV**] B. de Saint Venant, *Mémoire sur la torsion des prismes, avec des considérations sur leur flexion ainsi que sur l'équilibre intérieur des solides élastiques en général, et de formules pratiques pour le calcul de leur résistance à divers efforts s'exerçant simultanément*, Mémoires présentés par divers savant à l'académie des sciences de l'institut impérial de France, 2ème Ser **14** (1856), 233-560.

[**S0**] J. Serrin, *On surfaces of constant mean curvature which span a given space curve*, Math. Zeitschr. **112** (1969), 77-88.

[**S1**] J. Serrin, *The problem of Dirichlet for quasilinear elliptic differential equations with many independent variables*, Phil. Trans. R. Soc. Lond., A **264** (1969), 413–496.

[**S2**] J. Serrin, *A symmetry problem in potential theory*, Arch. Ration. Mech. Anal. **43** (1961), 304–318.

[**T**] D.F. Tepper, *Free boundary problem*, SIAM J. Math. Anal. **5** (1974), 841–846.

[**W**] M. Wiegner, *On the asymptotic behaviour of solutions of nonlinear parabolic equations*, Math. Z. **188** (1984), 3–22.

On Existence of Solutions for Non-coercive Problems

Philip Korman

I. A Model Non-coercive Problem Associated with Water Waves.

In its simplest form the model has the following form: $u(x,y,z)$ is 2π periodic in x and z solution of

$$
(1)\qquad
\begin{aligned}
u_y + F u_{xx} &= g(x,z) & y&=1\\
\Delta u &= f(x,y,z) & 0<y&<1\\
u &= 0 & y&=0.
\end{aligned}
$$

Here $F=$ const <0, and f and g are given functions, 2π periodic in x and z.

This problem is non-coercive, i.e. it does not satisfy the Lopatinski-Schapiro condition. On the other hand it has most of the properties of coercive elliptic problems, in particular, the maximum principle: if $f \le 0$ and $g \ge 0$ then $u \ge 0$. (Indeed, the point of minimum P cannot lie on $y=1$ since then $u_y(P)<0$, $-u_{xx}(P)\le 0$, contradicting $g(P)\ge 0$. Hence P must lie on $y=0$). The maximum principle easily implies the estimate $|u|_{L^\infty} \le c\left(|f|_{L^\infty} + |g|_{L^\infty}\right)$.

By $\|\cdot\|_m$ we denote the m-th Sobolev norm for the domain V, $V=\{0<y<1, 0\le x,z\le 2\pi\}$. Then ([**4**])

$$\|u\|_{m+2} \le c\left(\|f\|_m + \|g\|_{m+1}\right), \quad m = \text{ integer } \ge 0.$$

We shall discuss nonlinear problems of the type

$$
(2)\qquad
\begin{aligned}
u_y &= \rho(u_{xx}) & y&=1\\
\Delta u &= \varepsilon f(x,y,z,D^2u) & 0<y&<1\\
u &= 0 & y&=0.
\end{aligned}
$$

Here $\rho(0)=0$ and f is 2π periodic in x,z and depends on u and its derivatives of order up to two. We are interested in "small" (or perturbation) solutions.

Let us examine (1) with the solvability of (2) in mind. Let $f \equiv 0$ and $g = \sum_{j,k=-\infty}^{\infty} g_{jk} e^{ijx+ikz}$. Then the solution of (1) is

$$(3) \quad u = \sum_{j,k=-\infty}^{\infty} \frac{g_{jk} \sinh \sqrt{j^2+k^2}\, y}{\sqrt{j^2+k^2} \cosh \sqrt{j^2+k^2} - F j^2 \sinh \sqrt{j^2+k^2}} e^{ijx+ikz}.$$

If one uses iterations to solve (2), then it is necessary to have an estimate of the type

$$(4) \qquad \|u_{xx}\|_m \leq c\,\|g\|_m \quad \text{at } y = 1,$$

i.e. a gain of two x derivatives. For this the sign of F is crucial.

Case (i): $F < 0$. Here one easily derives (4), and so if $\dot{\rho}(0) > 0$ then (2) is solvable by Picard iterations. Indeed, we have the following result [**4**].

THEOREM. *Consider the problem*

$$(5) \qquad \begin{aligned} u_y &= \rho(u_{xx}) && y = 1 \\ \Delta u &= \varepsilon f(x, y, z, u, Du) && 0 < y < 1 \\ u &= 0 && y = 0. \end{aligned}$$

Assume that f is 2π periodic in x and z, and

(i) $\rho(0) = 0$, $\dot{\rho}(0) > 0$
(ii) $\rho, f \in C^4$ *(in all arguments).*

Then for ε sufficiently small the problem (5) has a C^2 solution, 2π periodic in x and z.

The main part of the proof involves the derivation of *a priori* estimates for the linearized problem.

Case (ii): $F = 0$. There is a gain of only one x derivative for u as compared to g in (3). Hence if one uses iterations to solve (2) in case $\rho(0) = \dot{\rho}(0) = 0$, then there is loss of one derivative at each step. Using the Nash-Moser technique we have proved the following theorem [**3**].

THEOREM. *Assume that*

(i) $\rho(0) = \dot{\rho}(0) = 0$, $\dot{\rho}(t) > 0$ *for $|t|$ small,*
(ii) $\rho, f \in C^{55}$ *in all variables.*

Then for ε sufficiently small the problem (2) has a C^2 solution, 2π periodic in x and z.

Case (iii): $F > 0$. This is the case of small divisors in (3). It is easy to show that the set of zero divisors,

$$F_{j,k} = \frac{\sqrt{j^2 + k^2}}{j^2} \coth \sqrt{j^2 + k^2},$$

is everywhere dense on the positive semiaxis. When F is restricted to be rational, then for $1 \leq F < \infty$ and $F \neq F_{j,k}$ the denominators are bounded away from zero, while for $0 < F < 1$ and $F \neq F_{j,k}$ there may be arbitrarily small divisors [**5**].

THEOREM. *For the problem (1) assume that F is rational, $F \geq 1$ $F \neq F_{j,k}$. Then we have the estimate ($m =$ integer ≥ 0)*

$$\|u\|_m \leq c\left(\|f\|_{m+1} + \|g\|_m\right).$$

If in addition $f \in H^5$, $g \in H^4$, then the problem (1) has a unique solution of class C^2.

Finally, we mention some other results. By using the degree theory, we have proved existence for (2) without requiring the solution to be small [**7**]. In the same paper we studied the distribution of eigenvalues for (1), which turned out to be similar to that for coercive problems. We have also considered problems involving higher order derivatives on the boundary and derived similar results. For example, the problem

$$\begin{aligned} u_y &= u_{xx}^2 - u_{xxxx} + u_{xzz} && y = 1 \\ \Delta u &= \varepsilon y \sin x \sin z && 0 < y < 1 \\ u &= 0 && y = 0, \end{aligned}$$

has a C^∞ solution, 2π periodic in x and z, provided ε is small enough [**5**].

II. Some Applications of the Domain Perturbation Method.

The boundary condition at $y = 1$ of the problem (1) arose in work of Shinbrot on the three dimensional water waves [**9**]. Starting with a fluid satisfying the Laplace equation in its domain and Bernoulli's equation on the free boundary, he maps (non-conformally) the unknown fluid domain onto a fixed one, the strip $0 \leq y \leq 1$. The equation of motion transforms into a nonlinear one, but when the bottom is assumed to be small, so are the nonlinear terms, and the problem can be solved by a contractive mapping argument. This approach is known as the domain perturbation method, apparently originated by D. Joseph [**2**]. The boundary condition Shinbrot gets at $y = 1$ has the following structure:

$$u_y - \tau(u_{yxx} + u_{yzz}) + Fu_{xx} = \varepsilon \cdot \text{ nonlinear terms.}$$

When the surface tension $\tau = 0$, we get our boundary condition in (1). Here F is the Froude number, $F = U^2/g$, with U denoting velocity of a stream and g the gravity. The case $F < 0$ corresponds to the gravity "pointing up" (and is the simplest one mathematically).

We used the domain perturbation method to attack the following problem.

PROBLEM: Let $r = B(\theta)$ be a closed curve in the plane. Find another closed curve $r = H(\theta)$ outside of $B(\theta)$, and a function u on the closed region between $B(\theta)$ and $H(\theta)$ with

$$\begin{aligned} &u = 0, \tfrac{\partial u}{\partial n} = -1 && r = H(\theta) \\ &\Delta u = 0 && B(\theta) < r < H(\theta) \\ &u = 0 && r = B(\theta), \end{aligned} \tag{6}$$

where $\frac{\partial}{\partial n}$ denotes the outward normal derivative.

This problem was considered by Hamilton [**1**] (and also earlier by Schaeffer [**8**] and Acker). Hamilton proved that for every smooth convex curve B there exists a unique solution of (6) (the curve $H(\theta)$ is also smooth and convex, and u is smooth). This result is strictly two-dimensional, since conformal mappings were used to derive *a priori* estimates.

We have attempted to apply the domain perturbation method in the dimensions greater than two, but difficulties appear already in the dimension three: the technique appears to break down because of the singularities of the Laplace operator in spherical coordinates. However, in two dimensions we have proved a result which complements that in [**1**], since it does not require convexity of the curve $B(\theta)$. We describe it next [**6**].

If $B(\theta) \equiv 1$ then the solution of (6) is $H(\theta) = h_0$, $u = -h_0 \log r + 1$, where $h_0 \log h_0 = 1$ $(h_0 \simeq 1.76)$. We then assume that $B = 1 + \varepsilon b(\theta)$, and look for the solution in the form

$$\begin{aligned} H &= h_0 + \varepsilon h(\theta) \\ u &= -h_0 \log r + 1 + \varepsilon v(r, \theta). \end{aligned} \tag{7}$$

THEOREM. *Assume that* $b(\theta) \in C^{m+\alpha}$, $\varepsilon|b|_{m+\alpha}$ *is sufficiently small,* $m =$ *integer* ≥ 2, $0 < \alpha < 1$. *Then there exists a pair of functions* $(v, h) \in C^{m+\alpha} \times C^{m+\alpha}$, *satisfying (6).*

PROOF: (Sketch) The change of variables $(r, \theta) \to (r', \theta')$, defined by

$$(8) \qquad \begin{aligned} r' &= (h_0 - 1)\frac{r - 1 - \varepsilon b}{\varepsilon(h - b) + h_0 - 1} + 1 \\ \theta' &= \theta \end{aligned}$$

maps the fluid domain $1 + \varepsilon b \leq r \leq h_0 + \varepsilon h$ onto the annulus $1 \leq r' \leq h_0$. Inserting the relations (7) and (8) in (6), we obtain (with $h(\theta) = v(h_0, \theta)$ and dropping the primes)

$$(9) \qquad \begin{aligned} v &= h + \varepsilon t(\varepsilon, h) & r &= h_0 \\ v_r &= -\tfrac{h}{h_0} + \varepsilon g(\varepsilon, Dv, Dh, Db) & r &= h_0 \\ \Delta v &= \varepsilon f(\varepsilon, D^2 v, D^2 h, D^2 b) & 1 &< r < h_0 \\ v &= h_0 b + \varepsilon q(\varepsilon, b) & r &= 1 \end{aligned}$$

with the functions t, g, f, q being smooth in their arguments. We define a map $T : (w, k) \to (v, h)$ from $C^{m+\alpha} \times C^{m+\alpha}$ to itself by solving (here $\bar{g} \equiv g + \frac{1}{h_0} t$)

$$(10) \qquad \begin{aligned} v_r + \tfrac{1}{h_0} v &= \varepsilon \bar{g}(\varepsilon, Dw, Dk, Db) & r &= h_0 \\ \Delta v &= \varepsilon f(\varepsilon, D^2 w, D^2 k, D^2 b) & 1 &< r < h_0 \\ v &= h_0 b + \varepsilon q(\varepsilon, b) & r &= 1, \end{aligned}$$

and then solving the first equation in (9) for $h(\theta)$. One verifies that T is a contraction by using the Schauder's estimates for the coercive problem (9).

The domain perturbation method can be used to attack rather general problems involving boundary conditions of arbitrary order with variable coefficients, and with nonlinear equations of motion. Next we discuss how one should approach non-coercive problems of the type (9), in particular the problem (with $F =$ const < 0, $h = u(x, 1, z)$)

$$(11) \qquad \begin{aligned} u_y + F u_{xx} &= \varepsilon g(\varepsilon, Du, Dh, Db) & y &= 1 \\ \Delta u &= \varepsilon f(\varepsilon, D^2 u, D^2 h, D^2 b) & 0 &< y < 1 \\ u &= 0 & y &= 0. \end{aligned}$$

Since the problem (11) is non-coercive, the Schauder's estimates are not applicable here. One can derive estimates in the Sobolev spaces (see [**5**])

$$(12) \qquad \|u\|_{m+2} + \overline{\|u\|}_{m+2} \leq c\left(\|f\|_m + \|g\|_{m+1}\right),$$

with $\overline{\|\cdot\|}_m$ denoting the norm on $y = 1$. However, the estimates (12) are not sufficient for solving (11) by using iterations, because of the loss of smoothness when taking traces. To solve (12) we use the spaces Λ^m, which are defined as the closure of the trigonometric polynomials with respect to the norm $\|\cdot\|_m$, defined as follows:

$$\|u(x,y,z)\|_0 = \sum_{j,k=-\infty}^{\infty} \max_y |u_{jk}(y)|, \|u\|_m = \sum_{|\alpha|\leq m} \|D^\alpha u\|_0,$$

see [**6**] for details.

ACKNOWLEDGEMENTS: I wish to thank L. Nirenberg for posing the problems (1) and (2), and for many useful discussions. Also, I wish to thank Professors B. Kawohl and P.L. Lions for useful comments after the talk, and the organizers for the opportunity to present my work.

Department of Mathematical Sciences
University of Cincinnati, Cincinnati OH 45221

Research at MSRI supported in part by NSF Grant DMS 812079-05.

References

1. R. Hamilton, *The inverse function theorem of Nash and Moser*, Bull. Amer. Math. Soc. **7** (1982), 65–222.
2. D. Joseph, *Domain perturbations: the higher order theory of infinitesimal water waves*, Arch. Ration. Mech. Anal. **51** (1973), 295–303.
3. P. Korman, *Existence of solutions for a class of nonlinear non-coercive problems*, Communs Partial Diff. Eqns. **8** (1983), 819–846.
4. P. Korman, *Existence of periodic solutions for a class on nonlinear problems*, Nonlinear Analysis TMA **7** (1983), 873–879.
5. P. Korman, *On existence of periodic solutions for a class of nonlinear, non-coercive problems*, preprint.
6. P. Korman, *On existence of solutions for several classes of free boundary problems*, preprint.
7. P. Korman, *Existence of solutions for a class of quasilinear non-coercive problems*, Nonlinear Analysis TMA **12** (1986), 1471–1476.
8. D. Schaeffer, *A stability theorem for the obstacle problem*, Adv. in Math. **17** (1975), 34–47.
9. M. Shinbrot, *Water waves over periodic bottoms in three dimensions*, J. Inst. Math. Applic. **25** (1980), 367–385.

Diffusion-Reaction Systems in Neutron-Fission Reactors and Ecology

ANTHONY W. LEUNG

1. Introduction.

This article presents a few recent results in the application of parabolic and elliptic systems to nuclear reactor dynamics and ecology. Many open questions concerning the existence and stability of positive solutions remain unanswered. For example, the existence of nontrivial positive steady state solutions in the temperature dependent case (non-prompt feedback) in Section 2 is unknown.

The domain in R^d will always be open, connected and bounded. $H^{2+\ell}(\bar{\Omega})$, $0 < \ell < 1$, denotes the Banach space of all real-valued functions u, Hölder continuous on $\bar{\Omega}$ with all first and second derivatives also Hölder continuous in $\bar{\Omega}$ and with finite value for the norm $|u|_{\Omega}^{2+\ell}$ (as described in [**4**]). We assume $\delta\Omega \in H^{2+\ell}$. For any $T > 0$, let $\Omega_T = \Omega \times (0, T)$. $H^{2+\ell,1+\ell/2}(\bar{\Omega}_T)$ denotes the Banach space of all real-valued functions u having all derivatives of the form $D^\alpha D_t^r u$, $2r + |\alpha| \leq 2$, Hölder continuous on $\bar{\Omega}_T$ and having finite norm $|u|_{\Omega_T}^{(2+\ell)}$ (see [**4**]).

2. Multigroup, Fission-Reactor Problem.

For the reactor problem, we consider the multigroup diffusion equations describing neutron-flux fission. The functions $u_i(x)$ or $\tilde{u}_i(x,t)$, $i = 1, \dots, n$, $x = (x_1, \dots, x_d) \in \Omega$ are the neturon flux of the ith energy group (decreasing energy for increasing i). $T(x)$ is the core temperature above average coolant temperature. We have an elliptic system for the steady state problem with temperature feedback:

$$(2.1)\qquad \begin{aligned} &\Delta u_i + \sum_{j=1}^{n} H_{ij}(x,T)u_j = 0, \quad i = 1, \dots, n \\ &\Delta T - c(x)T + \sum_{j=1}^{n} G_j(x,T)u_j = 0, \quad \text{in } \Omega. \end{aligned}$$

Here $\Delta \equiv \sum_{i=1}^{d} \frac{\partial^2}{\partial x_i{}^2}$, $c(x)$ represents the cooling function which is positive and continuous in $\bar{\Omega}$. The functions H_{ij}, G_j are functions of space and temperature, continuous for $x \in \bar{\Omega}$, $T \geq 0$, and Lipschitz in T uniformly in x. In nuclear engineering terminologies, H_{ij} describe fission, removal, group-transfer and absorption "cross sections", taking into account the parameters of diffusion, neutron and energy release. Details can be found in [**10**]. The advisability of a temperature dependent nonlinear feedback model has been proposed and studied in [**3**] and [**12**].

Let $\tilde{h}_{ij} = \inf\{H_{ij}(x,T) \mid x \in \bar{\Omega}, T \geq 0\}$, $\bar{h}_{ij} = \sup\{H_{ij}(x,T) \mid x \in \bar{\Omega}, T \geq 0\}$, $i,j = 1,\ldots,n$. Similarly, define $\tilde{g}_i, \bar{g}_i$ to be the corresponding inf and sup of G_i, $i = 1,\ldots,n$. We will always assume that:

$$(C1) \qquad -\infty < \tilde{h}_{ij} \leq \bar{h}_{ij} < \infty, \quad 0 \leq \tilde{g}_i \leq \bar{g}_i < \infty$$

for $i,j = 1,\ldots,n$. For $i \neq j$, H_{ij} describe group transfer and fissions of neutrons from other groups; while H_{ii} is affected by control rods and absorptions. Consequently, we always assume that:

$$(C2) \qquad \begin{aligned} &0 \leq \tilde{h}_{ij} \leq \bar{h}_{ij} < \infty \text{ each } i,j \text{ with } i \neq j, \\ &-\infty < \tilde{h}_{ii} < \bar{h}_{ii} < \infty, \quad i = 1,\ldots,n. \end{aligned}$$

For convenience, we define $\tilde{H}$ and $\bar{H}$ to be $n \times n$ square matrices whose (i,j)-th entries are respectively $\tilde{h}_{ij}$ and $\bar{h}_{ij}$ for $1 \leq i \leq j \leq n$. Hypotheses (C1) and (C2) are always assumed in Theorem 2.1–2.3 below. The following theorem gives criteria when certain components of a nonnegative steady state must be identically zero. Let $\lambda = \lambda_1$ be the first eigenvalue for $\Delta u + \lambda u = 0$ in Ω, $u = 0$ on $\partial\Omega$; and let $\omega(x)$ be a positive eigenfunction in Ω.

THEOREM 2.1. *Suppose that the $n \times n$ matrix $\tilde{H} - \lambda_1 I$ has a square $m \times m$ submatrix $A = (a_{ij})$, (formed by deleting the $k_1, \ldots, k_{n-m}$-th rows and columns of $\tilde{H} - \lambda_1 I$, $1 \leq k_1 < k_2 < \cdots < k_{n-m} \leq n$, $1 \leq m \leq n$) with the properties that:*

(I) *$A\vec{c} > 0$ for some positive m vector $\vec{c} > 0$; and*
(II) *For each $i = 2,\ldots,m$ at least one of $a_{i1}, a_{i2}, \ldots, a_{ii-1}$ is positive.*

Then equation (2.1) has no solution $(\hat{u}_1(x), \ldots, \hat{u}_n(x), \hat{T}(x))$ satisfying all the following three properties

(i) *each component is $C^2(\Omega) \cap C^1(\bar{\Omega})$*

(ii) $\hat{u}_i(x) \geq 0$, $i = 1, \ldots, n$, $\hat{T}(x) \geq 0$ *in* $\bar{\Omega}$,

(iii) $\hat{u}_{s_i}(x) \not\equiv 0$ *in* $\bar{\Omega}$, *where* s_i *is the first positive integer not included in* $k_i, \ldots, k_{n-m}$.

(That is, any solution of (2.1) satisfying (i) and (ii) must have its s_1-th component identically zero. When $m = n$, $\tilde{H} - \lambda_1 I = A$. A vector $\vec{v} > 0$ means each component of $\vec{v}$ is positive.)

PROOF: Let $1 \leq s_1 < s_2 < \cdots < s_m \leq n$ be integers not in $\{k_1, \ldots, k_{n-m}\}$. Assume that $(\hat{u}_1(x), \ldots \hat{u}_n(x), \hat{T}(x))$ exists as described with properties (i)–(iii). We construct a family of lower bounds for $\hat{u}_{s_i}(x)$, $i = 1, \ldots, m$ by letting $u^{\delta}_{s_i}(x) = c_i \delta \omega(x)$, where $\vec{c} = (c_1, \ldots, c_m)$ is as described in (I) above,and $\delta > 0$ is a parameter for the family. Define $u^{\delta}_{k_i}(x) \equiv 0$, $i = 1, \ldots, n-m$, and $T^{\delta}(x) \equiv 0$ for $x \in \bar{\Omega}$. For each $i = 1, \ldots, m$, when $u_j(x) \geq u^{\delta}_j(x)$ for all $j \neq s_i$ and $T(x) \geq T^{\delta}(x)$, we have

$$
\begin{aligned}
&\Delta u^{\delta}_{s_i}(x) + H_{s_i s_i}(x, T(x)) u^{\delta}_{s_i}(x) + \sum_{j=1, j\neq s_i}^{n} H_{s_i j}(x, T(x)) u_j(x) \\
&\geq \left[-\lambda_1 + \tilde{h}_{s_i s_i}\right] c_i \delta \omega(x) + \sum_{j=1, j\neq i}^{m} \tilde{H}_{s_i s_j} u^{\delta}_{s_j} \qquad (2.2)\\
&= \left[-\lambda_1 + \tilde{h}_{s_i s_i}\right] c_i \delta \omega(x) + \sum_{j=1, j\neq i}^{m} a_{ij} c_j \delta \omega = (A\vec{c})_i \delta \omega(x) > 0.
\end{aligned}
$$

Using properties (i)–(iii) and the first equation in (2.1), we deduce from the sign of $\Delta \hat{u}_{s_i} - C \hat{u}_{s_i}(x)$ for large $C > 0$, and maximum principle that $\hat{u}_{s_i}(x) > 0$ for $x \in \Omega$. Similarly, we deduce that $\hat{u}_{s_i}(x) > 0$ or $\hat{u}_{s_i}(x) \equiv 0$ in Ω. However, assumption (II) implies that the zero function is not a solution of the s_i-th equation successively, and we must have $\hat{u}_{s_i}(x) > 0$, $x \in \Omega$, $i = 1, 2, \ldots, m$.

Let $J \equiv \{\tau \mid \hat{u}_{s_i}(x) > u^{\delta}_{s_i}(x),\ i = 1, \ldots, m \text{ for all } 0 \leq \delta \leq \tau,\ x \in \Omega\}$. The above shows that J is nonempty. If there is a l.u.b. $\bar{\delta}$ for J, we consider $\Delta\big(\hat{u}_{s_i} - u^{\bar{\delta}}_{s_i}\big) - P\big(\hat{u}_{s_i} - u^{\bar{\delta}}_{s_i}\big)$ for large positive P and use (2.1) and maximum principles at the boundary and interior to deduce $\hat{u}_{s_i} \equiv u^{\bar{\delta}}_{s_i}$ in $\bar{\Omega}$ for some i. For this i, we have the contradiction:

$$
\begin{aligned}
0 &= \Delta \hat{u}_{s_i} + H_{s_i s_i}(x, \hat{T}) \hat{u}_{s_i} + \sum_{j=1, j\neq s_i}^{n} H_{s_i j}(x, \hat{T}) \hat{u}_j \\
&= \Delta u^{\bar{\delta}}_{s_i} + H_{s_i s_i}(x, \hat{T}) u^{\bar{\delta}}_{s_i} + \sum_{j=1, j\neq s_i}^{n} H_{s_i j}(x, \hat{T}) \hat{u}_j > 0.
\end{aligned}
$$

The last inequality is due to (2.2) with $\delta = \bar{\delta}$. Thus J is unbounded. By letting $\delta \to +\infty$, we have $u^{\delta}_{s_i}(x) \to +\infty$, implying the nonexistence of $(\hat{u}_1(x), \ldots, \hat{u}_n(x), \hat{T}(x))$.

We have also the following analog for *blow-up* in the parabolic case.

THEOREM 2.2. *Suppose that the $n \times n$ matrix $\tilde{H} - \lambda_1 I$ has a square submatrix $A = (a_{ij})$, (formed as described in Theorem 2.1) with the property that:*

(I) *$A\vec{c} > 0$ for some positive m vector $\vec{c} > 0$.*

Let $(\tilde{u}_1(x,t), \ldots, \tilde{u}_n(x,t), \tilde{T}(x,t))$ be a solution of:

$$
(2.3)\qquad
\begin{aligned}
\frac{\partial \tilde{u}_i}{\partial t} &= \Delta \tilde{u}_i + \sum_{j=1}^{n} H_{ij}(x,\tilde{T})\tilde{u}_j, \quad i = 1, \ldots, n \\
\frac{\partial \tilde{T}}{\partial t} &= \Delta \tilde{T} - c(x)\tilde{T} + \sum_{j=1}^{n} G_j(x,\tilde{T})\tilde{u}_j
\end{aligned}
$$

for $(x,t) \in \Omega \times (0,\infty)$, with each component function in $C^2(\Omega \times (0,\infty)) \cap C^1(\bar{\Omega} \times [0,\infty))$ and initial conditions satisfying:

$$
\begin{aligned}
&\tilde{u}_i(x,0) > 0 \text{ for } x \in \Omega \text{ and} \\
&\frac{\partial \tilde{u}_i}{\partial \eta}(x,0) < 0 \text{ for } x \in \partial\Omega, \quad i = s_i, \ldots, s_m
\end{aligned}
$$

(Recall that $s_1, \ldots, s_m$ are those integers between 1 and n not included in $\{k_1, \ldots, k_{n-m}\}$ described in Theorem 2.1), and

$$
\hat{u}_i(x,t) \geq 0, \quad i = 1, \ldots, n \quad \tilde{T}(x,t) \geq 0 \qquad \text{for } (x,t) \in (\bar{\Omega} \times \{0\}) \cup (\delta\Omega \times (0,\infty)).
$$

Then $\tilde{u}_i(x,t) \to +\infty$ for all $x \in \Omega$, $i = s_1, \ldots, s_m$ as $t \to +\infty$.

(More precisely, for such x and i, we have $\tilde{u}_i(x,t) \geq \varepsilon_1 \omega(x) e^{\varepsilon_2 t}$ for some positive constants $\varepsilon_1, \varepsilon_2$ and all $t \in [0,\infty)$).

The last theorem is proved by means of differential inequalities, and so is the following theorem concerning *decay* to zero.

THEOREM 2.3. *Suppose that the $n \times n$ matrix $\bar{H} - \lambda_1 I$ has the property that:*

$$(\bar{H} - \lambda_1 I)\vec{y} < 0 \text{ for some positive } n \text{ vector } \vec{y} > 0.$$

Then equation (2.1) with boundary conditions

$$u_i(x) = 0, \quad i = 1, \dots, n, \quad T(x) = 0 \text{ for } x \in \delta\Omega$$

has the solution $(0, \dots, 0)$ *as the only solution with the properties that each component is in* $C^2(\Omega) \cap C^1(\bar{\Omega})$ *and nonnegative in* $\bar{\Omega}$.

The last three theorems do not illustrate any existence of positive steady states. When we make a simplifying assumption that feedback is "prompt" — the reaction coefficients are functions of the neutron fluxes u_i directly — we can find conditions for the existence of positive steady states in the next theorem. We consider the Dirichlet problem:

$$(2.4) \qquad \begin{aligned} &\Delta u_i + \sum_{j=1}^{n} H_{ij}(u_1, \dots, u_n) u_j = 0 \text{ in } \Omega, \quad i = 1, \dots, n \\ &u_i(x) = 0, \quad x \in \delta\Omega, \quad i = 1, \dots, n \end{aligned}$$

The functions $H_{ij}(u_1, \dots, u_n)$ are assumed to be in the class C^1. Define $h'_{ij} = \inf\{H_{ij}(u_1, \dots, u_n) \mid u_k \geq 0, k = 1, \dots, n\}$, $h''_{ij} = \sup\{H_{ij}(u_1, \dots, u_n) \mid u_k \geq 0, k = 1, \dots, n\}$. Analogous to (C1), (C2), we assume that:

$$(P1) \qquad \begin{aligned} &-\infty < h'_{ii} \leq h''_{ii} < \infty, \quad i = 1, \dots, n; \\ &0 < h'_{1j}, \quad j = 2, \dots, n; \\ &0 \leq h'_{ij} \leq h''_{ij} < \infty, \quad 1 \leq i, \quad j \leq n, \quad i \neq j; \\ &\text{for each } i = 2, \dots, n \text{ at least one of } h'_{i1}, h'_{i2}, \dots, h'_{ii-1} \text{ is positive.} \end{aligned}$$

When u_1 is large, fast neutrons have to be removed by activating control rods to prevent overheating, we thus assume that

(P2) There exist positive constants p and $U^\star$ such that $H_{11}(u_1, \dots, u_n) \leq -p$ for all $u_1 \geq U^\star$ in $\{(u_1, \dots, u_n) \mid u_k \geq 0, \quad k = 1, \dots, n\}$.

For simplicity we also assume that:

(P3) $H_{11}(u_1, \dots, u_n)$ is continuously differentiable with respect to $u_2, \dots, u_n$ in $\{(u_1, \dots, u_n) \mid u_k \geq 0, k = 1, \dots, n\}$ with $\left| \dfrac{\partial H_{11}}{\partial u_j} \right| < K$ for $j = 2, \dots, n$, where K is some positive constant.

THEOREM 2.4. *Let* $\hat{H} = (\hat{H}_{ij})$ *be an* $n \times n$ *matrix:*

$$\hat{H}_{ij} = \begin{cases} h''_{ij} & \text{for all } (i,j) \neq (1,1) \\ -p & \text{for } (i,j) = (1,1). \end{cases}$$

Suppose that (P1)–(P3) are satisfied. Further, let

$$H_{11}(0,\dots,0) > \lambda_1, \text{ and } \hat{H}\vec{c} \leq 0 \tag{2.5}$$

for some positive n vector $\vec{c} > 0$. Then the Dirichlet problem (2.4) has a solution $(\hat{u}_1(x),\dots,\hat{u}_n(x))$ with components in $H^{2+\ell}(\bar{\Omega})$ and $\hat{u}_i(x) > 0$ in Ω, $i = 1,\dots,n$.

($\hat{H}\vec{c} \leq 0$ means each of its components is ≤ 0.)

PROOF: By (2.5), there exists a small $\alpha > 0$ so that $H_{11}(u, 0,\dots,0) > \lambda_1$ for $0 \leq u < \alpha$. Choose $0 < \varepsilon < \min\{\alpha, h'_{12}/K,\dots,h'_{1n}/K\}$. For each $i = 3,\dots,n$, let $1 \leq \hat{\imath} \leq i-1$ so that $h_{i\hat{\imath}} > 0$ (such $\hat{\imath}$ exists by (P1)). Choose $0 < \delta_2 < h'_{21}\varepsilon/|h'_{22} - \lambda_1|$, $\delta_1 = \varepsilon$, $0 < \delta_i < h'_{i\hat{\imath}}\delta_{\hat{\imath}}/|h'_{ii} - \lambda_1|$, for $i = 3,\dots,n$. (For $i = 2,\dots,n$, if $h'_{ii} - \lambda_1 = 0$ let $\delta_i > 0$ be arbitrary). Define $v_1(x) = \varepsilon\omega(x)$, $v_i(x) = \delta_i\omega(x)$, $i = 2,\dots,n$ for $x \in \bar{\Omega}$; and also define $w_i(x) \equiv c_i$ for $i = 1,\dots,n$ in $\bar{\Omega}$ where $\vec{c} = (c_1,\dots,c_n)$ is given in (2.5). From (P1) to (P3) and (2.5) we can deduce that v_i, w_i are, respectively, lower and upper solutions in the sense that:

$$\begin{aligned} &\Delta v_i(x) + \sum_{j=1}^{n} H_{ij}(z_1,\dots,z_{i-1},v_i,z_{i+1},\dots,z_n)v_i \geq 0 \\ &\Delta w_i(x) + \sum_{j=1}^{n} H_{ij}(z_1,\dots,z_{i-1},w_i,z_{i+1},\dots,z_n)w_i \leq 0 \end{aligned} \tag{2.6}$$

for each $i = 1,\dots,n$, all z_j satisfying $v_j(x) \leq z_j \leq w_j(x)$, $j \neq 1$, $x \in \Omega$. By [**13**], (2.6) implies that there exists a solution $(\hat{u}_1(x),\dots,\hat{u}_n(x))$ as described in the statement of the theorem with $v_i(x) \leq \hat{u}_i(x) \leq w_i(x)$, $i = 1,\dots,n$, $x \in \bar{\Omega}$. From the construction of v_i, we have $\hat{u}_i(x) > 0$ in Ω, $i = 1,\dots,n$.

The details for the proof of all the above theorems can be found in [**11**].

3. Ecological, Population problem.

In this section, we consider some recent results concerning population reaction-diffusion in ecology. In many biological studies, it was found that one should include the effect of density-dependent diffusion rates, drift terms and spatially varying growth rates. We consider the following initial

Dirichlet boundary value problem for n competing species (note conditions (3.2)):

$$
\begin{aligned}
&\frac{\partial u_i}{\partial t} = \operatorname{div}(\sigma_i(u_i)\nabla u_i) + u_i\big[a_i(x) + f_i(u_1,\dots,u_n)\big] \\
(3.1)\quad &\text{for } (x,t) = (x_1,\dots,x_d,t) \in \Omega \times (0,T], \quad T>0, \quad i=1,\dots,n; \\
&u_i(x,0) = \phi_i(x), \quad x \in \bar{\Omega}; \quad u_i(x,t) = \Phi_i(x), \quad (x,t) \in \delta\Omega \times [0,T].
\end{aligned}
$$

The functions $f_i : R^n \to R$ have Hölder continuous partial derivatives up to second order in compact sets, $i = 1,\dots,n$, and further satisfy:

$$
\begin{aligned}
&\frac{\partial f_i}{\partial u_j} < 0, \quad i,j = 1,\dots,n \text{ in } \{(u_1,\dots,u_n) \mid u_i \geq 0, \quad i=1,\dots,n\}, \\
(3.2)\quad &f_i(0,0) = 0, \\
&\sup_{s\geq 0} \frac{\partial f_i}{\partial u_i}(0,\dots,0,s,0,\dots,0) \stackrel{\text{def}}{=} r_i < 0,
\end{aligned}
$$

where $s \geq 0$ occurs in the ith component, $i = 1,\dots,n$.

The instrinsic growth rate function of the ith species, $a_i(x)$, is in $H^{1+\ell}(\bar{\Omega})$ and

$$
(3.3)\qquad a_i(x) \geq 0 \text{ in } \bar{\Omega}, \quad i = 1,\dots,n.
$$

The diffusivity function is nondecreasing, satisfying:

$$
(3.4)\quad \sigma_i(0) > 0, \quad \sigma_i(s) \text{ in } H^{\ell}(R), \quad \sigma_i'(s) \geq 0 \text{ in } [0,\infty), \quad \sigma_i''(s) \text{ is continuous in } [0,\infty), i = 1,\dots,n.
$$

These assumptions and equations are biologically plausible and sufficiently general, while the smoothness conditions are made convenient enough so that excessive technicalities do not arise. The initial-boundary value problem (3.1) under appropriate smoothness conditions for the initial boundary functions possess a solution in $H^{2+\ell,1+\ell/2}(\bar{\Omega}_T)$, each $T > 0$.

We assume that the intrinsic growth rate, $a_k(x)$, of a particular kth species is locally high in a subdomain Ω' of Ω. We will obtain a criterion on $a_k(x)$ which ensures that the population $u_k(x,t)$ will be bounded below by positive constants in compact subsets of Ω' for all $t > 0$. Such criteria can thus be interpreted as a survival condition for the kth species. Note that the growth rate does not have to be large in the entire domain.

THEOREM 3.1. *Let k be an integer, $1 \leq k \leq n$. Let $u = (u_1, \ldots, u_n)$ be a solution of (3.1) in the class $H^{2+\ell, 1+\ell/2}(\bar{\Omega}_T)$, $T > 0$, initially satisfying:*

$$0 \leq u_i(x,0) \leq b_i, \quad x \in \bar{\Omega}, \quad i = 1, \ldots, n$$

where b_i are positive numbers satisfying $b_i \geq \left|r_i^{-1}\right| \cdot \max\{a_i(x) \mid x \in \bar{\Omega}\}$. Suppose that there exists a subdomain $\Omega' \subseteq \Omega$ (with principal eigenvalues λ', i.e. $\lambda = \lambda' > 0$ is the first eigenvalue for the problem $\Delta\phi + \lambda\phi = 0$ in Ω', $\phi = 0$ on $\delta\Omega'$) with the properties:

(3.5)

$$0 < u_k(x,0), \quad x \in \bar{\Omega}'; a_k(x) - \sigma_k(0)\lambda' + f_k(b_1, \ldots, b_{k-1}, 0, b_{k+1}, \ldots, b_n) > 0$$

for all $x \in \bar{\Omega}'$. Then the solution u satisfies:

$$0 < u_k(x,t) \text{ for } (x,t) \in \Omega' \times [0,T].$$

Moreover, $u_k(x,t) \geq \delta > 0$ for all x in any compact set contained in Ω', $0 \leq t \leq T$ (where δ is some constant depending on the compact set, independent of T); and

$$0 \leq u_i(x,t) \leq b_i \text{ for } (x,t) \in \bar{\Omega}_T, i = 1, \ldots, n.$$

PROOF: Let $\theta(x)$ be a positive eigenfunction in Ω', associated with the principal eigenvalue $\lambda = \lambda'$. Define $v_i(x,t) \equiv 0$ in $\bar{\Omega}_T$ for $i \neq k$, $1 \leq i \leq n$; and

$$v_k(x,t) \begin{cases} \varepsilon\theta(x) & \text{if } x \in \Omega' \\ 0 & \text{if } x \in \bar{\Omega} \backslash \Omega' \end{cases}$$

in $\bar{\Omega}_T$. (Here ε is a sufficiently small positive constant to be determined). Define $w_i(x,t) \equiv b_i$ in $\bar{\Omega}_T$, $i = 1, \ldots, n$. For $i \neq k$, clearly we have

$$\text{(3.6)} \quad \operatorname{div}(\sigma_i(v_i)\nabla v_i) + v_i\big[a_i(x) + f_i(w_1, \ldots, w_{i-1}, v_i, w_{i+1}, \ldots, w_n)\big] - \frac{\partial v_i}{\partial t} \geq 0$$

for $(x,t) \in \Omega \times [0,T]$. From the assumptions on σ_i and (3.5), one can also obtain inequality (3.6) for $i = k$, $(x,t) \in (\Omega \backslash \delta\Omega') \times [0,T]$, provided $\varepsilon > 0$ is sufficiently small. From (3.2) we can also deduce that for $i = 1, \ldots, n$:

$$\text{(3.7)} \quad \operatorname{div}(\sigma_i(w_i)\nabla w_i) + w_i\big[a_i(x) + f_i(v_1, \ldots, v_{i-1}, w_i, v_{i+1}, \ldots, v_n)\big] - \frac{\partial w_i}{\partial t} \leq 0$$

for $(x,t) \in \Omega \times [0,T]$. From inequalities (3.6), (3.7) and initial conditions, one can deduce that:

$$0 \le v_i \le u_i(x,t) \le w_i = b_i, \quad i = 1,\dots,n$$

for $(x,t) \in \bar{\Omega} \times [0,T]$. From the definition of v_k, we obtain the strict positivity of u_k in compact subsets of Ω' as stated. Details of proof can be found in [**9**].

The next theorem considers criteria for extinction or decay of the particular kth species. Consider equations (3.1) under conditions (3.2)–(3.4), with the special homogeneous boundary condition on the kth component.

$$\Phi_k(x) \equiv 0, \quad x \in \delta\Omega. \tag{3.8}$$

(Moreover, we assume compatibility conditions $\phi_i(x) = \Phi_i(x)$ if $x \in \delta\Omega$, $\phi_i \in H^{2+\ell}(\bar{\Omega})$, and

$$\left\{\operatorname{div}(\sigma_i(\phi_i)\nabla\phi_i) + \phi_i\left[a_i(x) + f_i(\phi_1,\dots,\phi_n)\right]\right\}\Big|_{x\in\delta\Omega} = 0 \text{ for } i = 1,\dots,n.)$$

THEOREM 3.2. *Let k be an integer, $1 \le k \le n$. Suppose that*

$$a_k(x) < \sigma_k(0)\lambda_1 \text{ for all } x \in \bar{\Omega},$$

(where λ_1 is the principal eigenvalue for Ω). Let $C_i > 0$, $i = 1,\dots,n$ be such that for $x \in \bar{\Omega}$

$$a_i(x) + f_i(0,\dots,0,C_i,0,\dots,0) \le 0$$

(here C_i appears in the ith component), and $\bar{C}_i = \max\{C_i, \sup_{x\in\bar{\Omega}} \phi_i(x)\}$. Then there is a constant $q > 0$ so that the property:

$$\sigma_k'(s) \le q, \text{ for all } 0 \le s \le \bar{C}_k$$

implies that any solution $(u_1,\dots,u_n)$ of (3.1) with homogeneous condition (3.8) and each component in $H^{2+\ell,1+\ell/2}(\bar{\Omega}_T)$, $T > 0$ must satisfy:

$$0 \le u_k(x,t) \le Ke^{-\varepsilon t} \text{ in } \bar{\Omega}_T$$

where K, ε are positive constants indepedent of T. Moreover, we have

$$0 \le u_i(x,t) \le \bar{C}_i, \quad i = 1,\dots,n \text{ in } \bar{\Omega}_T.$$

REMARK: C_i exists by hypothesis (3.2); the size of q can be found in terms of the principal eigenfunction of a domain $\hat{\Omega} \supset \bar{\Omega}$. Details can be found in [9].

For situations more restrictive than (3.1), positive steady states and their stabilities are found in [6] and [7]. We consider

$$
\text{(3.9)} \qquad \begin{aligned} \frac{\partial u_1}{\partial t} &= \sigma_1 \Delta u_1 + u_1\big[a + f_1(u_1, u_2)\big] \\ \frac{\partial u_2}{\partial t} &= \sigma_2 \Delta u_2 + u_2\big[b + f_2(u_1, u_2)\big] \end{aligned} \qquad x \in \Omega, \quad t > 0
$$

where a, b, σ_1, σ_2 are positive constants, $f_i : R^2 \to R$ have Hölder continuous partial derivatives up to second order in compact sets, $i = 1,2$. Further

$$
\text{(3.10)} \qquad f_i(0,0) = 0, \quad i = 1,2; \quad \frac{\partial f_i}{\partial u_j} < 0, \text{ each } i,j = 1,2.
$$

for (u_1, u_2) in the first open quadrant.

THEOREM 3.3. *Suppose that* $a > \sigma_1 \lambda_1$, $b > \sigma_2 \lambda_1$ *and there are positive constants* $\bar{k}_1, \bar{k}_2$ *such that:*

$$
\begin{aligned} &a - \sigma_1 \lambda_1 + f_1(0, \bar{k}_2) > 0, \quad && b + f_2(0, \bar{k}_2) < 0, \\ &a + f_1(\bar{k}_1, 0) < 0, && b - \sigma_2 \lambda_1 + f_2(\bar{k}_1, 0) > 0, \end{aligned}
$$

then the system (3.9) under Dirichlet boundary conditions $u_1(x,t) = 0$, $x \in \delta\Omega$, $t \geq 0$, $i = 1,2$, *has a steady state solution* $(\bar{u}_1(x), \bar{u}_2(x))$ *with* $\bar{u}_i(x) \in H^{2+\ell}(\bar{\Omega})$, $\bar{u}_i(x) > 0$ *for* $x \in \Omega$, $i = 1,2$.

The stability of such a steady state is considered by the following theorem.

THEOREM 3.4. *Under the conditions of Theorem 3.3, suppose further that the steady state solution satisfies* $\frac{\partial \bar{u}_i}{\partial \eta} < 0$ *on* $\delta\Omega$, $i = 1,2$, *and*

$$
\sup_{x \in \Omega} \left| \frac{\bar{u}_i(x)}{\bar{u}_j(x)} \cdot \frac{(\partial f_j / \partial u_i)(\bar{u}_1(x), \bar{u}_2(x))}{(\partial f_j / \partial u_j)(\bar{u}_1(x), \bar{u}_2(x))} \right| < \inf_{x \in \Omega} \left| \frac{\bar{u}_i(x)}{\bar{u}_j(x)} \cdot \frac{(\partial f_i / \partial u_i)(\bar{u}_1(x), \bar{u}_2(x))}{(\partial f_i / \partial u_j)(\bar{u}_1(x), \bar{u}_2(x))} \right| < \infty
$$

for each $1 \leq i, j \leq 2$, $i \neq j$, *then* $(\bar{u}_1(x), \bar{u}_2(x))$ *is asymptotically stable.*

(Here asymptotic stability is interpreted to mean that for any solution $(u_1(x,t), u_2(x,t))$ with $u_i \in H^{2+\ell, 1+\ell/2}(\bar{\Omega} \times [0,T])$ each $T > 0$, $i = 1,2$ of system (3.9) with boundary conditions $u_i(x,t) = 0$ and initial conditions $u_i(x,0)$ whose values and first partial derivatives are close enough to $\bar{u}_i(x)$ respectively for all $x \in \bar{\Omega}$, $i = 1,2$, one has $u_i(x,t) \to \bar{u}_i(x)$ uniformly as $t \to +\infty$, $i = 1,2$.)

Theorem 3.3 is proved as Theorem 2.4 by using a Theorem in [**13**]. Theorem 3.4 is proved by using differential inequalities for parabolic systems. The upper and lower solutions are constructed by making use of the steady state solutions. However, the homogeneous boundary conditions make it technically cumbersome. For example, the upper solution for the second equation in (3.9) is constructed in the form $\left[1 + (\delta - \varepsilon\bar{u}_2(x))e^{-\sigma t}\right]\bar{u}_2(x)$, where $\delta, \varepsilon, \sigma$ are appropriately chosen positive constants. Details and examples can be found in [**7**].

For prey-predator systems with nonlinear boundary conditions some results are obtained in [**8**]. We consider

$$\begin{aligned} \frac{\partial u_1}{\partial t} &= \sigma_1 \Delta u_1 + u_1\left[a + f_1(u_1,u_2)\right] \\ \frac{\partial u_2}{\partial t} &= \sigma_2 \Delta u_2 + u_2\left[-r + f_2(u_1,u_2)\right] \end{aligned} \qquad x \in \Omega, \quad t > 0 \tag{3.11}$$

where a, r, σ_1, σ_2 are positive constants, $f_i : R^2 \to R$ have Hölder continuous partial derivatives up to second order in compact sets, $i = 1,2$. Further

$$\begin{gathered} f_i(0,0) = 0, \quad i = 1,2; \\ \frac{\partial f_1}{\partial u_1} < 0, \quad \frac{\partial f_1}{\partial u_2} < 0, \quad \frac{\partial f_2}{\partial u_1} > 0, \quad \frac{\partial f_2}{\partial u_2} < 0 \end{gathered} \tag{3.12}$$

for (u_1, u_2) in the first open quadrant, and for each $m \geq 0$,

$$\begin{aligned} \lim_{u_2 \to +\infty} f_2(m, u_2) &= -\infty, \\ \lim_{u_1 \to +\infty} f_2(u_1, m) &= +\infty, \\ \lim_{u_1 \to +\infty} f_1(u_1, 0) &= -\infty. \end{aligned} \tag{3.13}$$

In realistic ecological situations, one expects that the boundary conditions are likely to be coupled, mixed and nonlinear. For example, the predator may be under control at the boundary, while prey cannot move across it. Diffusion of predators inward across the boundary may be further enhanced by higher concentrations of prey and predator. We consider a simple boundary condition:

$$\frac{\partial u_1}{\partial \eta} = 0, \quad \frac{\partial u_2}{\partial \eta} - p(u_1) - q(u_2) = 0 \text{ on } \delta\Omega \tag{3.14}$$

with p,q having Hölder continuous first derivatives in compact subsets of $[0,\infty)$, and having the following monotonic and concavity properties:

(3.15)

$$p(0) \geq 0, \quad p'(u) > 0, p''(u) < 0 \text{ for all } u > 0, \quad \lim_{u\to+\infty} p(u) = P < \infty;$$

$$q(0) \geq 0, \quad q'(u) > 0, q''(u) < 0 \text{ for all } u > 0, \quad \lim_{u\to+\infty} q(u) = Q < \infty;$$

$$p(0) + q(0) > 0.$$

THEOREM 3.5. *Suppose* $\left|\frac{\partial f_2}{\partial u_2}\right|$ *is uniformly bounded in the first open quadrant, and there is an* $R > 0$*, such that for* $\alpha, \beta > R$:

$$\left|\frac{\partial f_1}{\partial u_2}(0,\beta)\frac{\partial f_2}{\partial u_1}(\alpha,\beta)\right| \leq \rho \left|\frac{\partial f_2}{\partial u_2}(\alpha,\beta)\frac{\partial f_1}{\partial u_1}(\alpha,0)\right|$$

for a constant $\rho < 1$*. Then for* $a > 0$ *sufficiently large the system (3.11) with boundary conditions (3.14) has a steady state solution* $(\bar{u}_1(x), \bar{u}_2(x))$*, such that* $\bar{u}_i(x) \in C^1(\bar{\Omega}) \cap C^2(\Omega)$ *and* $\bar{u}_i(x) > 0$ *for all* $x \in \bar{\Omega}$*. (Here, we assume properties (3.12), (3.13) and (3.15) for* f_i, p, q*).*

Further study of the stability of the steady state above as $t \to +\infty$ in (3.11) can be found in [8]. Results are similar to Theorem 3.4, although the interaction and boundary conditions are changed.

Department of Mathematical Sciences, University of Cincinnati, Cincinnati OH 45221.

REFERENCES

1. G. S. Chen and A. Leung, *Nonlinear multigroup neutron-flux systems: blow-up, decay, and steady states*, J. Math. Phys. **26** (1985), 1553–1559.
2. P. C. Fife and M. M. Tang, *Comparison principles for reaction-diffusion systems: irregular comparison functions and applications to questions of stability and speed of propagation of disturbances*, J. Diff. Eqs. **40** (1981), 168–185.
3. W. E. Kastenberg, *A stability criterion for space-dependent nuclear-reactor systems with variable temperature feedback*, Nuc. Sci. Eng. **37** (1969), 19–29.
4. O. A. Ladyzhenskaya, V. A. Solonikov and N. N. Ural'ceva, "Linear and Quasilinear Equations of Parabolic Type," Trans. of Math. Monographs 23, AMS, Providence, 1968.
5. A. Leung and D. Clark, *Bifurcations and large-time asymptotic behavior for prey-predator reaction-diffusion equations with Dirichlet boundary data*, J. Diff. Eqs. **35** (1980), 113–127.
6. A. Leung, *Equilibria and stabilities for competing-species reaction-diffusion equations with Dirichlet boundary data*, Math. Anal. Appl. **73** (1980), 204–218.
7. A. Leung, *Stabilities for equilibria of competing-species reaction-diffusion equations with homogeneous Dirichlet conditions*, Funk. Ekv. (Ser. Interna.) **24** (1981), 201–210.
8. A. Leung, *A semilinear reaction-diffusion prey-predator system with nonlinear coupled boundary conditions: equilibrium and stability*, Indiana Univ. Math. J. **31** (1982), 223–241.
9. A. Leung, *Nonlinear density-dependent diffusion for competing species interactions: large-time asymptotic behavior*, Proc. of the Edinburgh Math. Soc. **27** (1984), 131–144.
10. A. Leung and G. S. Chen, *Positive solutions for temperature-dependent two group neutron flux equations: equilibria and stabilities*, SIAM Math. Anal. **15** (1984), 490–499.
11. A. Leung and G. S. Chen, *Elliptic and parabolic systems for neutron fission and diffusion*, J. Math. Anal. Appl. **120** (1986), 655–699.
12. I. Stakgold and L. Payne, *Nonlinear problems in nuclear reactor analysis*, in "Proc. Conference on Nonlinear Problems in Physical Sciences and Biology," Springer Lecture Notes in Math, vol. 322, Springer-Verlag, New York, 1972, pp. 298–307.
13. L. Y. Tsai, *Nonlinear boundary value problems for systems of second order elliptic differential equations*, Bull. Inst. Math. Acad. Sinica **5** (1977), 157–165.

Numerical Searches for Ground State Solutions of a Modified Capillary Equation and for Solutions of the Charge Balance Equation

HOWARD A. LEVINE

In this talk we shall present some numerical results for solutions of two degenerate elliptic boundary value problems. While quite different in origin they have some common features which make their numerical solution quite interesting.

The first problem is

$$\vec{\nabla} \cdot \left((1 + |\vec{\nabla} u|^2)^{-1/2} \vec{\nabla} u \right) + f(u) = 0 \qquad x \in \mathbb{R}^n \tag{1.1}$$

$$\lim_{|x| \to \infty} u(x) = 0 \tag{1.2}$$

where f is one of

$$f_1(u) = \lambda |u|^{q-1} u - u$$

or

$$f_2(u) = \lambda |u|^q - u$$

with $\lambda > 0$, $q > 1$ and where u is a real valued function. We call (1.1) a modified capillary equation. Equation (1.1) is degenerate at $|\vec{\nabla} u| = \infty$. Concus and Finn have studied (1.1) when $\lambda = 0$ (see [**2**]). We present some numerical solutions of (1.1)–(1.2) in the case $\lambda > 0$.

A similar degenerate problem occurs in plasma physics. Among the transport equations describing the particle dynamics in a tandem mirror magnetic fusion apparatus, is the so-called charge balance equation. It is basically a conservation law balancing the radial loss of electrons against axial loss [**5**,**7**]. The z or axially averaged form of this equation may be written, after appropriate scaling,

$$\frac{d}{dy}(G(y, \frac{d\phi}{dy})) - H(y, \phi(y)) = 0, \quad 0 < y < Y \tag{1.3}$$

where the boundary conditions are

$$\phi(Y) = \phi_1 \tag{1.4a}$$

$$\phi \text{ is analytic at } y = 0, \tag{1.4b}$$

and where y denotes the square of the radial variable. The second boundary condition is necessary because G has a geometric degeneracy associated with radial problems, namely

$$\lim_{y\to 0^+} \frac{G(y,\phi')}{y} \neq 0 \tag{1.5}$$

for $\phi' \neq 0$.

It is also assumed that

$$\frac{\partial H}{\partial \phi} > 0. \tag{1.6a}$$

In addition to the degeneracy in (1.5), G also has the profile given in Figure 12. Solutions of (1.3), (1.4a,b) are required to be such that except at a finite number of values of $\phi'(y)$,

$$\frac{\partial G}{\partial \phi'}(y,\phi'(y)) > 0, \tag{1.6b}$$

an inequality based on physical considerations.

The study of (1.1)–(1.2) was undertaken by Ni and Serrin [**8**], who showed that if $n > 2$, there are no positive radial solutions (ground states) if

$$q > q(n) \equiv (n+2)/(n-2).$$

Franchi, Lanconelli and Serrin [**4**] also showed that if $1 < q < q(n)$ and

$$\lambda < \lambda(q) \equiv \left[\frac{1}{2}\frac{(q-1)}{(q+1)}\right]^{\frac{1}{2}(q-1)},$$

then there are no ground states (Figure 1). The arguments are based on the Pohozaev identity, the necessary condition for solvability being

$$\max_{u\in[0,\bar{u}]} \left| \int_0^u f(\eta)\, d\eta \right| < 1$$

where $\bar{u}$ is the largest root of $\int_0^u f(\eta)\, d\eta$.

In consequence of these results, Serrin asked us to numerically investigate the radial case when $q < q(n)$ and $\lambda > \lambda(q)$. Thomas Evers, a student of mine, carried out these computations in his MS. paper [**3**], which we present here in summary form. (Some of the numerical results presented here were obtained after Evers completed his work by John Axtell and the author using modifications and improvements of Evers's code.)

The radial problem is

(1.7)
$$\left[(u'(r))^2+1\right]^{-3/2}u''(r)+(n-1)r^{-1}\left[1+(u'(r))^2\right]^{-1/2}u'(r)+f(u(r))=0$$
$$0<r<\infty$$

with

$$u'(0)=u(\infty)=0. \tag{1.8}$$

An initial attempt to solve (1.7)–(1.8) via the shooting method by writing (1.7) in the form of a two-dimensional system with $y_1(r) = u(r)$, $y_2(r) = u'(r)$ failed because, as we shall see numerically, solutions of the initial value problem for (1.7) are not always global.

Therefore, following Concus and Finn [**2**] in the case $\lambda = 0$, arc length was introduced as a parameter and the equation (1.7) recast in the form

$$\begin{aligned}\frac{dr}{ds}&=\cos(\psi(s))\\ \frac{d\psi}{ds}&=-\frac{(n-1)}{r(s)}\sin(\psi(s))-f(u(s))\\ \frac{du}{ds}&=\sin(\psi(s))\end{aligned} \tag{1.9}$$

with the initial conditions

$$\begin{aligned}r(0)&=0\\ \psi(0)&=0\\ u(0)&=u_0 \quad (u_0 \text{ is the shooting parameter})\end{aligned} \tag{1.10}$$

where ψ is the angle between the r axis and the curve (Figure 2). A solution of (1.9) gives rise to u as a single valued differentiable function of r if and only if ψ satisfies

$$|\psi(s)|<\pi/2.$$

The numerical integration of (1.9) is complicated by the presence of a zero divisor. Therefore the following approximations were used near $s = 0$:

$$\begin{aligned}r&=s\\ u(r)&=u(s)=u_0+\tfrac{1}{2}u_0''s^2\end{aligned}$$

for $0 < s < h$ where

$$u_0'' = -f(u_0)/n \tag{1.11}$$

which is found from (1.7). One then solves (1.9) beginning at $s = h$ with starting values determined as follows:

$$\begin{aligned} r(h) &= h \\ \psi(h) &= \sin^{-1}(u_0''h) \\ u(h) &= u_0 + \tfrac{1}{2}u_0''h^2. \end{aligned} \tag{1.12}$$

Naturally, improved starting values can be found by taking more terms in the Taylor series about $r = 0$.

The triple $(r(s), \psi(s), u(s))$ can then be found numerically and u plotted against r.

In order to verify that we have a solution of (1.7)–(1.8), various subcases of the Pohozaev identity were employed. For example, if (1.7) is multiplied by $u'(r)r^{n-1}$, the resulting expression integrated by parts and referred to the variable s, then any solution of (1.7)–(1.8) must satisfy

$$\begin{aligned} (n-1)\int_0^\infty & r^{n-2}(s)\cos(\psi(s))F(u(s))\,ds \\ &= \lim_{s\to\infty}\left[(n-1)\int_0^s r^{n-2}(\eta)\,d\eta - r^{n-1}(s)\right]. \end{aligned} \tag{1.13}$$

We also must have

$$\int_0^\infty r^{n-1}(s)u(s)f(u(s))\cos(\psi(s))\,ds = \int_0^\infty r^{n-1}(s)\sin^2(\psi(s))\,ds \tag{1.14}$$

and

$$\begin{aligned} \int_0^\infty r^{n-1}(s)f(u(s))\cos(\psi(s))\,ds &= \lim_{s\to\infty} r^{n-1}(s)\sin(\psi(s)) \\ &= 0. \end{aligned} \tag{1.15}$$

We summarize the essential observations of the calculations. Evers took $n = 2$ in all of his computations.

(1) The curves in some of the Figures do not appear to start at $R = 0$. This is due to the failure of the code to include the solution at grid

values in $[0, h]$. Since on this interval, the solution is approximately given by

$$u(r) = u_0 - \tfrac{1}{2} f(u_0) r^2 / n;$$

we did not feel the need to include the graph of u on this interval.

(2) In Figures 4A,B we see that a multivalued curve starts at the point $(0, 2)$ and cascades down toward the $u = 0$ axis. In this computation $q = 2$ and $f(u) = \lambda |u|^2 - u$. After crossing the axis, the curve becomes a spiral in the fourth quadrant. Detail plots near $r = 0$ are given in the insets of each panel.

Several observations are to be noted. First, with $f(u) = \lambda u|u| - u$, such spirals do not occur (Figure 5). Secondly, we see that as the dimension increases, the number of oscillations near $r = 0$ decreases. (For this value of λ, with $n = 4$, the curve fails to be multivalued near $r = 0$ (plot not shown).) Third, between $n = 2.9$ and $n = 3$, the spiral does not occur at the indicated values of (λ, u_0). However, by increasing λ (Figure 4B.6), the spiral reappears. This is consistent with Serrin's observation [**10**] that the first value of u_0 beyond which oscillation occurs near $r = 0$ increases with dimension.

The spiral will not appear in the case that f has three real roots, $0, \pm\lambda^{-1/(q-1)}$. In the case that f has only the real roots $0, \lambda^{-1/(q-1)}$, if $f(-u) = -f(u)$, we may predict the spiral from (1.9). When $n = 1$, the system decouples and, from the second and third equations, we see that

$$-\cos(\psi(s)) + F(u(s)) = -1 + F(u_0)$$

where $F'(u) = f(u)$, $F(0) = 0$. This conservation law leads to a spiral in the (r, u) plane. For example, with $f(u) = \lambda u^2 - u$ and $u_0 = -1$, $\psi_0 = 0$, we have

$$H(-u) = \tfrac{1}{3}\lambda + \tfrac{3}{2} - \cos\psi$$

where $H(x) \equiv \tfrac{1}{3}\lambda x^3 + \tfrac{1}{2}x^2$ for $x \geq 0$. Then

$$-H^{-1}(\tfrac{1}{3}(\lambda + \tfrac{15}{2})) \leq u \leq -H^{-1}(\tfrac{1}{3}(\lambda + \tfrac{3}{2})) \leq 0.$$

See Figure 4C, first panel. With $h(x) \equiv H'(x)$, we find that

$$r(\psi) = \int_{\psi}^{0} \frac{\cos\eta \, d\eta}{h(H^{-1}(\tfrac{1}{3}(\lambda + \tfrac{9}{2}) - \cos\eta))}.$$

We see that, as ψ decreases, $r(\psi)$ oscillates as it becomes larger, i.e. the curve $(r(s), u(s))$ is a spiral. The pitch of the spiral is controlled by λ, becoming tighter as λ increases (Figures 4E.1, 4C.1, and 4E.2). Also, because the integrand is periodic, $r(\psi) \to +\infty$ as $\psi \to -\infty$.

When $n > 1$, we cannot solve (1.9) explicitly. However, we observe that as $r \to \infty$, we obtain the same system asymptotically as that for $n = 1$. Thus we expect spiral-like solutions for $n > 1$ and $u_0 < 0$. (See Figures 4C.2–4C.6 and Figures 4D.2–4D.6) However, as n increases, the angle the axis of the spiral makes with the r axis decreases from zero to a limiting angle of $-\frac{\pi}{2}$. Again the larger λ is, the more tightly wound the spiral becomes (See Figures 4E.2–4E.6). Also, for fixed λ, the spiral moves off to $r = \infty$ as n increases. Further experimentation showed that for fixed $u_0 (= -1)$, the limiting angle appeared to be achieved for some finite value of n and that the distance between the spiral axis and the u axis increased like

$$R(n) = 0.31907 \cdot n^{0.554057}$$

for large n (See Table 2). However, we do not as yet have a proof of these observations.

(3) The condition for the nonexistence of ground states, $\lambda \leq \lambda(q)$, is not optimal (Figure 3). Here $\lambda(q) = 0.408$. In Figure 5, with $q = 2$ and $\lambda = 0.908$, we see that positive solutions, when they exist, are asymptotic to $\lambda^{-1/(q-1)}$ as $r \to +\infty$.

(4) On the other hand, for λ large, $\lambda = 5.408$, ground state solutions do appear to exist. From Figures 6, we infer that there must be a ground state solution with $0.443514 < u(0) < 0.443517$. In Figures 7 and 8 with $q = 3$, $f(u) = \lambda u|u|^{q-1} - u$, we again infer the existence of a positive ground state with $u(0)$ in $(0.49486, 0.49487)$ for $\lambda = 20.25$ and with $u(0)$ in $(0.2207, 0.2208)$ with $\lambda = 100.25$.

(5) In Figures 9, we see that there appears to be a second radial solution which changes signs once. For this solution $u(0) \in (0.3426, 0.3427)$.

(6) From Figures 6, 7, 8, and 9 we see that, at least for some numbers $R > 0$, the Dirichlet problem for (1.1) has positive radial solutions in $\{|x| < R\}$ with $u = 0$ on $|x| = R$.

(7) In Figures 10, with $q = 2$, $u_0 = 6.9$, and $\lambda = 1.408$, we have given two close-ups of the multivalues nature of the projection of the solution trajectories onto the $(r, 0, u)$ plane. Again only a portion of the

curves are shown, the graph in the immediate neighborhood of $R = 0$ being omitted. Figure 10.2 is a detail of Figure 10.1.

(8) In Figure 11.2, we observe a phenomenon Evers called "backspiraling." We observe a multivalued trajectory for which $r(s)$ is negative and oscillatory while u increases.

In order to eliminate backspiraling, and, at the same time improve the quality of Evers's graphs, we rewrote his code to pay more careful attention to the behavior of the ratio

$$\frac{\sin\psi(s)}{r(s)}$$

when both numerator and denominator are small. This ratio can be ill-defined (or nearly so) at points other than $s = 0$ as we can see from Figure 10. However, it is easy to see that if, as $s \to s_0$, we have

$$\lim_{s\to s_0} r(s) = \lim_{s\to s_0} \sin\psi(s) = 0,$$

then from L'Hopital's rule,

$$\lim_{s\to s_0} \frac{\sin\psi(s)}{r(s)} = -\tfrac{1}{n} f(u(s_0)),$$

(as one might expect from (1.11)).

To see how backspiraling can arise if the incremental steps beyond $r = h$ are not small relative to h, we consider a simple argument based on the Euler polygon method. Let (r_i, ψ_i, u_i), denote, for $i = 2, 3, \ldots$, the approximate solution of (1.9) obtained by the Euler polygon method. We have starting values

$$r_1 = h, \quad \psi_1 = -hf(u_0)/n, \quad u_1 = u_0 - \tfrac{1}{2}h^2 f(u_0)/n.$$

If ℓ is the step size for the method, then

$$\begin{aligned} r_2 &= h + \ell\cos\psi_1 \\ \psi_2 &= \psi_1 + \ell[-(n-1)(\sin\psi_1)/h - f(u_1)] \\ u_2 &= u_1 + \ell\sin\psi_1. \end{aligned}$$

Thus, if $\ell > h$ and for some integer p,

$$\psi_1 \cong \pi \pmod{2p\pi}$$

we will have $r_2 < 0$. If ψ_1 is slightly smaller than π mod $(2n\pi)$, then $u_2 > u_1$, and, for h sufficiently small $u_2 > u_0$ also.

In this connection, it is interesting to observe that there is a reciprocal relation between λ and h when this instability is most likely to occur. We have, from the definition of ψ_1 that

$$\lambda h \cong (n\pi \text{ mod } (2p\pi) - u_0 h) u_0^{-q}$$

which says that the larger the initial step, the smaller λ must be in order to avoid this kind of instability for large incremental steps of size ℓ (= INCR).

There is also an approximate reciprocal relationship between λ and ℓ which can be demonstrated numerically. In Figures 12A.1–12A.6 we have fixed h and ℓ (TOUT and INCR) and allowed λ to take values as indicated. In Figures 12B.1–12B.6 we have fixed λ and decreased ℓ, the incremental step size. The graphs become increasingly "noisy" as λ increases for fixed ℓ and as ℓ increases for fixed λ. Although the range of values taken (the projections of the graphs on the R, U axes) are not the same in the pairs of graphs with roughly equal $\ell\lambda$ products (for which equal numbers of points were plotted), one cannot but be impressed with the striking similarity between such graphs.

The quadrature schemes used in LSODE, the ODE solver used in [**3**], are of course much more sophisticated than the Euler polygon scheme above. Nevertheless, similar principles apply to them.

(9) If one looks for a "singular" solution near $r = 0$ in the form $u(r) \approx \alpha r^{-\beta}$, then one finds that, to leading order

$$u(r) \approx \left(\frac{n-1}{\lambda r}\right)^{1/q} \equiv v(r)$$

as $r \to 0^+$. The multivalued solutions appear to oscillate above $v(r)$. In the upper graph of Figure 10 and in Figures 13A,B $v(r)$ is plotted against r. We conjecture that, for fixed λ, as $u_0 \to +\infty$, the multivalued solution converges to an appropriate singular solution with $v(r)$ as the lead term in its expansion. In the case $f(u) = \lambda u$, Concus and Finn have obtained existence and uniqueness of such a singular solution [**1**]. Elsewhere in this volume, Atkinson examines the question of singular solutions when $f(u) = \lambda u^q$.

In Table 1, we present some numerical data based on (1.13)–(1.15). The integrals were evaluated using Simpson's fourth order rule.

The results of these computations led Serrin and Peletier to prove the following:

THEOREM [**8**]. *For $1 < q < (n+2)/(n-2)$ and all λ sufficiently large, (1.1) has a radially symmetric ground state.*

Their proof is based on the shooting method. However, the question of the existence of higher order modes as indicated in Remark 5 above and in Figure 9, has yet to be answered.

More recently, motivated by these computations, Serrin [**10**] has shown that for $1 < q < (n+2)/(n-2)$, for all sufficiently large λ, there is $R_0 > 0$ such that for all $R > R_0 = R_0(\lambda, q, n)$, the Dirichlet problem has a positive radial solution in $\{x \mid |x| < R\}$.

We next turn to an abbreviated discussion of (1.3)–(1.4).

In order to solve (1.3)–(1.4) numerically, the authors in [**7**] also tried to solve the second order differential equation via the shooting method by writing $u_1 = \phi$, $u_2 = \phi'$ and solving the resulting second order system. In view of the requirement (1.6b), it is necessary to impose a jump in ϕ' as ϕ' approaches a root of $G_{\phi'}(y, \phi')$. Proceeding in this manner yields unsatisfactory results because the roots of $G_{\phi'}$ are dependent on y and because the equation is therefore singular (loses order) at an unknown value of ϕ', (which becomes known when the ODE solver returns a flag warning that the number of allowed predictor-corrector iterations per step has been exceeded and the calculation cannot proceed, because the error tolerance cannot be met in the required number of iterations).

Again, we introduce a stretching variable, namely s, and examine the system

$$\begin{aligned} \frac{dy}{ds} &= \frac{\partial G}{\partial \phi'}(y, \phi') \\ \frac{d\phi}{ds} &= \phi' \frac{\partial G}{\partial \phi'}(y, \phi') \\ \frac{d\phi'}{ds} &= H(y, \phi) - \frac{\partial G}{\partial y}(y, \phi'). \end{aligned} \tag{1.16}$$

This system is then solved with initial data determined from the compatibility condition

$$H(0, \phi_0) = G_y(0, \phi_0') \tag{1.17}$$

where now ϕ_0' is the shooting parameter. This condition is the analog of the first of the boundary conditions (1.8). To obtain a parabolic approximation to ϕ near $y = 0$, one solves for ϕ_0'' from (1.3) in the same manner as was done for (1.11). Again there are no zero divisors on the right-hand side of (1.16) and one encounters no difficulties in the use of a standard package to solve (1.16). However, as before, the projection of $(y(s), \phi(s), \phi'(s))$ into the $(y, \phi, 0)$ plane is still required to define ϕ as a well-defined function of y. This is contained in the requirement (1.6b) which in turn leads to the requirement that ϕ' must be discontinuous whenever it changes sign. To track this, one examines the sign of

$$G_{\phi'\phi'}(y(s_k), \phi'(s_k))$$

at those points where $\phi'(s_k) \neq 0$ and where $G_{\phi'}(y(s_k), \phi(s_k))$ is either zero or has sign opposite from that of $G_{\phi'}(y(s_{k-1}), \phi'(s_{k-1}))$. There are four cases in which this discontinuity must be introduced and an algebraic equation solved for the value of ϕ' on the opposite side of the discontinuity.

Of course, (1.3) can, and does, possess solutions that do not change sign at all. In this case, there will not be any jump in ϕ', although ϕ'' may undergo a rapid, but smooth, sign change. In Figures 15, the former case is shown while in Figures 16, the latter case is shown. For these computations, we have used

(1.18)
$$G(y, \phi') = y(y+1)n(y)(\phi')^2(1 + (\phi')^2)^{-1}(\phi' - \omega_0(y))$$

(1.19)
$$\omega_0(y) = -n'(y)/n(y)$$

(1.20)
$$H(y, \phi) = n^2(y)\left\{\exp(y-1) - \exp\left[-(1 + \phi/(1 - \tfrac{1}{2}y))\right]\right\}$$

where

(1.21)
$$n(y) = n_0 e^{-y}.$$

Here $0 < y < 2$.

The function $n(y)$ represents the axially averaged plasma density. The choice above is just one possible model for it. A second choice, with an inverted density profile is $n(y) = n_0(\exp(-y))(1 + \alpha y)$. This gives a non-constant $\omega_0(y)$.

In Figure 17, we indicate a few solutions in the (y, ϕ) plane which are found by choosing initial values satisfying (1.17) and then solving the system numerically, plotting ϕ versus y. We find that there is a region R, points of which cannot be on any such solution curve. It can be shown, computationally at least, and theoretically, in the autonomous case, that there is an unbounded region $\tilde{R} \subset R$, such that if we start at $(Y, \phi(Y))$ and vary $\phi'(Y)$, we cannot solve (1.16) backward in s to reach a point $(0, \phi_0)$ and have the solution be analytic in a right open neighborhood of $y = 0$, and hence satisfy (1.17) and its higher order analogs. We refer the reader to [**6**] for a complete theoretical analysis of the autonomous case as well as the details of the computations we have briefly discussed here.

ACKNOWLEDGEMENT: The author thanks John Axtell for computing Figures 10 and 11 from the programs of [**3**].

Department of Mathematics, Iowa State University, Ames IA 50011

This research was supported in part by AFOSR 84-0252 and in part by the Lawrence Livermore National Laboratory, Livermore CA 94550

Research at MSRI supported in part by NSF Grant DMS 812079-05.

Table 1.

		Figure 6		Figure 7		Figure 8		Figure 9	
	u_0	0.443514	0.443517	0.49486	0.49487	0.2207	0.2208	0.3426	0.3427
	λ	5.40825		20.25		100.25		100.25	
(1)	$(n-1)\int_0^\infty r^{n-1}(s)\cos(\psi(s))F(u(s))\,ds$	0.030706	0.030708	0.55390	0.55398	0.010930	0.010865	0.071749	0.071821
(2)	$\lim_{s\to\infty}\left[(n-1)\int_0^s r^{n-2}(\eta)\,d\eta - r^{n-1}(s)\right]$	0.030706	0.030707	0.55381	0.55387	0.010832	0.010857	0.056644	0.056705
(3)	$\int_0^\infty r^{n-1}(s)u(s)f(u(s))\cos(\psi(s))\,ds$	0.082822	0.083827	0.090491	0.090507	0.018499	0.018557	0.055896	0.05592
(4)	$\int_0^\infty r^{n-1}(s)\sin^2(\psi(s))\,ds$	0.082821	0.082826	0.090458	0.090472	0.018496	0.018556	0.061196	0.06125
(5)	$\int_0^\infty r^{n-1}(s)f(u(s))\cos(\psi(s))\,ds$	-1.9×10^{-4}	2.44×10^{-3}	-6.4×10^{-5}	1.5×10^{-3}	-4.0×10^{-4}	8.1×10^{-3}	0.2040	0.2041
(6)	$-\lim_{s\to\infty} r^{n-1}(s)\sin(\psi(s))$	-1.9×10^{-4}	2.44×10^{-3}	-2.1×10^{-4}	1.4×10^{-4}	-4.1×10^{-4}	8.1×10^{-3}	0.2038	0.2039
		q=2 n=2		q=3 n=2		q=3 n=2		q=3 n=2	

Note: Lines (1) and (2) correspond to the left and right-hand sides of (1.13) respectively. Similarly, lines (3) and (4) correspond to (1.14) and (5), (6) to (1.15). The integrals and the limits were evaluated up to the first value of s where $u'(s) = 0$ or where $u(s) = 0$, which value of s is then defined as infinity. The integrals were evaluated using Simpson's rule.

Typeset by $\mathcal{A}_{\mathcal{M}}\mathcal{S}$-TEX

Table 2.

$\lambda = 5.408 \qquad u(0) = -1 \qquad q = 2$

Dimension (N)	(A) Measured Value of $R(N)$	(B) $R(N) = 0.31907N^{0.554057}$
200	5.90	6.01
300	7.50	7.52
400	8.80	8.82
500	10.0	9.98
600	11.1	11.0
700	12.1	12.0
800	13.0	13.0
900	14.0	13.8
1000	14.8	14.7
1200	16.3	16.2
1400	17.8	17.7
1600	19.3	19.0
1800	20.5	20.3
2000	21.8	21.5
2500	24.5	24.4
3000	26.9	26.9
3500	29.4	29.3
4000	31.6	31.6
4500	33.5	33.7
5000	35.5	35.8
5500	37.5	37.7
6000	39.5	39.6
6500	41.0	41.4
7000	42.5	43.1
7500	44.0	44.8
8000	45.8	46.4

Note: The formula for $R(N)$ in the third column is obtained from a least squares fit of the data in the first two columns (plots are not shown). The standard deviation of the differences between columns (A) and (B) is 0.14.

Explanation of Figures.

Figure 1. (λ, q) plane. There are no ground states for (λ, q) in the hatched region ([**4,7**]).

Figure 2. The change of variables leading to (1.9).

Figure 3. (λ, q) plane. There are no ground states for some λ's, $\lambda > \lambda(q)$ but there are ground states for $\lambda \gg \lambda(q)$.

Figures 4A. $u_0 = 2.0$, λ (- LAM) $= 5.408$, ℓ (= INC) $= 10^{-2}$. Panels are labelled by dimension n (= DIMEN) increasing from left to right and from top to bottom. The insets show the local (near $r = 0$) behavior of the solutions ($n = 1.1, 1.3, 1.7, 2.0$, for Figures 4A.1–4, respectively).

Figures 4B. Same description as for Figures 4A ($n = 2.3, 2.9, 3$, for Figures 4B.1–3). Spiral disappears when $n = 3$, $\lambda = 5.408$, returns for $n = 3$ at $\lambda = 50.408$ (for Figure 4B.4).

Figures 4C,D. $u_0 = -1.0$, $\lambda = 5.408$, $\ell = 10^{-2}$. Panels are labelled by dimension n (= DIMEN). (Figure 4C: $n = 1, 1.03, 1.10, 1.25, 2.00, 2.50$ for Figures 4C.1–6 respectively. Figure 4D: $n = 3.00, 4.00, 6.00, 10.00, 25.00$, 100.00 for Figures 4D.1–6 respectively.)

Figures 4E. Dependence of spirals on λ. For Figures 4E.1, 4E.3, and 4E.5, $\lambda = 0.00$; for Figures 4E.2, 4E.4, and 4E.6, $\lambda = 20.408$. Compare also with Figures 4C.1, 4C.3, and 4C.5.

Figures 5. Plots of u versus r, $f(u) = f_2(u) = \lambda u^2 - u$. $\lambda = 0.908$. There are no ground states for this value of λ. (Figure 5.1, $u_0 = 0.25$; Figure 5.2, $u_0 = 2.0$)

Figures 6. Plots of u versus r, $f(u) = f_2(u) = \lambda u^2 - u$, $\lambda = 5.408$, $q = 2$. There is a positive solution of (1.7)–(1.8) with $u_0 \in (0.443514, 0.443517)$. (Figure 6.1, $u_0 = 0.443514$; Figure 6.2, $u_0 = 0.443517$)

Figures 7. Plots of u versus r, $f(u) = f_1(u) = \lambda |u|^2 u - u$, $\lambda = 20.25$, $q = 3$. There is a positive solution of (1.7)–(1.8) with $u_0 \in (0.49486, 0.49487)$. (Figure 7.1, $u_0 = 0.3427$; Figure 7.2, $u_0 = 0.3426$)

Figures 8. Plots of u versus r, $f(u) = f_1(u) = \lambda|u|^2u - u$, $\lambda = 100.25$, $q = 3$. There is a positive solution of (1.7)–(1.8) with $u_0 \in (0.2207, 0.2208)$. (Figure 8.1, $u_0 = 0.2207$; Figure 8.2, $u_0 = 0.2208$)

Figures 9. Plots of u versus r, $f(u) = f_1(u) = \lambda|u|^2u - u$, $\lambda = 20.25$, $q = 3$. There appears to be a second solution of (1.7)–(1.8) for some $u_0 \in (0.3426, 0.3427)$ which changes sign once. (Figure 9.1, $u_0 = 0.49486$; Figure 9.2, $u_0 = 0.49487$)

Figures 10. Multivalued solution of (1.1) in a neighborhood of $r = 0$ for $u_0 = 6.9$ and $\lambda = 1.408$ (small), $q = 2$. The dashed curve is a plot of $v(r) = (\lambda r)^{-1/q}$. Scaling: Figure 10.1, $R : 10^{-1}$; Figure 10.2, $R : 10^{-2}$, $u : 10^{-1}$.

Figures 11. (Figure 11.1) Multivalued solution of (1.1) in a neighborhood of $r = 0$ with $u_0 = 6.9$, $q = 2$ and $\lambda = 10.408$. (Figure 11.2) Backspiraling with $u_0 = 6.9$ and $\lambda = 10.408$. Scaling: $R : 10^{-3}$, $u : 10^{-1}$.

Figures 12A. $\ell = 10^{-4}$, $u_0 = 6.9$, $R : 10^{-4}$, $u : 10^{-2}$. Panels are labelled by λ (= LAM): (50.408, 100.408, 200.408, 400.408, 800.408, 1600.408 for Figures 12A.1–6 respectively).

Figures 12B. $u_0 = 6.9$, $\lambda = 1600.408$, $R : 10^{-4}$, $u : 10^{-3}$. Panels are labelled by ℓ (= INC): (1.5625×10^{-6}, 3.125×10^{-6}, 6.25×10^{-6}, 1.25×10^{-5}, 2.5×10^{-5}, 5×10^{-5} for Figures 12B.1–6, respectively.

Figure 13A. Plot of multivalued solution when $n = 2$, $q = 3$, $u_0 = 2.7$ and $\lambda = 100.25$. The dashed curve is a plot of $v(r) = (\lambda r)^{-1/q}$. Scaling: $R : 10^{-2}$. (Insert: $R : 10^{-3}$, $u : 10^{-1}$.)

Figure 13B. Detail of insert. Scaling $R : 10^{-3}$, $u : 10^{-2}$.

Figure 14. Qualitative shape of $G(y, \phi')$ for fixed y.

Figures 15. Solution of (1.16)–(1.21) with a discontinuous derivative.

Figures 16. Solution of (1.16)–(1.21) with a continuous derivative.

Figure 17. Envelope of solutions of (1.3) obtained by solving (1.16) with shooting parameter $\phi_0' = v_0$, shooting from $y = 0$ in the positive y direction.

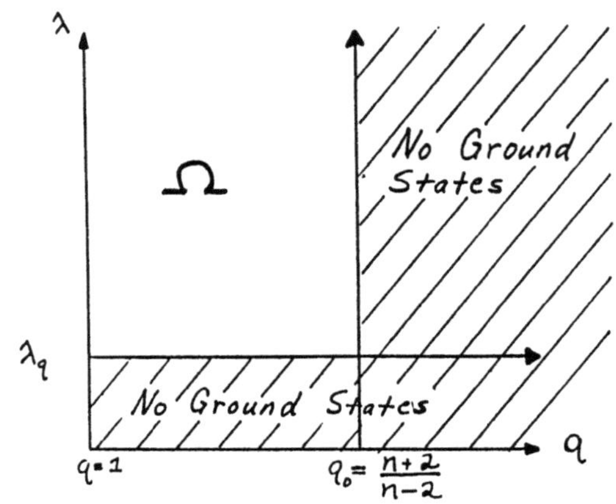

Figure 1

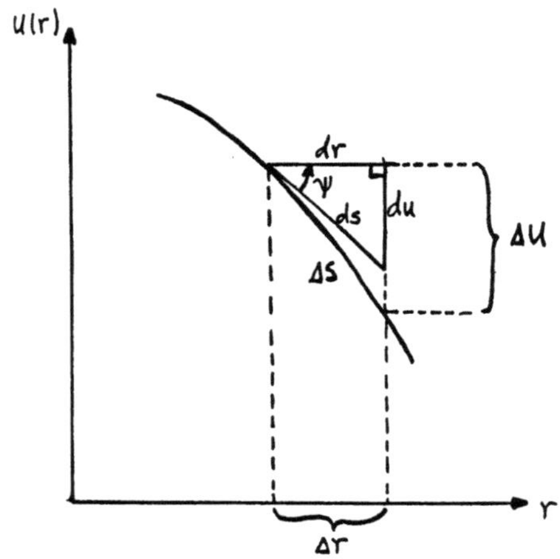

Figure 2

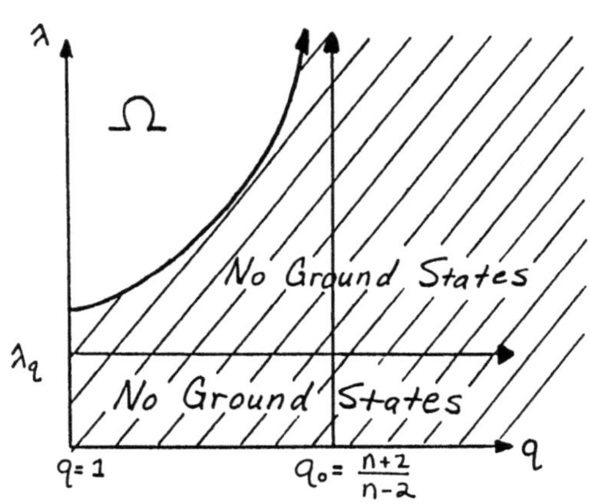

Figure 3

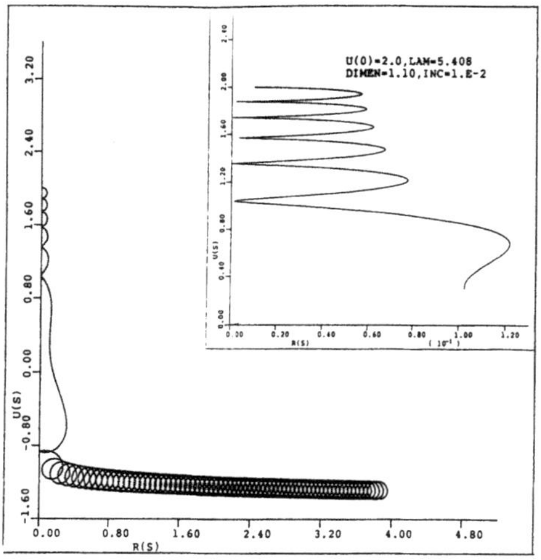

Figure 4A.1

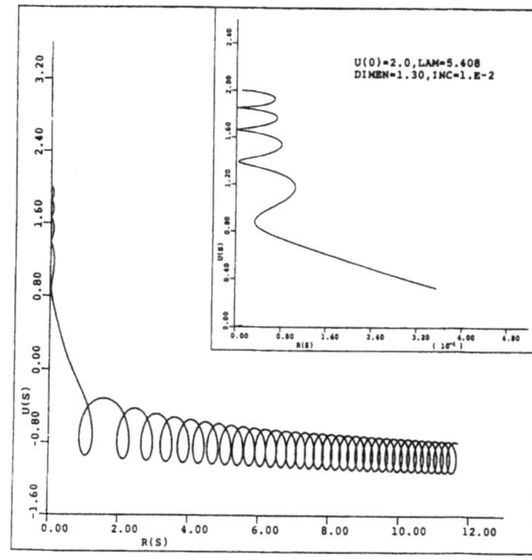

Figure 4A.2

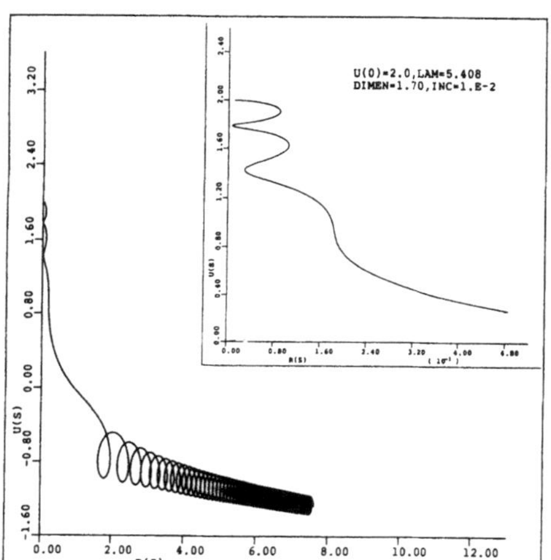

Figure 4A.3

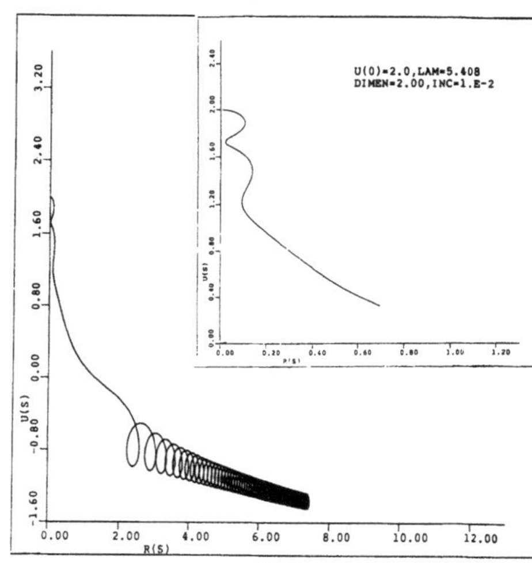

Figure 4A.4

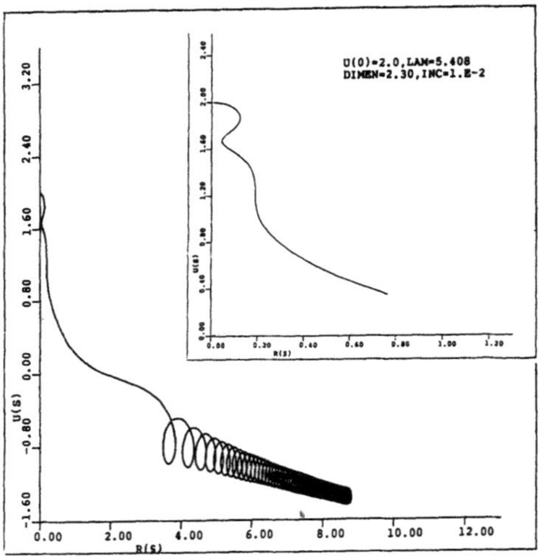

Figure 4B.1

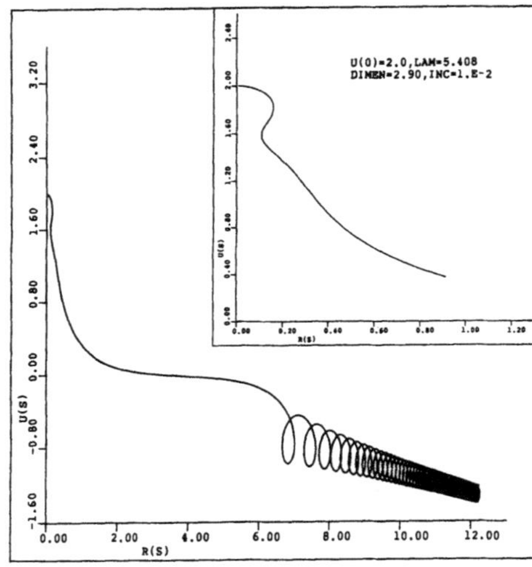

Figure 4B.2

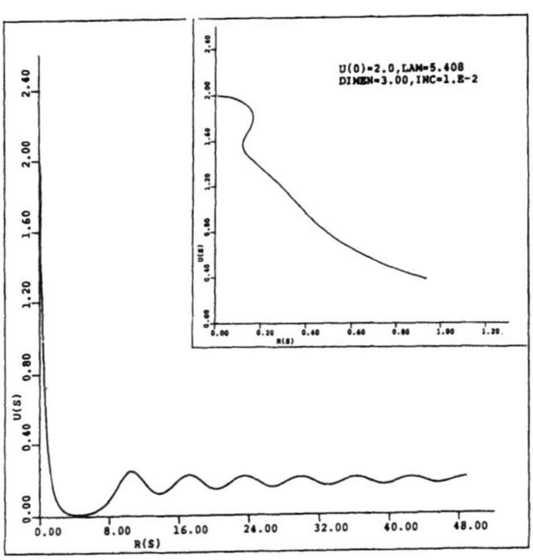

Figure 4B.3

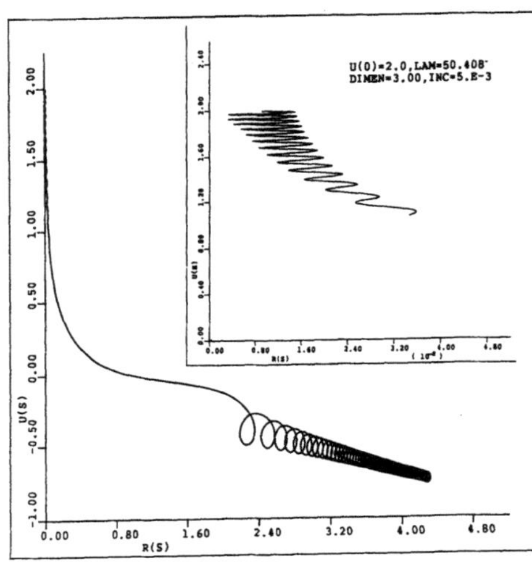

Figure 4B.4

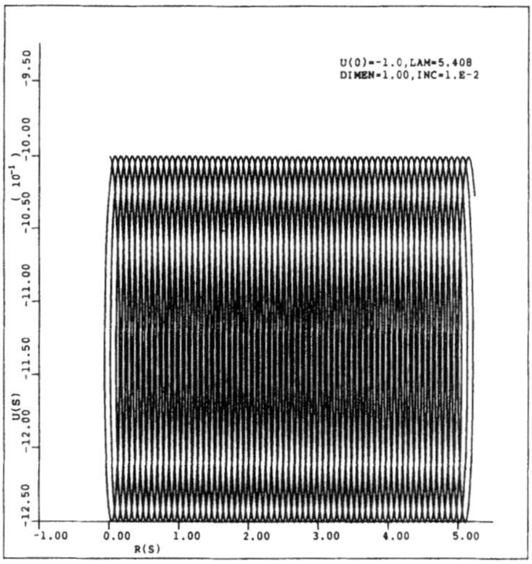

Figure 4C.1

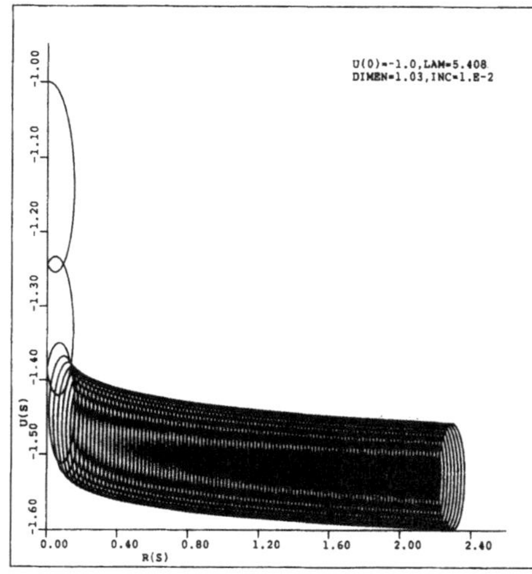

Figure 4C.2

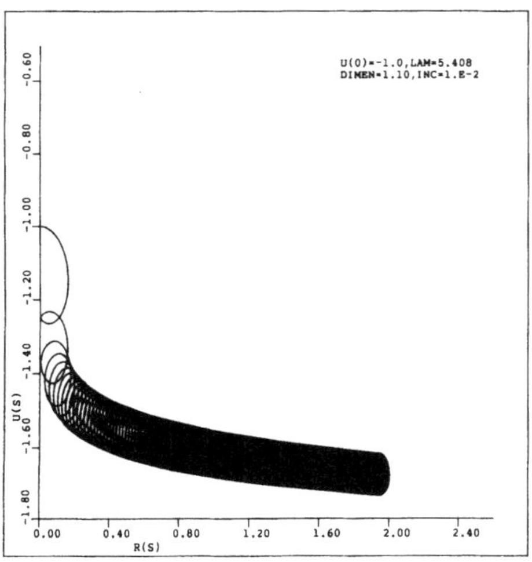

Figure 4C.3

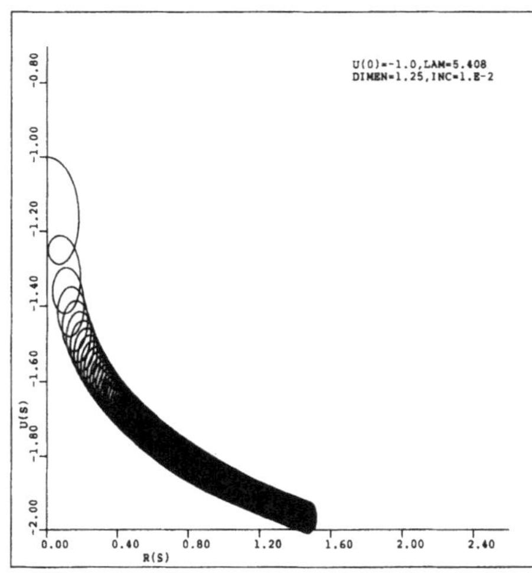

Figure 4C.4

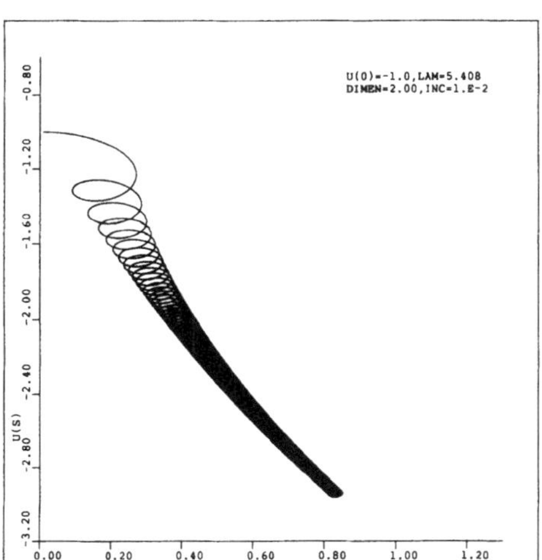

Figure 4C.5

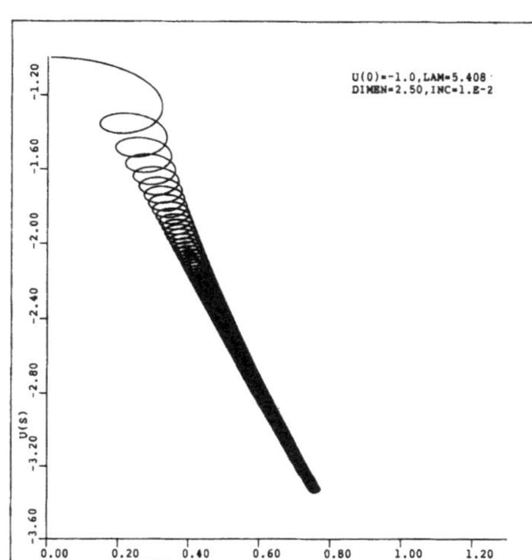

Figure 4C.6

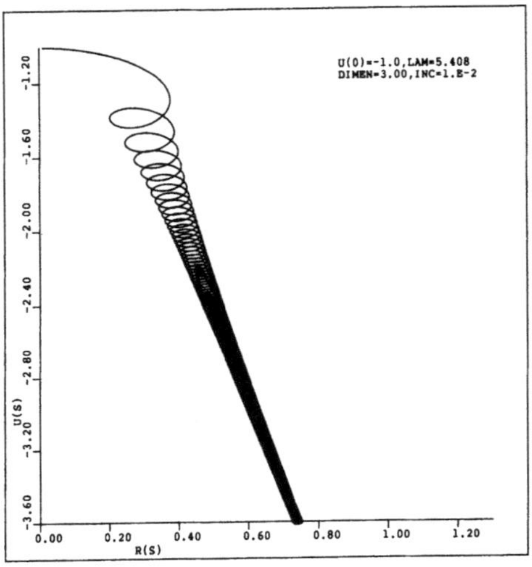

Figure 4D.1

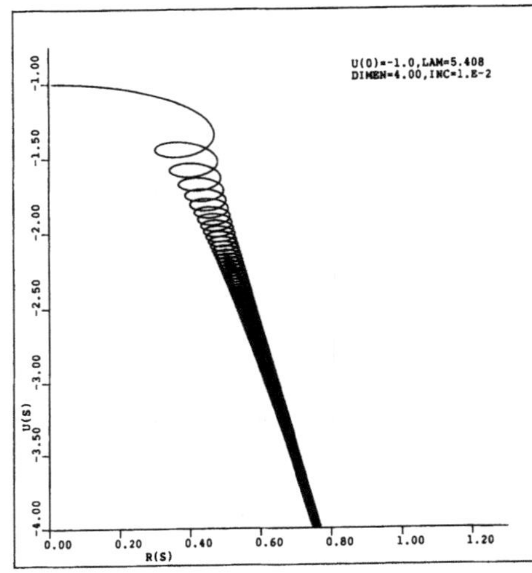

Figure 4D.2

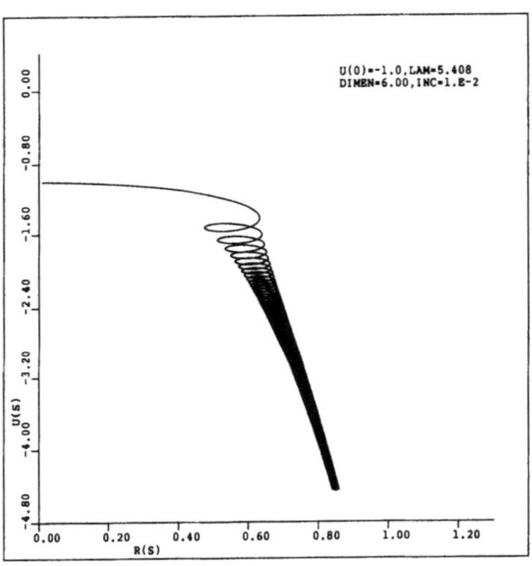

Figure 4D.3

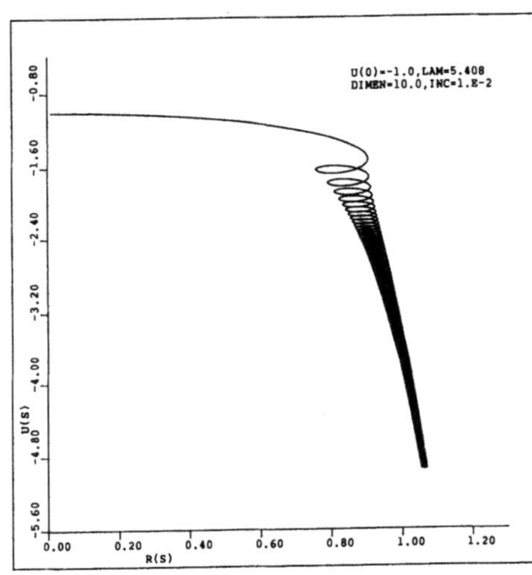

Figure 4D.4

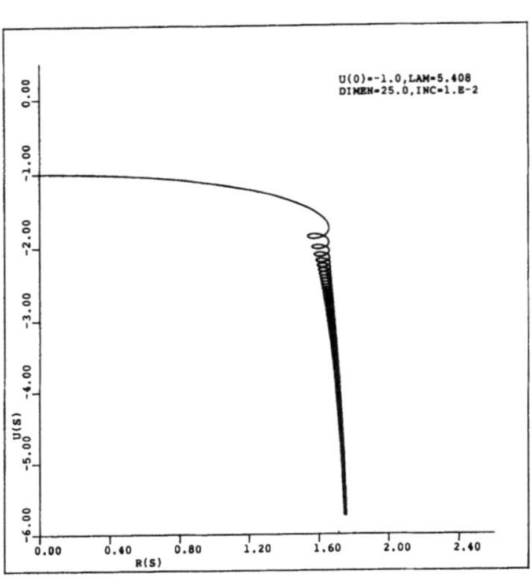

Figure 4D.5

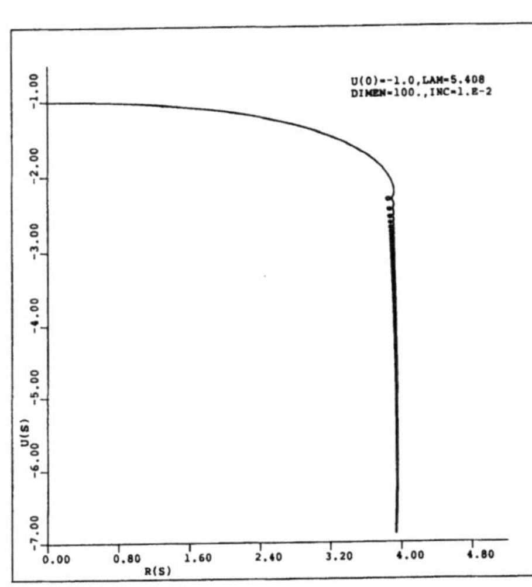

Figure 4D.6

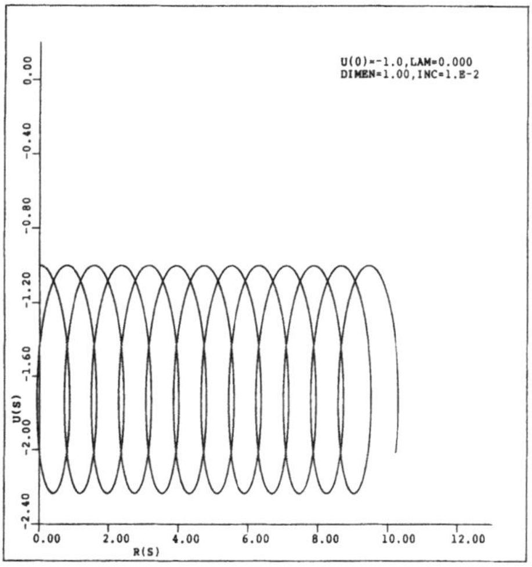

Figure 4E.1

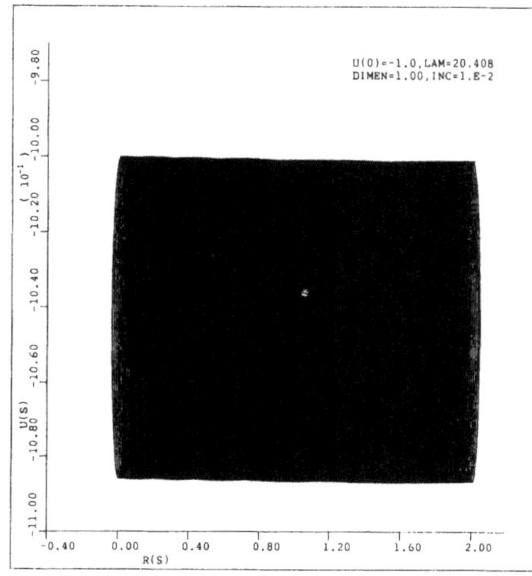

Figure 4E.2

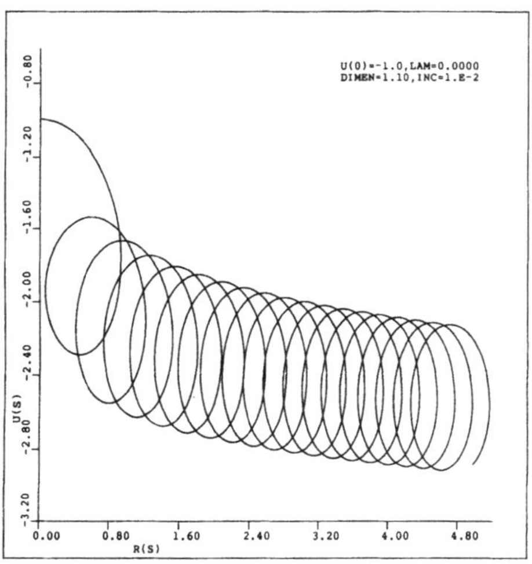

Figure 4E.3

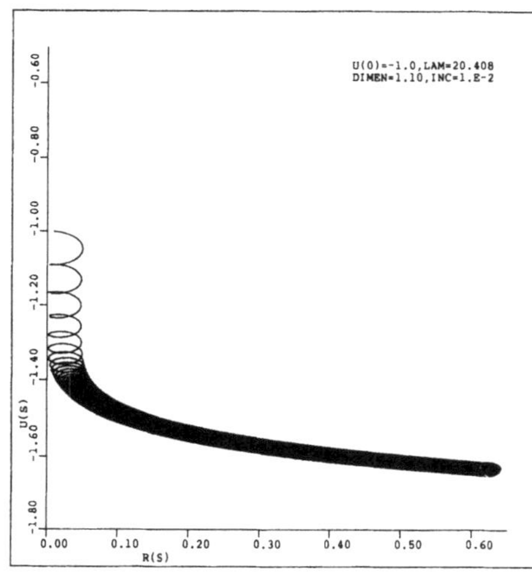

Figure 4E.4

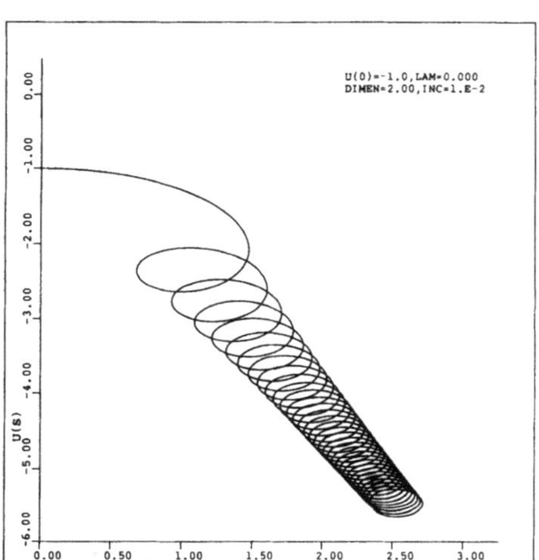

Figure 4E.5

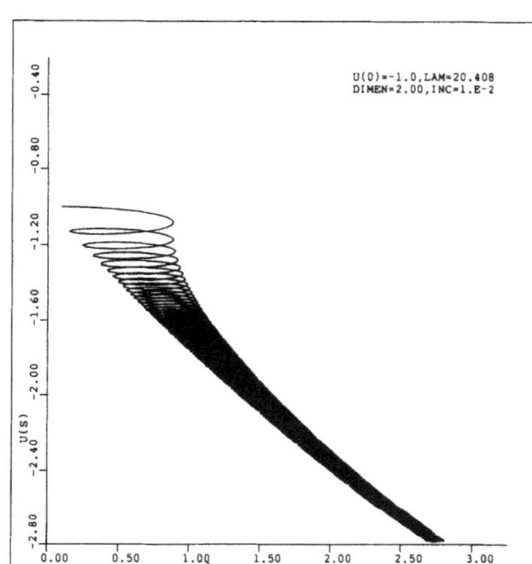

Figure 4E.6

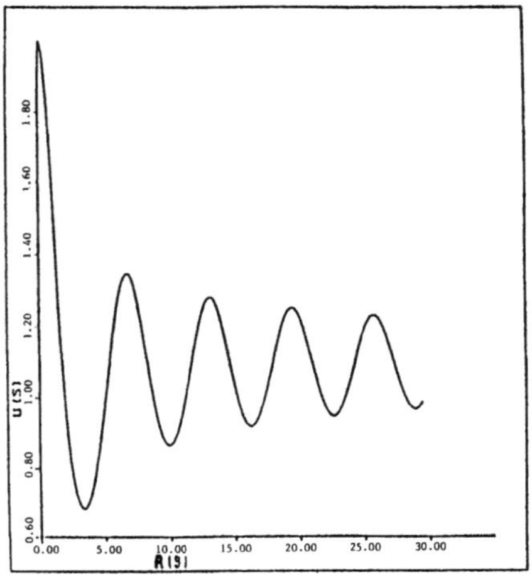

Figure 5.1

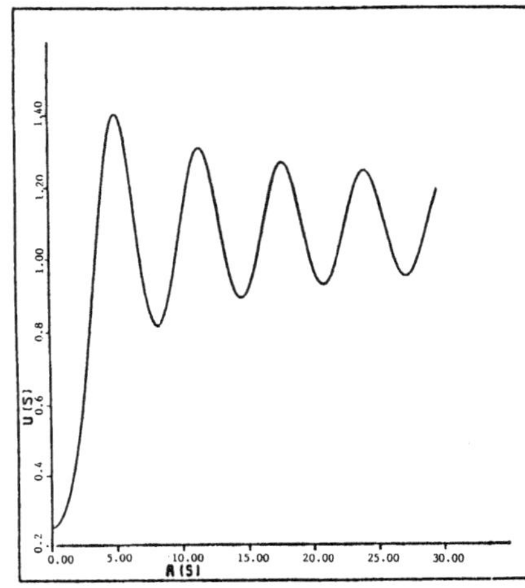

Figure 5.2

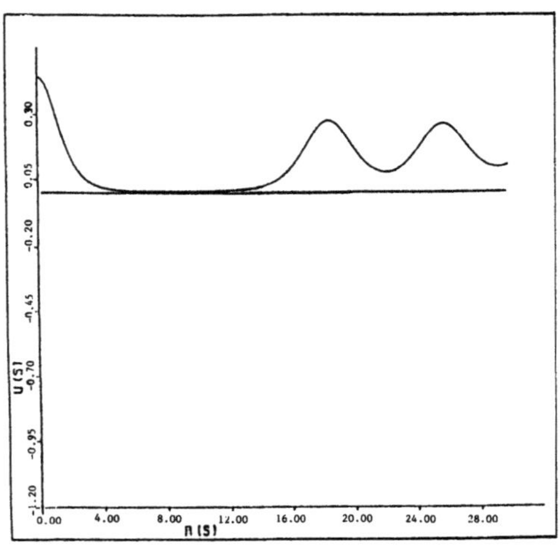

Figure 6.1

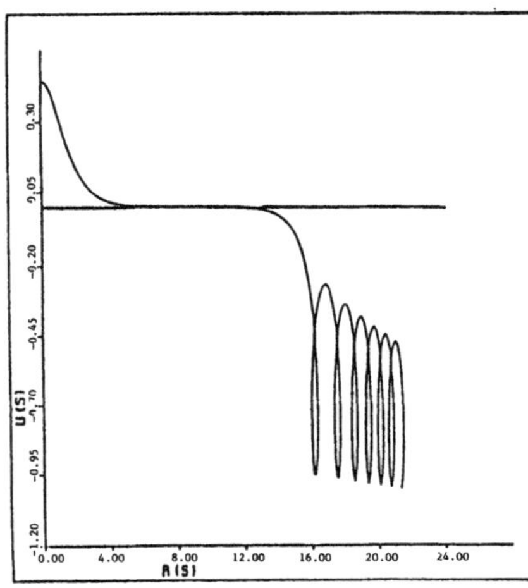

Figure 6.2

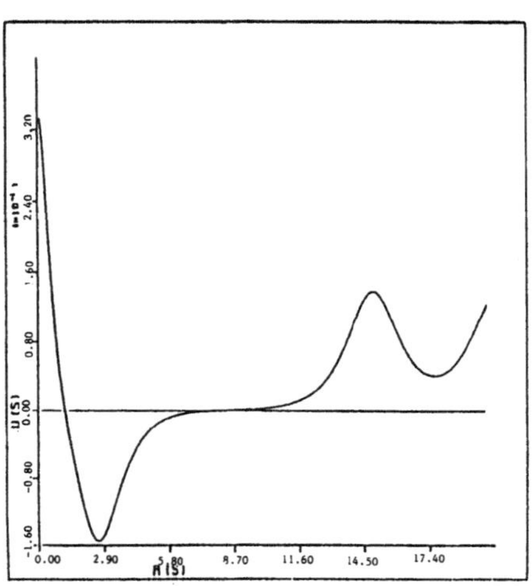

Figure 7.1

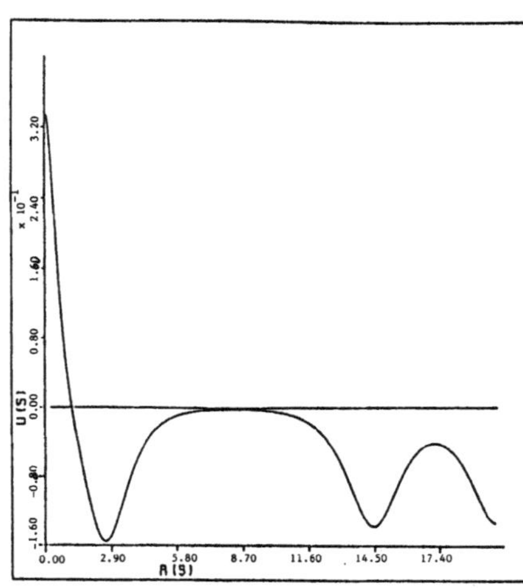

Figure 7.2

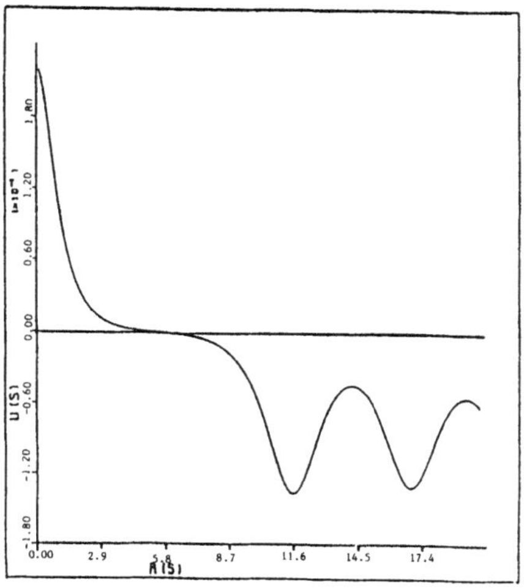

Figure 8.1

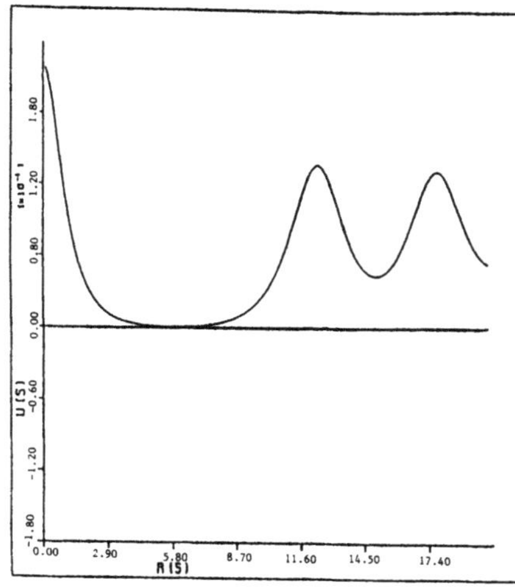

Figure 8.2

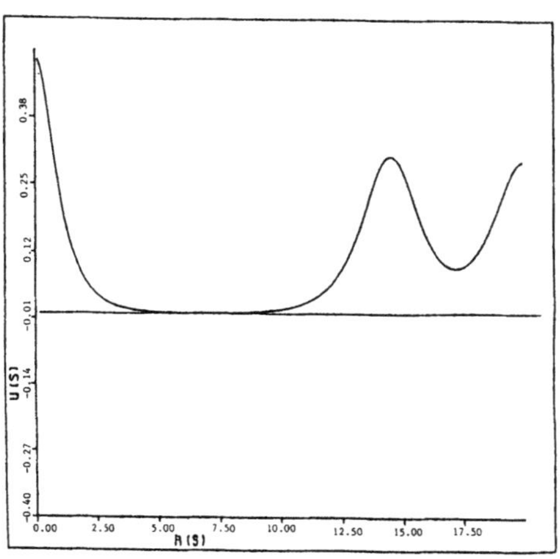

Figure 9.1

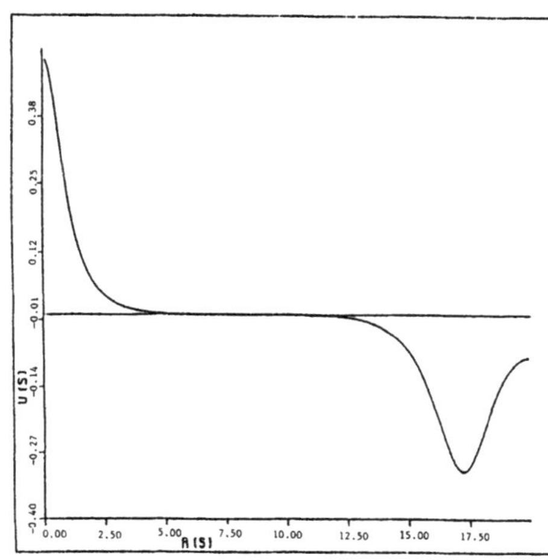

Figure 9.2

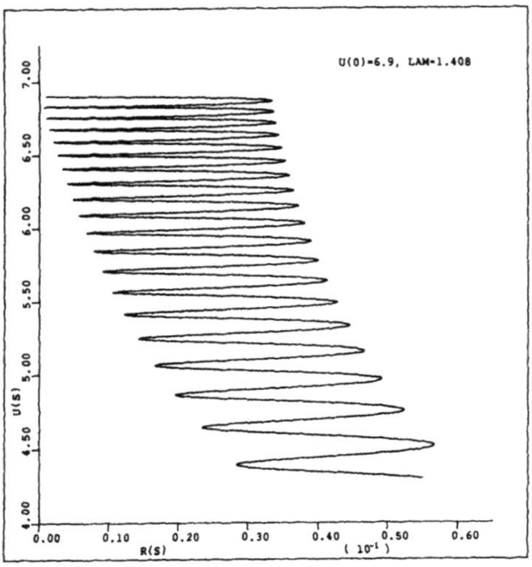

Figure 10.1

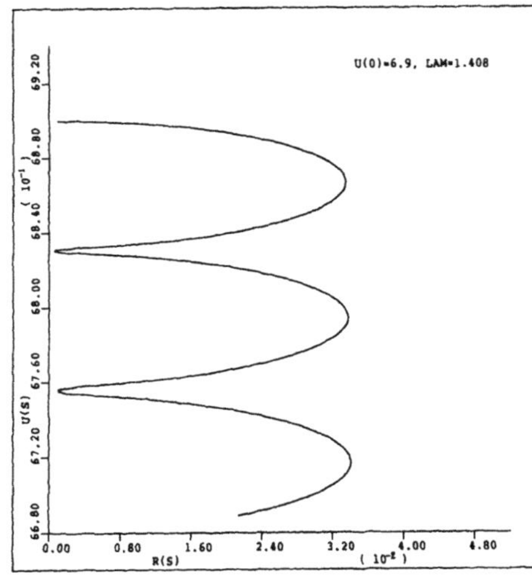

Figure 10.2

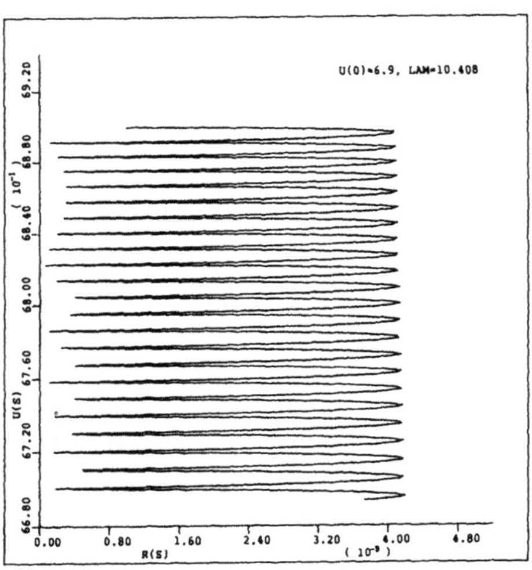

Figure 11.1

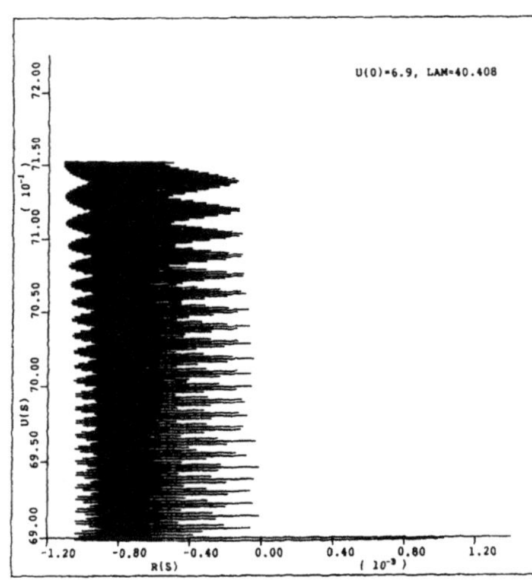

Figure 11.2

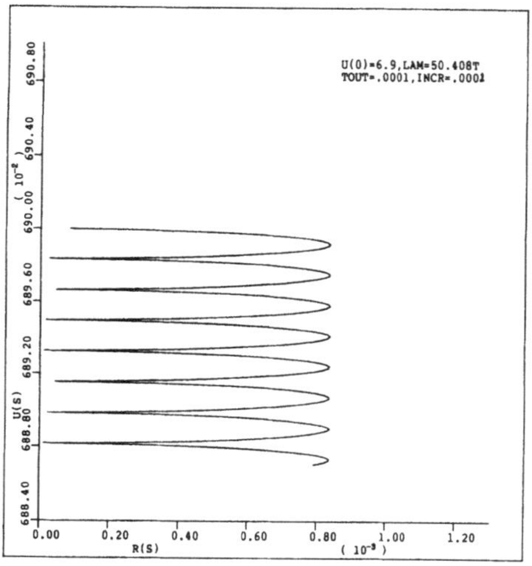

Figure 12A.1

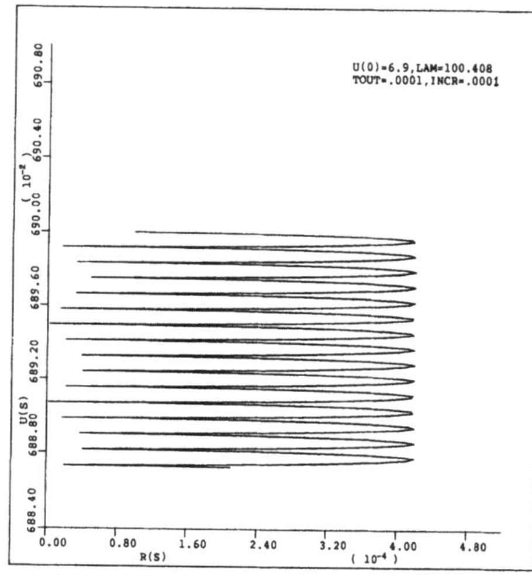

Figure 12A.2

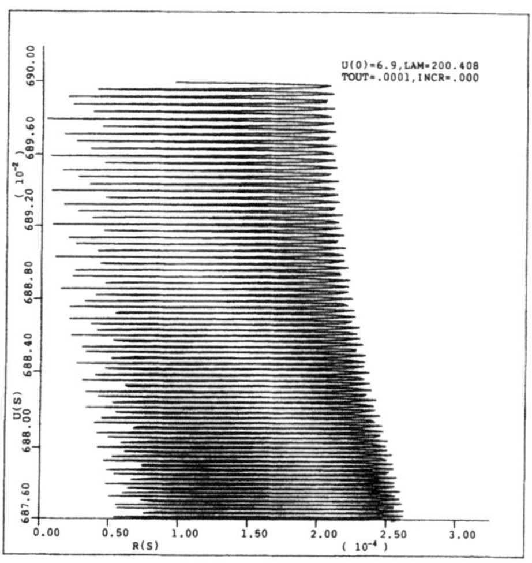

Figure 12A.3

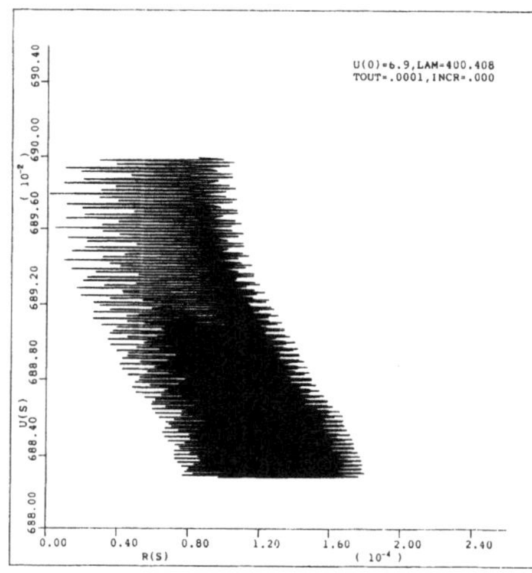

Figure 12A.4

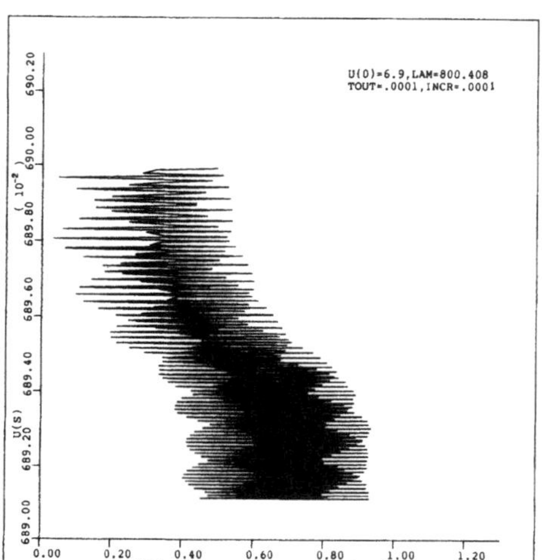

Figure 12A.5

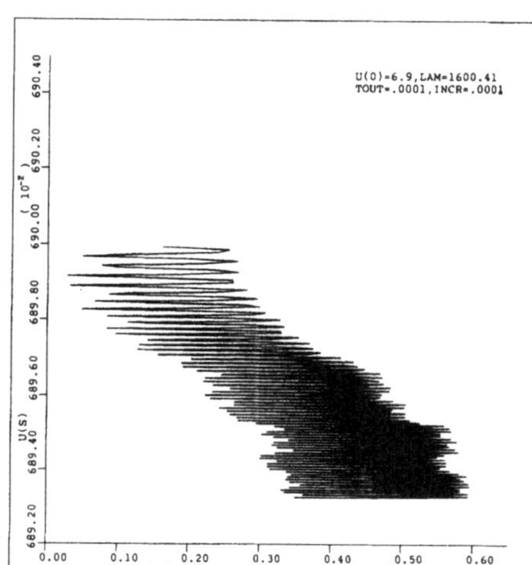

Figure 12A.6

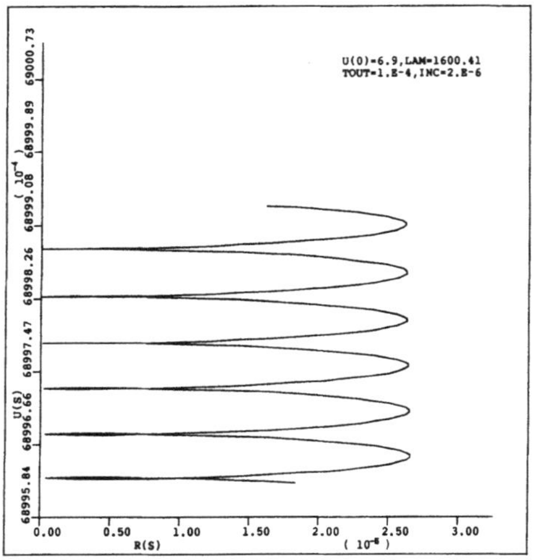

Figure 12B.1

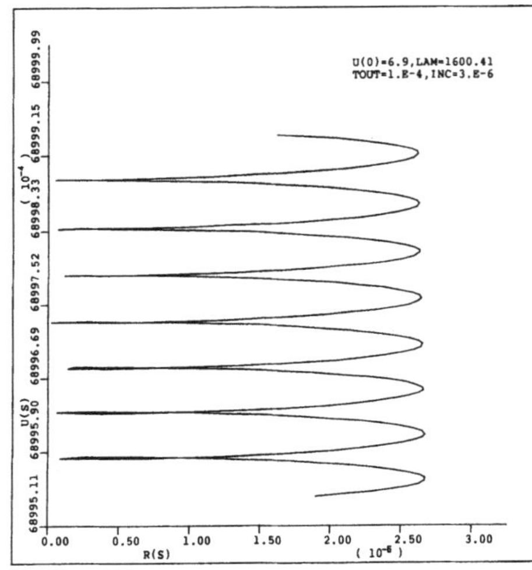

Figure 12B.2

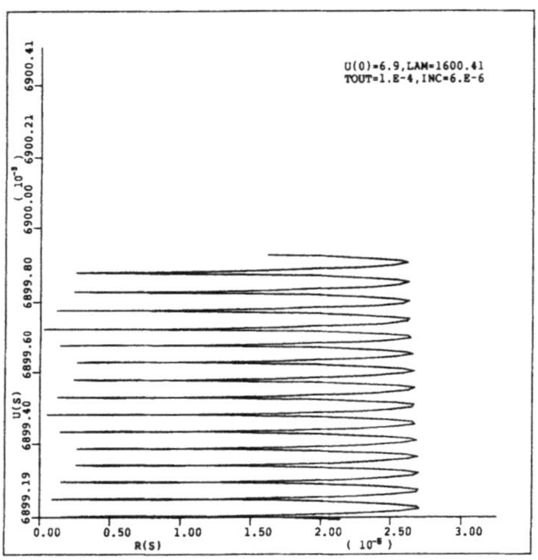

Figure 12B.3

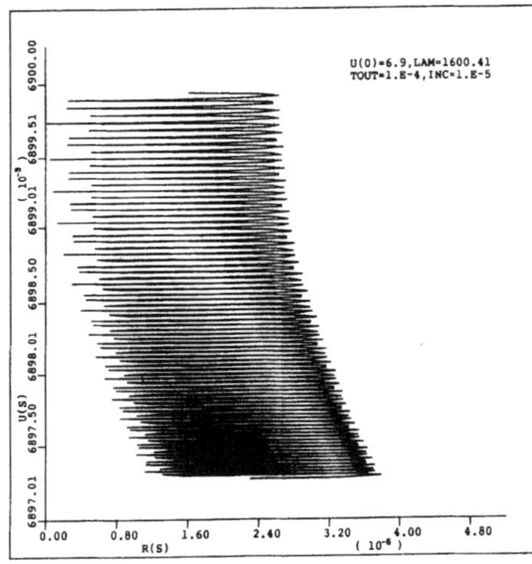

Figure 12B.4

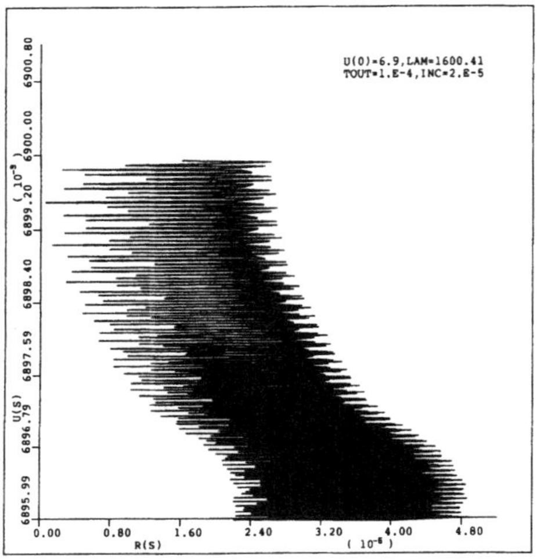

Figure 12B.5

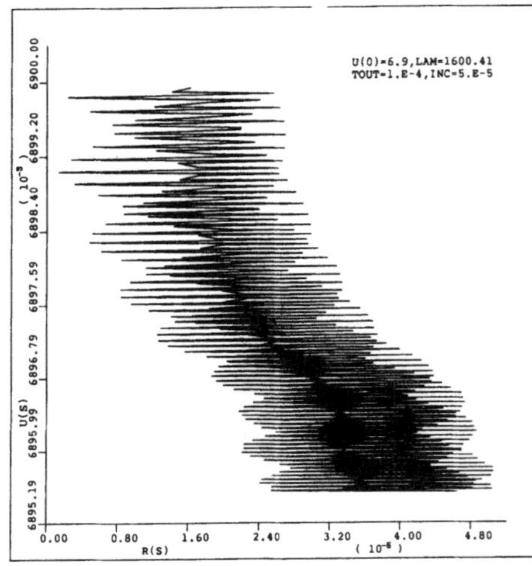

Figure 12B.6

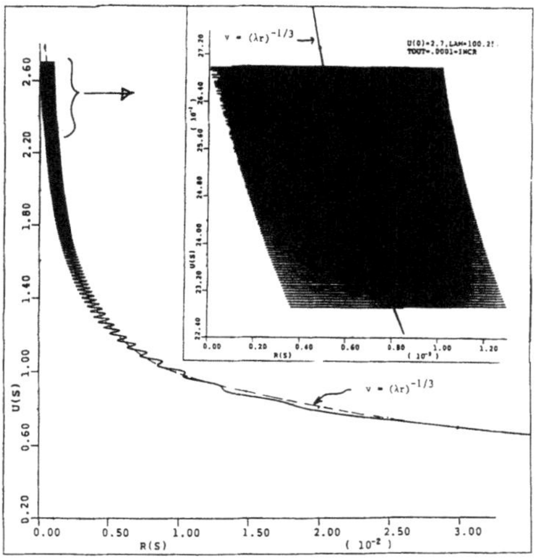

Figure 13A

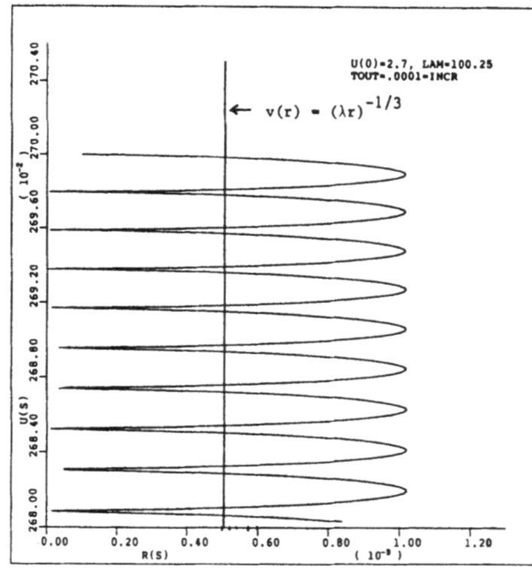

Figure 13B

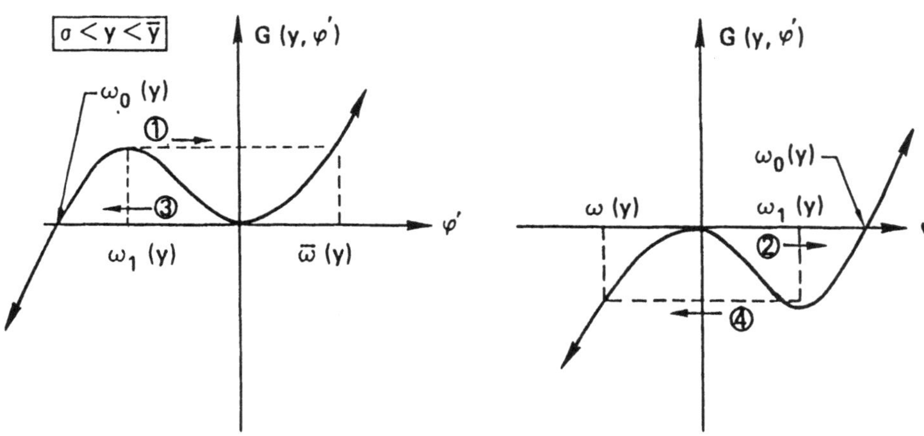

Qualitative shape of G (y, φ')

Figure 14

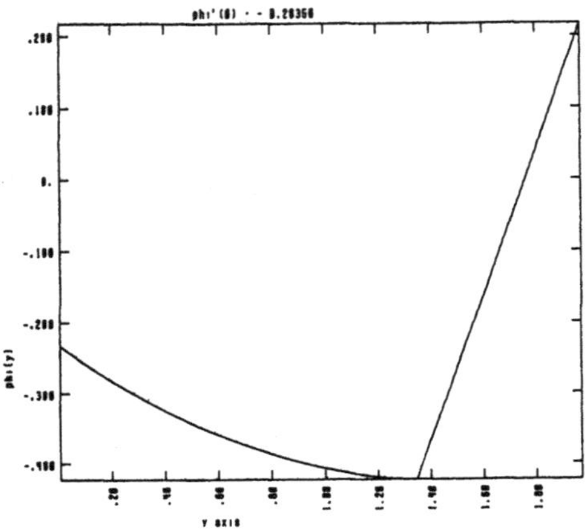

Figure 15.1

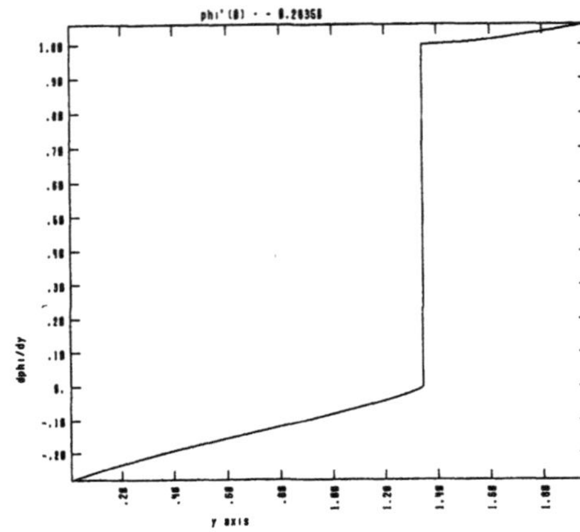

Figure 15.2

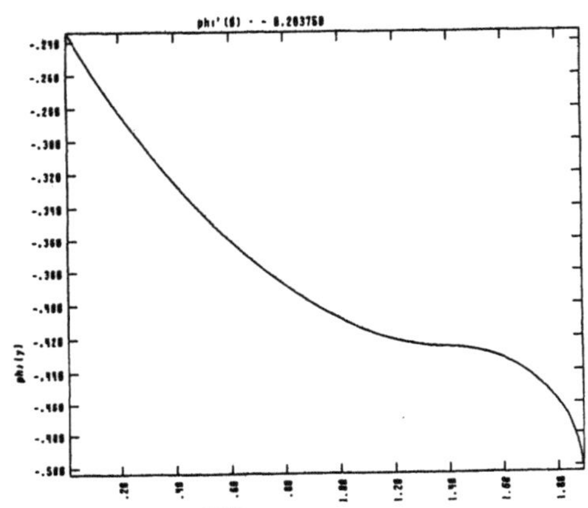

Figure 16.1

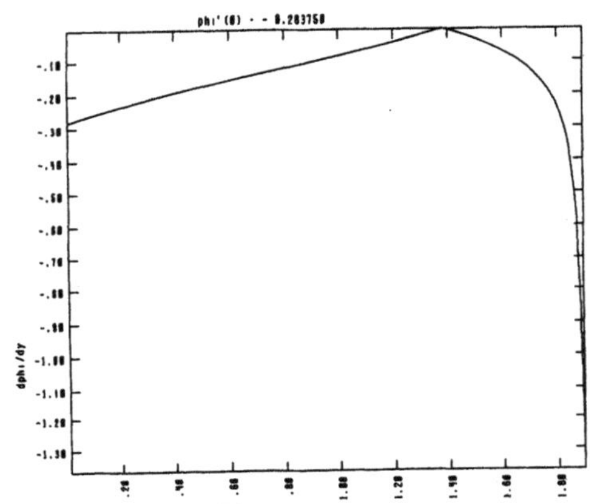

Figure 16.2

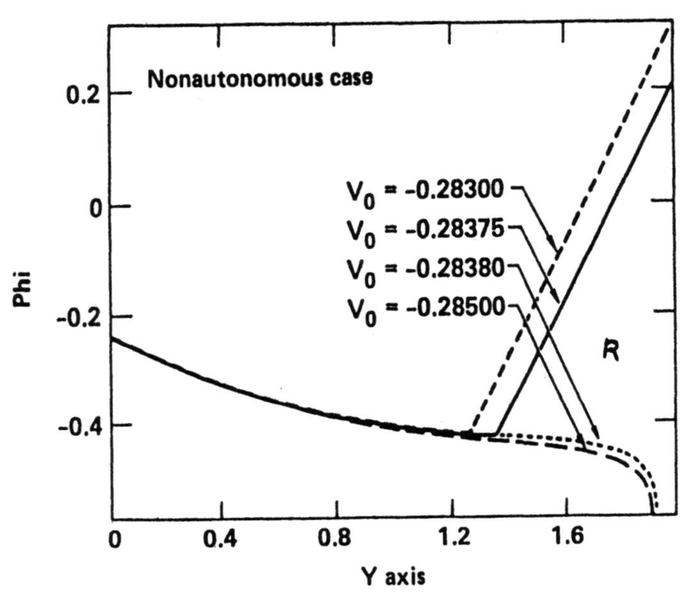

Figure 17

REFERENCES

1. P. Concus and R. Finn, *A singular solution of a capillary problem (I. Existence, II. Uniqueness)*, Inventiones Math **19** (1975), 143–159.
2. P. Concus and R. Finn, *The shape of a pendant liquid drop*, Phil. Trans. Royal Soc. London, Ser. A292 (1979), 307–342.
3. T. K. Evers, *Numerical search for ground state solutions of a modified capillary equation*, M.S. paper, Department of Mathematics, Iowa State University (1985).
4. B. Franchi, E. Lanconelli and J. Serrin, *Esistenze e unicita degli stati fondamentali per equazione ellicitì quasilineare*, Rend. Acad. Naz. Lincei (1986).
5. D. E. Hastings and S. P. Auerbach, *Additive noise and the radial electric field in systems with convective and diffusive losses*, Physics of Fluids **28** (1985), 2219–2221.
6. H. A. Levine and S. P. Auerbach, *The charge balance equation*, Submitted to J. Math. Phys..
7. A. A. Mirin, S. P. Auerbach, R. H. Cohen, J. M. Gilmore, L. D. Pearlstein and M. E. Rensink, *Radial transport calculations for tandem mirrors*, Nuclear Fusion **23** (1983), 703–738.
8. W.-M. Ni and J. Serrin, *Nonexistence theorems for quasilinear partial differential equations*, Rend. Circolo. Math. Palermo, Suppl. **8** (1985), 171–185.
9. L. A. Peletier and J. Serrin, *Ground states for the prescribed mean curvature equation*, Proc. AMS (in press).
10. J. Serrin, *Positive solution of a prescribed mean curvature problem*, (in press).

On Positive Solutions of Semilinear Elliptic Equations in Unbounded Domains

P. L. LIONS

Abstract. We give a short survey of some existence results concerning positive solutions of semilinear elliptic equations in unbounded domains. We consider only solutions which vanish at infinity. We also investigate the existence of "minimum-type" solutions as well as solutions given by "higher" critical points.

Summary.

I. Introduction.

We present here some existence results concerning the equation

$$(1) \qquad -\sum_{i,j=1}^{N} \frac{\partial}{\partial x_i}\left(a_{ij}(x)\frac{\partial u}{\partial x_j}\right) = g(x,u) \text{in } \mathbb{R}^N$$

and we will look for solutions u satisfying

$$(2) \qquad u > 0 \text{ in } \mathbb{R}^N, \quad u(x) \to 0 \text{ as } |x| \to \infty.$$

We will always assume that $a_{ij} \equiv a_{ji} \quad (\forall i,j)$, that (1) is uniformly elliptic, i.e.

$$(3) \qquad \exists \nu > 0, \forall \xi \in \mathbb{R}^N, \quad \sum_{i,j=1}^{N} a_{ij}(x)\xi_i\xi_j \geq \nu|\xi|^2, \text{ a.e. } x \in \mathbb{R}^N$$

and that the nonlinearity is subcritical which roughly speaking means that $g(x,u)|u|^{-(N+2)/(N-2)}$ goes to 0 as $|u| \to \infty$ (at least if $N \geq 3$). In addition, we will assume that there exists a trivial solution, namely 0, i.e. we assume that $g(x,0) = 0$ a.e. Typical examples of problems we wish to study are:

Example 1: a_{ij}, g do not depend on x and thus the problem is translation invariant, i.e. if u is a solution of (1)–(2) then $u(\cdot + \varsigma)$ is also a solution of (1)–(2) for all $\varsigma \in \mathbb{R}^N$.

Example 2: $g(x,t) = b(x)|t|^{p-1}t - a(x)t$ on $\mathbb{R}^N \times \mathbb{R}$, with $1 < p < \frac{N+2}{N-2}$ if $N \geq 3$, $1 < p < \infty$ if $N \leq 2$. And even if we consider more general classes of a, b, a typical example containing most of the difficulties we want to address is the case when

$$(4) \qquad 0 < \underline{a} \leq a \leq \bar{a}, \qquad 0 < \underline{b} \leq b \leq \bar{b} \qquad \text{a.e. in } \mathbb{R}^N$$

for some positive constants $\underline{a}, \bar{a}, \underline{b}, \bar{b}$.

Let us also mention that we consider similar problems set in general unbounded domains. In all cases, we see from (2) that we are looking for solutions which vanish at infinity: in fact, we are looking for solutions with "finite energy," i.e.

$$(5) \qquad \nabla u \in L^2(\mathbb{R}^N), \qquad G(x,u) \in L^1(\mathbb{R}^N)$$

where $G(x,t) = \int_0^t g(x,s)\, ds$.

Such problems in bounded domains may be studied and solved by variational techniques as in A. Ambrosetti and P. H. Rabinowitz [**2**], or by L^∞ a priori estimates as in H. Brezis and R. E. L. Turner [**16**], D. G. de Figueiredo, P. L. Lions and R. D. Nussbaum [**24**], B. Gidas and J. Spruck [**32**], or by a combination of these arguments (see also [**24**]). However, in unbounded domains, the situation is in general quite different and the above methods fail: L^∞ a priori estimates are not enough to ensure the right behavior at infinity or to avoid the trivial solution while the usual variational arguments fail because of some lack of compactness we describe now. Indeed, in unbounded domains such as $\mathbb{R}^N$, one way to see that some lack of compactness appears is to recall that the Rellich-Kondrakov theorem on the compactness of the embedding H_0^1 (or H^1) $\hookrightarrow L^2$ is no more true when dealing with general unbounded domains. This can also be seen in the context of Example 1 since the problem in this case is translation invariant and thus as soon as there exists a solution, the set of solutions is not compact in any Sobolev space like $H^1(\mathbb{R}^N)$.

These difficulties are not purely technical, as it can be seen from various nonexistence results taken from M. J. Esteban and P. L. Lions [**28**] that we recall in Section IV. Furthermore, such problems are not academic ones since they occur in various branches of mathematical physics, such as classical fields equations, propagation of laser beams, fluid mechanics (solitary waves, vortex rings...), Hartree-Fock theory in quantum mechanics (atomic physics, nuclear physics) — see, for instance, H. Berestycki and P. L. Lions [**9**], D. Gogny and P. L. Lions [**33**], P. L. Lions [**36**] and the references therein.

A natural attempt in order to solve such problems is to take opportunity of the variational structure of the problem and to try to obtain the existence of solutions through the simplest variational approaches, namely (constrained) minimization or min-max principles of the Mountain Pass Lemma type. These approaches lead in fact to solutions that could be called (and often are) ground state solutions by analogy with quantum mechanics. Mathematically, a correct definition of a ground state solution could be either a solution with the lowest "energy" among all nontrivial solutions, or a solution having at most 1 as Morse index (number of negative eigenvalues of the linearized operator). However, we will not expand further on this rather technical question even if it is possible to complete the statements of our existence results in Sections II–III by similar informations. In Section II, we essentially consider the case of Example 2 while in Section III, we consider more general nonlinearities. At this stage, let us mention that we can treat these questions by the use of the concentration-compactness method (see P. L. Lions [**37**]) which was precisely devised to study "ground state solutions" and we will mention how the results of [**37**], W. Y. Ding and W. M. Ni [**27**], and M. J. Esteban and P. L. Lions [**29**] fit in this approach.

In fact, the concentration-compactness method also makes precise the lack of compactness in these questions and this combined with the idea of A. Bahri on the study of critical points at infinity (see A. Bahri [**5**], A. Bahri and J. M. Coron [**6**]) leads to existence results of "higher critical points". Similar existence results in special cases may be obtained directly. We give a few examples of results obtained by these approaches in Section IV.

We conclude this introduction by a few remarks: first of all, even if our assumptions include those cases, particular cases of the above problems are standard modifications of the case of bounded domains. Indeed, if we consider the case of Example 2 and if, for instance, we assume $b \in L^\infty_+$, $a \to +\infty$ as $|x| \to \infty$, or $a \geq \underline{a} > 0$ and $b \in L^\alpha_+ + L^\beta_+$ with $\frac{2N}{(N-2)} \frac{1}{p+1} \leq \alpha, \beta < \infty$, or $a \geq 0$ and $b \in L^\alpha_+$ with $\alpha = \frac{2N}{(N-2)} \frac{1}{p+1} (\dots)$ then existence follows from the weak continuity of the functional $\left(\int_{\mathbb{R}^N} b|u|^{p+1}\, dx\right)$ on sequences $(u_n)_n$ satisfying: ∇u_n is bounded in $L^2(\mathbb{R}^N)$, au_n^2 is bounded in L^1. One possible result which can be obtained along those lines is given in E. Brüning [**17**]. Next, we want to emphasize the fact that we will not consider the case of the limit Sobolev exponent, i.e. $p = \frac{N+2}{N-2}$ for $N \geq 3$ and we refer instead to F. V. Atkinson and L. A. Peletier [**3**], P. L. Lions [**38**], A. Bahri and J. M. Coron [**6**] and W. Y. Ding [**26**].

Finally, we would like to mention that there are several related aspects of semilinear elliptic equations that we do not consider here, such as (for example)

(i) the symmetries of solutions with the fundamental results on positive solutions due to B. Gidas, W. M. Ni and L. Nirenberg [**30,31**] and some existence results of nonsymmetric solutions (complex solutions in P. L. Lions [**39**], real solutions in the limit case in W. Y. Ding [**26**])

(ii) vector-valued solutions (see H. Berestycki and P. L. Lions [**9,10**], P. L. Lions [**37**], H. Brezis and E. H. Lieb [**14**])

(iii) multiplicity results (see H. Berestycki and P. L. Lions [**9**], H. Berestycki and C. H. Taubes [**11**]).

II. Homogeneous nonlinearities in $\mathbb{R}^N$.

We will basically consider the case of Example 2, i.e.

$$g(x,t) = b(x)|t|^{p-1}t - a(x)t \quad \text{on } \mathbb{R}^N \times \mathbb{R},$$

with $1 < p < \frac{N+2}{N-2}$ if $N \geq 3$, $1 < p < \infty$ if $N \leq 2$. And we will assume

$$(6) \quad \begin{cases} \exists \nu > 0, \forall \varphi \in \mathcal{D}(\mathbb{R}^N), \quad \displaystyle\int_{\mathbb{R}^N} \sum_{i,j=1}^N a_{ij}(x) \frac{\partial \varphi}{\partial x_i} \frac{\partial \varphi}{\partial x_j} + \int_{\mathbb{R}^N} a\varphi^2\, dx \geq \\ \qquad \geq \nu \displaystyle\int_{\mathbb{R}^N} |\nabla \varphi|^2 + \varphi^2\, dx, \end{cases}$$

$$(7) \qquad a_{ij} \in L^\infty(\mathbb{R}^N)(1 \le i,j \le N), \quad \sup_{y\in\mathbb{R}^N} \int_{y+B} |a|^\alpha \, dx < \infty$$

with $\alpha = \frac{N}{2}$ if $N \ge 3$, $\alpha > 1$ if $N = 2$, $\alpha = 1$ if $N = 1$,

$$(8) \qquad b \ge 0 \text{ a.e. in } \mathbb{R}^N, \quad b \not\equiv 0, \quad \sup_{y\in\mathbb{R}^N} \int_{y+B} |b|^\beta \, dx < \infty$$

with $\beta = \frac{2N}{N-2}\frac{1}{p+1}$ if $N \ge 3$, $\beta > 1$ if $N = 2$, $\beta = 1$ if $N = 1$.

Observe that (6) implies (3) (choose $\varphi = \cos(\lambda(x \cdot e))\psi$ with $\lambda > 0$, $e \in \mathbb{R}^N$, $\psi \in \mathcal{D}(\mathbb{R}^N)$, let $\lambda \to +\infty$, and then $\psi^2 \to \delta_{x_0}$ for almost all x_0).

Next, we observe that in order to solve (1)–(2) it is enough to build a solution of

$$(9) \qquad -\sum_{i,j=1}^N \frac{\partial}{\partial x_i}\left(a_{ij}\frac{\partial u}{\partial x_j}\right) + au = \theta b u^p \text{ in } \mathbb{R}^N, \quad u > 0 \text{ in } \mathbb{R}^N$$

for some $\theta > 0$, where $u \in H^1(\mathbb{R}^N)$. And this is the case if we find a minimum of

$$(10) \qquad I = \inf\left\{\mathcal{E}(u)/u \in H^1(\mathbb{R}^N), \quad J(u) = 1\right\}$$

with

$$\mathcal{E}(u) = \int_{\mathbb{R}^N} \tfrac{1}{2}\sum_{i,j=1}^N a_{ij}\frac{\partial u}{\partial x_i}\frac{\partial u}{\partial x_j} + \tfrac{1}{2}au^2 \, dx,$$

$$J(u) = \int_{\mathbb{R}^N} b\frac{|u|^{p+1}}{p+1} \, dx.$$

Before going further in the analysis of the minimization problem, we have to check that the functional J is meaningful over the space $H^1(\mathbb{R}^N)$. To this end, we just observe that if $(B_i)_{i\ge 1}$ denotes a collection of unit balls such that every point in $\mathbb{R}^N$ belongs to at least one and at most m (a fixed number) of these balls, then for some $C \ge 0$ independent of u

$$J(u) \le C\sum_i \int_{B_i} b|u|^{p+1} \, dx \le C\sum_i \sup_j \left(\int_{B_j} b^\beta \, dx\right)^{1/\beta}\left(\int_{B_i} |u|^{\frac{2N}{N-2}} \, dx\right)^\gamma$$

if $N \geq 3$ (if $N = 1,2$ we modify easily this argument), where $\gamma = \frac{N-2}{2N}(p+1)$. Hence, we have denoting by C various constants independent of u

$$J(u) \leq C \sum_i \left(\int_{B_i} |u|^{\frac{2N}{N-2}} \, dx \right)^{\gamma} \leq C \sum_i \left(\int_{B_i} |\nabla u|^2 + u^2 \, dx \right)^{\frac{p+1}{2}}$$
$$\leq C \left\{ \sup_j \left(\int_{B_j} |\nabla u|^2 + u^2 \, dx \right)^{\frac{p-1}{2}} \right\} \left(\sum_i \int_{B_i} |\nabla u|^2 + u^2 \, dx \right)$$
$$\leq C \left(\int_{\mathbb{R}^N} |\nabla u|^2 + u^2 \, dx \right)^{\frac{p+1}{2}},$$

thus proving our claim. One shows similarly that $\left(\int_{\mathbb{R}^N} au^2 \, dx\right)$ is well-defined on $H^1(\mathbb{R}^N)$.

Recalling that a minimizing sequence $(u_n)_n$ is a sequence such that $J(u_n) = 1$ for all $n \geq 1$, $\mathcal{E}(u_n) \underset{n}{\to} I$, we may now state the following general result taken from [**37,29**] (it is not stated in [**37**] even if the proofs given in [**37**] immediately yield it as explained in [**29**]).

THEOREM II.1. *For any minimizing sequence $(u_n)_n$ of (10), there exists a sequence $(y_n)_n$ in $\mathbb{R}^N$ such that $(u_n(\cdot + y_n))_n$ is relatively compact in $L^{\bar{p}}(\mathbb{R}^N)$ (for all $2 \leq \bar{p} \leq \frac{2N}{N-2}$ if $N \geq 3$, $2 \leq \bar{p} < \infty$ if $N \leq 2$).*

Of course, if for some sequence $(u_n)_n$, the associated sequence $(y_n)_n$ is bounded then there exists a minimum of (10), since $\mathcal{E}(u)$ and $J(u)$ are continuous on $H^1(\mathbb{R}^N)$ by the above arguments.

As we will see, all the existence results of minima of (10) we know follow from Theorem II.1. In order to apply this result, it is convenient to study what happens if $|y_n|$ (or a subsequence) goes to ∞ as n goes to ∞.

We will not prove the above result and refer instead to [**37,29**]. Let us only mention that the concentration-compactness method allows a complete classification of the possible losses of compactness of, say, the embedding of $H^1(\mathbb{R}^N) \hookrightarrow L^q(\mathbb{R}^N)$ with $2 < q < \frac{2N}{N-2}$ if $N \geq 3$, $2 < q < \infty$ if $N \leq 2$. Roughly speaking, neglecting in particular the extraction of subsequences, the "only possible loss" of compactness consists in the splitting of the original sequence of functions into (essentially) a finite number of sequences of functions with disjoint supports which go infinitely away from each other and such that each of those subsequences is relatively compact in L^q (or in $L^2 \ldots$). For the above particular minimization problem, the fact that the minimizing sequences have only "one piece" in the above scenario is deduced from the following remark based upon the homogeneity of the nonlinearity

$$I_\lambda = \lambda^{\frac{2}{p+1}} I \quad \text{for all } \lambda > 0$$

where I_λ stands for the same infimum as in (10) with 1 replaced by λ, and thus $I = I_1 < I_\lambda + I_{1-\lambda}$, for all $\lambda \in (0,1)$. And this strict subadditivity is enough to ensure the compactness up to a translation stated in Theorem II.1.

Now, if $(y_n)_n$ goes to ∞ and if $v_n = u_n(\cdot + y_n)$ is relatively compact in $L^q(\mathbb{R}^N)$, we consider the following set of coefficients

$$\{(a_{ij}(\cdot + y_n))_{i,j}, a(\cdot + y_n), b(\cdot + y_n)\} = \mathcal{C}(y_n).$$

Next, we claim that, extracting a subsequence if necessary, these coefficients converge weakly to some set of coefficients $\mathcal{C} = \{(\tilde{a}_{ij})_{i,j}, \tilde{a}, \tilde{b}\}$ satisfying also (6)–(7)–(8). The convergence for $a(\cdot + y_n)$ (respectively $b(\cdot + y_n)$), is in L^α_{loc} weak (respectively L^β_{loc} weak) while for $(a_{ij}(\cdot + y_n))_{i,j}$ the right convergence to use is the so-called G-convergence (or equivalently, here Γ-convergence). See, for instance, E. De Giorgi [**25**], H. Attouch [**4**], F. Murat and L. Tartar [**43**] and the references therein.

We will not detail this technical point here: we just recall that if $a_{ij}(x) \to \tilde{a}_{ij}$ as $|x| \to \infty$ for all $1 \le i,j \le N$ then $(a_{ij})_{i,j}$ G-converges to $(\tilde{a}_{ij})_{i,j}$. To simplify notations, if the above convergences hold we will say that $\mathcal{C}(y_n)$ converges to $\mathcal{C}$ and we then introduce

$$I(\mathcal{C}) = \inf\left\{\tilde{\mathcal{E}}(u)/u \in H^1(\mathbb{R}^N), \tilde{J}(u) = 1\right\} \tag{11}$$

where

$$\tilde{\mathcal{E}}(u) = \int_{\mathbb{R}^N} \frac{1}{2}\sum_{i,j=1}^N \tilde{a}_{ij}\frac{\partial u}{\partial x_i}\frac{\partial u}{\partial x_j} + \tfrac{1}{2}\tilde{a}u^2\,dx,$$

$$\tilde{J}(u) = \int_{\mathbb{R}^N} \tilde{b}\frac{|u|}{p+1}^{p+1}.$$

If $\tilde{b} \equiv 0$, we agree that $I(\mathcal{C}) = +\infty$.

And we finally set

$$I^\infty = \inf_{\mathcal{C}} I(\mathcal{C}) \tag{12}$$

where the infimum is taken over all sets of coefficients $\mathcal{C}$ obtained as the limit of $\mathcal{C}(y_n)$ for some sequence $(y_n)_n$ in $\mathbb{R}^N$ such that $|y_n| \underset{n}{\to} \infty$.

The infimum (12) is called the problem at infinity. It is not very difficult to deduce from Theorem II.1 the following (see [**37,29**])

THEOREM II.2.

1. *Every minimizing sequence of (10) is relatively compact in $L^q(\mathbb{R}^N)$ (for all $2 \leq q \leq \frac{2N}{N-2}$ if $N \geq 3$, $2 \leq q < \infty$ if $N \leq 2$) if and only if $I < I^\infty$; one always has $I \leq I^\infty$. In particular, if $I < I^\infty$ there exists a minimum of (10) and all minimizing sequences are relatively compact in $H^1(\mathbb{R}^N)$.*
2. *Any minimizing sequence $(u_n)_n$ of (10) is relatively compact in $H^1(\mathbb{R}^N)$ up to a translation, i.e. there exists $(y_n)_n$ in $\mathbb{R}^N$ such that $(u_n(\cdot + y_n))_n$ is relatively compact in $L^q(\mathbb{R}^N)$. In addition, if $(\tilde{u}_n)_n$ is a minimizing sequence such that $\tilde{u}_n(\cdot + y_n)$ converges to some u in $L^q(\mathbb{R}^N)$ where $|y_n| \underset{n}{\to} \infty$ and $C(y_n)$ converges to C then u is a minimum of $I(C)$ and $I = I^\infty = I(C)$. Furthermore, if $a_{ij}(\cdot + y_n) \underset{n}{\to} \tilde{a}_{ij}(\cdot)$ a.e. $(\forall\, i, j)$, then $\tilde{u}_n(\cdot + y_n) \underset{n}{\to} \tilde{u}$ in $H^1(\mathbb{R}^N)$.*

REMARKS:

1. In addition, we know that $|\nabla \tilde{u}_n(\cdot + y_n)|^2$ is a tight sequence.
2. In particular, we may apply the above result to each $I(C_0)$. We then have to decide what is the corresponding problem at infinity $I^\infty(C_0)$.

For a general C_0, the class of sets of coefficients C' we may obtain from C_0 as we did from the original problem is contained in the class of C we described above and thus $I^\infty(C_0) \geq I(C_0) \geq I^\infty$ and all inequalities may be strict. This remark (and the Corollary which follows) is based upon a few observations concerning the class of coefficients sets C we are studying.

(i) The set of coefficients satisfying (6)–(7)–(8) with fixed constants is "relatively compact" (the topology may be defined precisely but we will not do so here); in particular we may extract converging subsequences from any sequence.

(ii) Denoting by $\tau_y C = \{(\tilde{a}_{ij}(\cdot + y))_{i,j}, \tilde{a}(\cdot + y), \tilde{b}(\cdot + y)\}$ for all $y \in \mathbb{R}^N$ if $C = \{(\tilde{a}_{ij})_{i,j}, \tilde{a}, \tilde{b}\}$, we observe that if C is a "limit set" of the original problem (10) then so is $\tau_y C$.

(iii) If $(C^n)_n$ is a sequence of "limit sets" of (10) which converges to some C then C is also in the same class.

With these observations, we may deduce the following result where we say that u is a minimum up to $\varepsilon > 0$ of a certain minimization problem such as (11) (i.e. $I(C)$) if

$$\tilde{J}(u) = 1, \qquad I(C) \leq \tilde{\mathcal{E}}(u) \leq I(C) + \varepsilon.$$

COROLLARY II.1.

1. *Let* $(a_{ij}^n)_{i,j}, a^n, b^n$ *be sequences of coefficients satisfying uniformly (6)–(8) and let* ε^n *be a sequence in* $(0,\infty)$ *converging to 0. For each* n, *we choose* u^n *a minimum up to* ε^n *of the corresponding infimum* I^n *(given by (10)). Then, there exists* $(y^n)_n$ *in* $\mathbb{R}^N$ *such that* $(u^n(\cdot + y^n))_n$ *is relatively compact in* $L^q(\mathbb{R}^N)$ *for all* $2 \leq q \leq \frac{2N}{N-2}$ *(if* $N \leq 3$, $2 \leq q < \infty$ *if* $N \leq 2$*).*
2. *Let* $(C^n)_n$ *be a minimizing sequence for the problem at infinity* I^∞ *and let* $(\varepsilon^n)_n$ *be a sequence in* $(0,\infty)$ *converging to 0. For each* n, *we choose* u^n *a minimum up to* ε^n *of* $I(C^n)$ *and we assume that* $u^n(\cdot + y^n)$ *converges in* $L^q(\mathbb{R}^N)$ *to some* $\tilde{u} \in L^1(\mathbb{R}^N)$, $C^n(y^n)$ *converges to some* C *which is thus an admissible set for* I^∞. *Then,* $I^\infty = I(C)$ *and* $I(C)$ *is achieved by* $\tilde{u}$.

We now present a list of applications of Theorem II.2 and Corollary II.1; we begin with the x-independent case which is an immediate application of Corollary II.1 ($I = I(C) = I^\infty, \forall C$). This case was already considered in the works of Z. Nehari [**44**], G. H. Ryder [**46**], C. V. Coffman [**20**], M. Berger [**12**], and W. Strauss [**48**] (at least as far as existence is concerned).

COROLLARY II.2. *Assume that* $(a_{ij})_{i,j}, a, b$ *are independent of* x, *i.e.* $a \in (0,\infty)$, $b \in (0,\infty)$, $(a_{ij})_{i,j}$ *is a positive definite matrix. Then, any minimizing sequence of (10) is relatively compact in* $H^1(\mathbb{R}^N)$ *up to a translation. In particular, there exists a minimum of (10).*

Similarly, one deduces the following result taken from M. J. Esteban and P. L. Lions [**29**]:

COROLLARY II.3. *Assume that, for each* $k \in \{1,\dots,N\}$, $(a_{ij})_{i,j}, a, b$ *are periodic with respect to* x_k *with a common period, say* T_k, *and satisfy (6)–(8). Then, any minimizing sequence of (10) is relatively compact in* $H^1(\mathbb{R}^N)$ *up to a translation (in* $\prod_{k=1}^N T_k\mathbb{Z}$*). In particular, there exists a minimum of (10).*

We next consider the natural situation when

$$\text{(13)} \quad \operatorname{meas}\left\{ \sum_{i,j=1}^N |a_{ij}(x) - a_{ij}^\infty| + |a(x) - a^\infty| + |b(x) - b^\infty| > \delta \right\} < \infty \quad \text{for all } \delta > 0$$

where $a^\infty, b^\infty, (a_{ij}^\infty)_{i,j}$ are thus respectively positive constants, a positive definite matrix. Then, obviously, any $\mathcal{C}$ is given by $\{(a_{ij}^\infty)_{i,j}, a^\infty, b^\infty\}$ and

$$I^\infty = I(\mathcal{C}) = \inf\left\{\mathcal{E}^\infty(u)/u \in H^1(\mathbb{R}^N), J^\infty(u) = 1\right\}$$

where

$$\mathcal{E}^\infty(u) = \int_{\mathbb{R}^N} \frac{1}{2}\sum_{i,j=1}^N a_{ij}^\infty \frac{\partial u}{\partial x_i}\frac{\partial u}{\partial x_j} + \tfrac{1}{2}a^\infty u^2\, dx,$$

$$J^\infty(u) = \int_{\mathbb{R}^N} b^\infty \frac{|u|^{p+1}}{p+1}\, dx.$$

Notice that we know by Corollary II.2 that I^∞ admits a minimum. We then have the following (see [**37,41**]):

COROLLARY II.4. *Assume (6)–(8) and (13). Then, any minimizing sequence of (10) is relatively compact in $H^1(\mathbb{R}^N)$ if and only if $I < I^\infty$. In particular, if $\mathcal{E}(\bar{u})J(\bar{u})^{-2/(p+1)} < I^\infty$ where $\bar{u}$ is a minimum of I^∞ (i.e. $\mathcal{E}^\infty(\bar{u}) = I^\infty$, $J^\infty(\bar{u}) = 1$), then any minimizing sequence of (10) is relatively compact in $H^1(\mathbb{R}^N)$ and there exists a minimum.*

Example II.1. Assume in addition to (6)–(8), (13)

$$(14)\qquad (a_{ij})_{i,j} \le (a_{ij}^\infty)_{i,j}, \quad a \le a^\infty, \quad b \ge b^\infty, \quad \text{a.e. in } \mathbb{R}^N$$

then as soon as these inequalities are not all equalities we deduce

$$\mathcal{E}(\bar{u})J(\bar{u})^{-\frac{2}{p+1}} < I^\infty$$

and we may apply Corollary II.4.

Example II.2. Assume now in addition to (6)–(8), (13)

$$(15)\qquad (a_{ij})_{i,j} \ge (a_{ij}^\infty)_{i,j}, \quad a \ge a^\infty, \quad b \le b^\infty, \quad \text{a.e. in } \mathbb{R}^N$$

then obviously $I = I^\infty$ and if there were a minimum of (10), this minimum would also be a minimum of I^∞ and this would imply that all inequalitites in (15) would be equalities.

Example II.3. We still assume (6)–(8), (13) and to simplify the presentation and notations, we take $a_{ij}(\cdot) \equiv \delta_{ij}$ for $1 \le i, j \le N$, $a \in L^\infty(\mathbb{R}^N)$, $b \in L^\infty(\mathbb{R}^N)$. In addition, we will assume either

$$(16)\qquad \begin{cases} a \equiv a^\infty, b \ge b^\infty \text{ a.e. } |x| \ge R_0, \\ \inf_{|x|\le R_n} \operatorname{ess}\left[b(x+y_n) - b^\infty\right] R_n^{(N-1)\frac{p-1}{2}} \\ \qquad\qquad \exp\left((p+1)(a^\infty)^{1/2}R_n\right) \underset{n}{\to} +\infty \end{cases}$$

or

$$\text{(17)}\qquad \begin{cases} b \equiv b^\infty, a \geq a^\infty \text{ a.e. } |x| \geq R_0, \\ \inf\limits_{|x|\leq R_n} \text{ess}\,[a^\infty - a(x+y_n)] \exp\left(2(a^\infty)^{1/2} R_n\right) \underset{n}{\to} +\infty \end{cases}$$

for some $R_0 < \infty$, $(y_n)_n$ in $\mathbb{R}^N$ such that $|y_n| \underset{n}{\to} \infty$, R_n in $(0, \infty)$ such that $R_n \underset{n}{\to} \infty$.

Then, we claim that we can find $\bar{u}$ minimum of I^∞ such that

$$\mathcal{E}(\bar{u}) J(\bar{u})^{-\frac{2}{p+1}} < I^\infty.$$

Indeed, we know there exists $u_1 > 0$ in $\mathbb{R}^N$ minimum of I^∞ solving

$$-\Delta u_1 + a^\infty u_1 = S b^\infty u_1^p \text{ in } \mathbb{R}^N, \quad u_1 \in H^1(\mathbb{R}^N)$$

where $S = 2(p+1)^{-1} I^\infty$. And by the results of B. Gidas, W. M. Ni and L. Nirenberg [**30,31**], we may assume that u_1 is radial. In addition, a straightforward argument (see [**48,9**]) shows that

$$u_1(x) \exp\left((a^\infty)^{1/2}|x|\right) |x|^{\frac{N-1}{2}} \to C_0$$

for some positive constant $C_0 > 0$.

Then, we choose $\bar{u}_n(x) = u_1(x - y_n)$ and we compute

$$\mathcal{E}(\bar{u}_n) = \int_{\mathbb{R}^N} \tfrac{1}{2}|\nabla u_1|^2 + \tfrac{1}{2} a(x+y_n) u_1^2 \, dx,$$

$$J(\bar{u}_n) = \int_{\mathbb{R}^N} \frac{b(x+y_n)}{p+1} u_1^{p+1} \, dx.$$

Hence, if (16) holds we deduce

$$\mathcal{E}(\bar{u}_n) J(\bar{u}_n)^{-\frac{2}{p+1}} = I^\infty \left\{ 1 + \int_{\mathbb{R}^N} \frac{1}{p+1} (b(x+y_n) - b^\infty) u_1^{p+1} \, dx \right\}^{-\frac{2}{p+1}}$$

while if (17) holds we obtain

$$\mathcal{E}(\bar{u}_n) J(\bar{u}_n)^{-\frac{2}{p+1}} = I^\infty - \tfrac{1}{2} \int_{\mathbb{R}^N} (a^\infty - a(x+y_n)) u_1^2 \, dx.$$

In the first case, we conclude observing that

$$\begin{aligned} &\int_{\mathbb{R}^N} (b(x+y_n) - b^\infty) u_1^{p+1} \, dx \\ &\geq \left(\int_{|x|\leq R_n} u_1^{p+1} \, dx \right) \inf_{|x|\leq R_n} \text{ess}(b(x+y_n) - b^\infty) - C \int_{|x|\geq R_n} u_1^{p+1} \, dx \\ &\geq \inf_{|x|\leq R_n} \text{ess}(b(x+y_n) - b^\infty) \frac{1}{b^\infty} - C \int_{|x|\geq R_n} u_1^{p+1} \, dx \\ &\geq \inf_{|x|\leq R_n} \text{ess}(b(x+y_n) - b^\infty) \frac{1}{b^\infty} - C \exp\left(-(p+1) a_\infty^{1/2} R_n\right) R_n^{-\frac{N-1}{2}(p-1)} \end{aligned}$$

for n large. While in the second case, we conclude similarly

$$\int_{\mathbb{R}^N} (a^\infty - a(x+y_n))u_1^2\,dx \geq$$
$$\left(\inf_{|x|\leq R_n} \text{ess}(a^\infty - a(x+y_n))\right) \int_{\mathbb{R}^N} u_1^2\,dx - C\exp(-2a_\infty^{1/2}R_n) > 0$$

for n large. Here and above, C denotes various constants independent of n.

Example II.4. We still assume (6)–(8), (13) and to simplify the presentation and notations, we take $a_{ij} \equiv a_{ij}^\infty$ for $1 \leq i,j \leq N$, $b \equiv b^\infty > 0$, and $a \in L^\infty$. We have seen in Example II.1 that if $a \leq a^\infty$ there exists a minimum of (10) while we saw in Example II.2 that if $a \geq a^\infty$ there does not exist a minimum of (10) (if $a \equiv a^\infty$). Hence, we now assume

$$\text{(18)} \qquad \inf \text{ess}(a - a^\infty) < 0.$$

In order to analyze the relation between I and I^∞, it is convenient to set $a = ta_1$ where $t > 0$, $a_1 \in L^\infty$, $a_1 \geq 0$ (for instance) $\text{meas}\{|a_1(x) - 1| > \delta\} < \infty$ for all $\delta > 0$ and a_1 satisfies (18) with a, a^∞ replaced by $a_1, 1$ so that $a^\infty = t$.

Then, we claim that there exists $t_0 \in [0,\infty)$ such that $I < I^\infty$ if $t > t_0$. Indeed, denoting by u_1 a minimum of I^∞ for $a = a_1$ (i.e. $a^\infty = 1$) then by a simple scaling argument $\lambda^{-N/(p+1)}u_1((\cdot - y/\lambda))$, for all $y \in \mathbb{R}^N$ and $\lambda > 0$, is a minimum of I^∞ for $a = \lambda^2 a_1$ (i.e. $a^\infty = \lambda^{-2}$). Thus, if we choose $\lambda > 0$ such that $\lambda^{-2} = t$, we may compute

$$\mathcal{E}\left(\lambda^{-\frac{N}{p+1}}u_1((\cdot - y)/\lambda)\right) = I^\infty + \lambda^{N\frac{p-1}{p+1}-2} \int_{\mathbb{R}^N} \{a_1(y+\lambda z) - 1\}u^2(z)\,dz$$
$$J\left(\lambda^{-\frac{N}{p+1}}u_1((\cdot - y)/\lambda)\right) = 1.$$

And we conclude provided we find $y \in \mathbb{R}^N$ such that

$$\int_{\mathbb{R}^N} \{a_1(y+\lambda z) - 1\}u^2(z)\,dz < 0$$

if $\lambda > 0$ is small enough. This is clearly possible if (18) holds.

We next present another application of our general results.

COROLLARY II.5. *Assume (6)–(8) and suppose in addition that for each sequence $(y_n)_n$ in $\mathbb{R}^N$ such that $|y_n| \underset{n}{\to} \infty$, $C(y_n)$ converges to $C = \{(\tilde{a}_{ij})_{i,j}, \tilde{a}, \tilde{b}\}$, there exists $h \in \mathbb{R}^N$ such that*

$$(a_{ij}(\cdot + h)) \leq (\tilde{a}_{ij})_{i,j}, \quad a(\cdot + h) \leq \tilde{a}, \quad b(\cdot + h) \geq \tilde{b} \text{ a.e. in } \mathbb{R}^N.$$

Then, there exists a minimum of (10).

REMARKS:

1. In fact, $I < I^\infty$ as soon as $(\tilde{a}, \tilde{b})$ does not coincide with (a, b) for each C.
2. This result extends a result due to W. Y. Ding and W. M. Ni [**27**] corresponding to a particular case when $a_{ij} \equiv \delta_{ij}$, $a \equiv$ constant > 0 and b is nonincreasing with respect to x_i for x_i large and nondecreasing for $-x_i$ large and additional assumptions are made that imply (19) in particular.

We conclude this section with a final remark inspired of [**27**].

COROLLARY II.6. *We assume (6)–(8) and we consider $(\hat{a}_{ij})_{i,j}, \hat{a}, \hat{b}$ also satisfying (6)–(8) such that*

$$\text{(20)}\quad \text{meas}\left\{ \sum_{i,j=1}^{N} |a_{ij} - \hat{a}_{ij}| + |a - \hat{a}| + \left|b - \hat{b}\right| \geq \delta \right\} < \infty, \text{ for all } \delta > 0.$$

We denote by $\hat{I}$ the infimum (10) corresponding to $(\hat{a}_{ij})_{i,j}, \hat{a}, \hat{b}$ and by $\hat{I}^\infty$ the corresponding problem at infinity. Then, $I^\infty = \hat{I}^\infty$. Thus, if $I < \hat{I}$, $I < I^\infty$ and in particular there exists a minimum of I.

III. More general nonlinearities in $\mathbb{R}^N$.

III.1. Reduction to some minimization problems.

We first explain how to build solutions of (1)–(2) under appropriate conditions on $g(x, t)$ by convenient minimization problems. The results which follow are minor extensions of results given in P. L. Lions [**37**] (along the lines of M. J. Esteban and P. L. Lions [**29**]) and they are based upon the reduction of (1)–(2) to a certain minimization problem involving an artificial constraint automatically satisfied by solutions of (1)–(2). This idea was introduced in C. V. Coffman [**18,19**], J. Hempel [**34**], and A. Ambrosetti [**1**]. We first observe that if u solves (1)–(2) "then"

$$\text{(21)}\qquad J(u) = 0$$

where

$$J(u) = \int_{\mathbb{R}^N} \sum_{i,j=1}^{N} a_{ij}(x) \frac{\partial u}{\partial x_i} \frac{\partial u}{\partial x_j} - g(x,u)\, u\, dx.$$

It is then "natural" to try to minimize over $M = \{u \in X - \{0\} / J(u) = 0\}$ the "action" functional

$$\mathcal{E}(u) = \int_{\mathbb{R}^N} \tfrac{1}{2} \sum_{i,j=1}^{N} a_{ij}(x) \frac{\partial u}{\partial x_i} \frac{\partial u}{\partial x_j} - G(x,u)\, dx \tag{22}$$

where $G(x,t) = \int_0^t g(x,s)\, ds$ and X will be either $H^1(\mathbb{R}^N)$ or $\mathcal{D}^{1,2}(\mathbb{R}^N)$ ($= \{u \in L^{2N/(N-2)}(\mathbb{R}^N),\ \nabla u \in L^2(\mathbb{R}^N)\}$ if $N \geq 3$, Hilbert space endowed of the scalar product $\int_{\mathbb{R}^N} (\nabla u \cdot \nabla v) dx$).

Next, if u is a minimum of $\mathcal{E}$ over M (and assuming that $\mathcal{E}, J$ are C^1 on X) we see there exists $\theta \in \mathbb{R}$ such that

$$- \sum_{i,j=1}^{N} \frac{\partial}{\partial x_i} \left(a_{ij} \frac{\partial u}{\partial x_j} \right) - g(x,u) =$$

$$= \theta \left\{ -2 \sum_{i,j=1}^{N} \frac{\partial}{\partial x_i} \left(a_{ij} \frac{\partial u}{\partial x_j} \right) - g(x,u) - \frac{\partial g}{\partial t}(x,u) u \right\} \text{ in } \mathcal{D}'(\mathbb{R}^N)$$

and multiplying by u and integrating by parts we find

$$0 = J(u) = \theta \left\{ J(u) + \int_{\mathbb{R}^N} \sum_{i,j=1}^{N} a_{ij} \frac{\partial u}{\partial x_i} \frac{\partial u}{\partial x_j} - \frac{\partial g}{\partial t}(x,u) u^2\, dx \right\}$$

or

$$0 = \theta \left\{ \int_{\mathbb{R}^N} \sum_{i,j=1}^{N} a_{ij} \frac{\partial u}{\partial x_i} \frac{\partial u}{\partial x_j} - \frac{\partial g}{\partial t}(x,u) u^2\, dx \right\}$$

while

$$0 = J(u) = \int_{\mathbb{R}^N} \sum_{i,j=1}^{N} a_{ij} \frac{\partial u}{\partial x_i} \frac{\partial u}{\partial x_j} - g(x,u) u\, dx.$$

Therefore, if we assume

$$g(x,t) = -a(x)t + f(x,t) \quad \forall t \in \mathbb{R}, \text{ a.e. } x \in \mathbb{R}^N \tag{23}$$

and

$$(24) \qquad f(x,t)|t|^{-\gamma}t^{-1} \quad \begin{array}{l} \text{is nondecreasing for } t \geq 0, \\ \text{is nonincreasing for } t \leq 0 \end{array}$$

a.e. $x \in \mathbb{R}^N$, for some $\gamma > 0$, we deduce easily that $\theta = 0$ and thus u is a solution in X of (1). To ensure the positivity it will be convenient to set

$$(25) \qquad f(x,t) = 0, \text{ for } t \leq 0, \text{ a.e. } x \in \mathbb{R}^N.$$

All throughout this section, we assume that (23), (25) hold, that $a(x) \in L^\infty(\mathbb{R}^N)$, $f(x,\cdot) \in C(\mathbb{R})$ uniformly in $x \in \mathbb{R}^N$, $a_{ij} \in L^\infty(\mathbb{R}^N)$, $(1 \leq i,j \leq N)$ and

$$(26) \quad \forall R < \infty, \exists C_R \geq 0, \quad |f(x,t)| \leq C_R \text{ if } |t| \leq R, \text{ a.e. } x \in \mathbb{R}^N.$$

$$(27) \quad \begin{cases} \lim_{t \to +\infty} f(x,t)t^{-\frac{N+2}{N-2}} = 0, & \text{uniformly in } x \in \mathbb{R}^N, \text{ if } N \geq 3 \\ \lim_{t \to +\infty} f(x,t)e^{-t^\beta} = 0, & \text{uniformly in } x \in \mathbb{R}^N, \text{ for some} \\ & \beta \in (0,2), \text{ if } N = 2 \end{cases}$$

$$(28) \quad f(x,t) \geq 0, \quad \forall t \in \mathbb{R}, \text{ a.e. } x \in \mathbb{R}^N,$$

$$(29) \quad \begin{cases} \lim_{t \to +\infty} f(x,t)t^{-1} = +\infty, \text{ uniformly for } x \in \mathcal{O} \text{ bounded} \\ \qquad \text{open set of } \mathbb{R}^N \end{cases}$$

We will also use the following assumptions

$$(30) \qquad \lim_{t \to 0+} f(x,t)t^{-1} = 0, \text{ uniformly in } x \in \mathbb{R}^N$$

$$(31) \qquad \lim_{t \to 0+} f(x,t)t^{-\frac{N+2}{N-2}} = 0, \text{ uniformly in } x \in \mathbb{R}^N, \text{ if } N \geq 3$$

We just saw that if (24) holds then we will build a solution of (1)–(2) (in X) if we solve

$$(32) \qquad I = \inf\{\mathcal{E}(u)/u \in X, u \neq 0, J(u) = 0\}.$$

To this end, we need to assume (6) or

$$\exists \nu > 0, \quad \forall \varphi \in \mathcal{D}(\mathbb{R}^N),$$

$$\int_{\mathbb{R}^N} \sum_{i,j=1}^N a_{ij} \frac{\partial \varphi}{\partial x_i} \frac{\partial \varphi}{\partial x_j} + a\varphi^2 \, dx \geq \nu \int_{\mathbb{R}^N} |\nabla \varphi|^2 \, dx.$$

Again, we have to introduce a notion of problem at infinity. In order to do so, we consider any sequence $(y_n)_n$ in $\mathbb{R}^N$ such that $|y_n| \underset{n}{\to} \infty$ and we set $\mathcal{C}(y_n) = \{(a_{ij}(\cdot + y_n))_{i,j},\ a(\cdot + y_n),\ f(\cdot + y_n, \cdot)\}$. Next, extracting a subsequence if necessary, we may assume that $(a_{ij}(\cdot + y_n))_{i,j}$ G-converges to some $(\tilde{a}_{ij})_{i,j}$, $a(\cdot + y_n)$ converges to some $\tilde{a}(w - L^\infty_\star)$, $f(\cdot + y_n, \cdot)$ converges to some $\tilde{f}(\cdot, \cdot)$ $(w - L^\infty(\mathbb{R}^N \times (-R, R))\star$, for all $R < \infty)$. In that case, we say that $\mathcal{C}(y_n)$ converges to $\mathcal{C} = \{(\tilde{a}_{ij})_{i,j}, \tilde{a}, \tilde{f}\}$. For any such limit system, we set

$$I(\mathcal{C}) = \inf \left\{ \tilde{\mathcal{E}}(u) / u \in X, u \neq 0, \quad \tilde{J}(u) = 0 \right\} \tag{34}$$

with

$$\tilde{\mathcal{E}}(u) = \int_{\mathbb{R}^N} \tfrac{1}{2} \sum_{i,j=1}^N \tilde{a}_{ij} \frac{\partial u}{\partial x_i} \frac{\partial u}{\partial x_j} + \tfrac{1}{2} \tilde{a} u^2 - \tilde{F}(x, u)\, dx,$$

$$\tilde{J}(u) = \int_{\mathbb{R}^N} \sum_{i,j=1}^N \tilde{a}_{ij} \frac{\partial u}{\partial x_i} \frac{\partial u}{\partial x_j} + \tilde{a} u^2 - \tilde{f}(x, u) u\, dx.$$

And if $\{u \in X, u \neq 0, \tilde{J}(u) = 0\} = \emptyset$, we set $I(\mathcal{C}) = +\infty$.

Finally, we introduce

$$I^\infty = \inf_{\mathcal{C}} I(\mathcal{C}) \tag{35}$$

where the infimum is taken over all limit systems $\mathcal{C}$. Then, we have the

THEOREM III.1. *We assume (24) and either (6), (30), or (31), (33), $a^- \in L^{N/2}(\mathbb{R}^N)$ and $N \geq 3$. In the first case, we choose $X = H^1(\mathbb{R}^N)$ while in the second case, we choose $X = \mathcal{D}^{1,2}(\mathbb{R}^N) \cap L^2(a^+)$ where $L^2(a^+) = \left\{\varphi / \int_{\mathbb{R}^N} a^+ \varphi^2\, dx < \infty\right\}$. Then, we have*

(i) $I = \inf\{\mathcal{E}(u) / u \in X, u \neq 0, J(u) \leq 0\} \leq I^\infty$

(ii) *Every minimizing sequence of I is relatively compact in $L^q(\mathbb{R}^N)$ (for all $2 \leq q \leq \frac{2N}{N-2}$ if $X = H^1$; and $q = \frac{2N}{N-2}$ if $N \geq 3$, $X = \mathcal{D}^{1,2} \cap L^2(a^+)$) if and only if $I < I^\infty$. In particular, if $I < I^\infty$, there exists a minimum of I and all minimizing sequences are relatively compact in X.*

(iii) *Let $(u_n)_n$ be a minimizing sequence of I; there exists $(y_n)_n$ in $\mathbb{R}^N$ such that $(u_n(\cdot + y_n))_n$ is relatively compact in $L^q(\mathbb{R}^N)$. In addition, if $(\tilde{u}_n)_n$ is a minimizing sequence such that $\tilde{u}_n(\cdot + y_n)$ converges* *to some* $\tilde{u}$ *in* $L^q(\mathbb{R}^N)$ *where* $|y_n| \underset{n}{\to} \infty$ *and* $C(y_n)$ *converges to* C, *then* $\tilde{u}$ *is a minimum of* $I(C)$ *and* $I = I^\infty = I(C)$. *Furthermore, if* $a_{ij}(\cdot + y_n) \underset{n}{\to} \tilde{a}_{ij}$ a.e. $(\forall i, j)$, *then* $\tilde{u}_n(\cdot + y_n) \underset{n}{\to} \tilde{u}$ *in* X.

(iv) *If* $I^\infty < \infty$, *let* $(C^n)_n$ *be a sequence of "limit systems" such that* $I(C^n) \underset{n}{\to} I^\infty$ *and let* $(\varepsilon^n)_n$ *be a sequence in* $(0, \infty)$ *converging to 0. For each* n, *we choose* u^n *a minimum up to* ε^n *of* $I(C^n)$, *then there exists* $(y^n)_n$ *in* $\mathbb{R}^N$ *such that* $(u^n(\cdot + y^n))_n$ *is relatively compact in* $L^q(\mathbb{R}^N)$. *In addition, if we assume that* $u^n(\cdot + y^n)$ *converges in* $L^q(\mathbb{R}^N)$ *to some* u *in* X, *and that* $C^n(y^n)$ *converges to some* C *which is therefore a "limit system," then* $I^\infty = I(C)$ *and* $I(C)$ *is achieved by* $\tilde{u}$.

REMARKS:

1. In addition, we know that $|\nabla u_n(\cdot + y_n)|^2$ or $|\nabla u^n(\cdot + y^n)|^2$ are tight.
2. If $\tilde{a}^n_{ij}(\cdot + y^n) \underset{n}{\to} \tilde{a}_{ij}$ a.e. $(\forall i, j)$ then $u^n(\cdot + y^n)$ converges to $\tilde{u}$ in X.

The above result has exactly the same type of applications as Theorem II.2; hence, we skip those straightforward adaptations.

We next want to conclude this section by another reduction to some minimization problem. We begin with the simple case when $(a_{ij})_{i,j}, g$ do not depend on x, i.e. we consider

$$\text{(36)} \qquad \begin{cases} -\displaystyle\sum_{i,j=1}^{N} a_{ij} \frac{\partial^2 u}{\partial x_i \, \partial x_j} = g(u) \text{ in } \mathbb{R}^N, \quad u > 0 \text{ in } \mathbb{R}^N, \\ \nabla u \in L^2(\mathbb{R}^N), \quad G(u) \in L^1(\mathbb{R}^N) \end{cases}$$

and $N \geq 2$, $(a_{ij})_{i,j}$ is a positive definite matrix, $g \in C(\mathbb{R})$ and we set $g(t) = 0$ if $t \leq 0$. Before stating the result and the assumptions, we first recall from [**9**] that if u solves (36) (and say $u \in L^\infty_{\mathrm{loc}}(\mathbb{R}^N)$) then the following identity holds:

$$\text{(37)} \qquad \frac{N-2}{2} \int_{\mathbb{R}^N} \sum_{i,j=1}^{N} a_{ij} \frac{\partial u}{\partial x_i} \frac{\partial u}{\partial x_j} \, dx = N \int_{\mathbb{R}^N} G(u) \, dx.$$

This identity is nothing but the conservation law associated with the "scaling" invariance of $\mathbb{R}^N$ and was obtained in the case of bounded domains by S. Pohozaev [**45**]. Again, in view of this identity, it is "natural" to consider

$$I = \inf \{ \mathcal{E}(u) / u \in X, u \neq 0, \quad J(u) = 0 \}$$

where

$$\mathcal{E}(u) = \int_{\mathbf{R}^N} \tfrac{1}{2} \sum_{i,j=1}^{N} a_{ij} \frac{\partial u}{\partial x_i} \frac{\partial u}{\partial x_j} - G(u)\, dx,$$

$$J(u) = \int_{\mathbf{R}^N} \frac{N-2}{2} \sum_{i,j=1}^{N} a_{ij} \frac{\partial u}{\partial x_i} \frac{\partial u}{\partial x_j} - NG(u)\, dx,$$

$$X = \{u / \nabla u \in L^2(\mathbf{R}^N), G(u) \in L^1(\mathbf{R}^N)\}.$$

Observe next that in view of the form of $\mathcal{E}$ and J, it is equivalent to consider

$$I = \inf \{E(u)/u \in X, u \neq 0, \quad J(u) = 0\} \tag{38}$$

where $E(u) = \frac{1}{2} \int_{\mathbf{R}^N} \sum_{i,j=1}^{N} a_{ij} \frac{\partial u}{\partial x_i} \frac{\partial u}{\partial x_j}\, dx$. Next, if u is a minimum of I then (neglecting differentiability technicalities) there exists $\theta \in \mathbf{R}$ such that

$$- \sum_{i,j=1}^{N} a_{ij} \frac{\partial^2 u}{\partial x_i \partial x_j} = \theta \left\{ (N-2) \sum_{i,j=1}^{N} a_{ij} \frac{\partial^2 u}{\partial x_i \partial x_j} + Ng(u) \right\}. \tag{39}$$

Applying now the analogue of (37) for this new equation, we deduce

$$[1 + \theta(N-2)] \frac{N-2}{2} \int_{\mathbf{R}^N} \sum_{i,j=1}^{N} a_{ij} \frac{\partial u}{\partial x_i} \frac{\partial u}{\partial x_j}\, dx = (N\theta) N \int_{\mathbf{R}^N} G(u)\, dx$$

and if $J(u) = 0$, we deduce $1 + \theta(N-2) = \theta N$, i.e. $\theta = \frac{1}{2}$ and (39) reduces to (36). Therefore, we only have to find a minimum of I where I is given by (38). The fact that minima of (38) yield solutions of (36) was used in M. Struwe [**49**] and is implicit in the works by Coleman, Glazer and Martin [**23**], H. Berestycki and P. L. Lions [9]. The result which follows is essentially the analogue of results taken from [**37**,**14**]. We will assume

$$\exists \varsigma > 0, \quad G(\varsigma) > 0 \tag{40}$$

$$\varlimsup_{t \to +\infty} G^+(t) t^{-\frac{2N}{N-2}} = 0, \varlimsup_{t \to 0+} G^+(t) t^{-\frac{2N}{N-2}} = 0 \text{ if } N \geq 3 \tag{41}$$

while

$$\varlimsup_{t \to +\infty} G^+(t) e^{-t^\beta} = 0 \text{ for some } \beta \in (0,2), G(t) < 0 \quad \text{for } t \in (0, \delta) \text{ if } N = 2 \tag{42}$$

for some $\delta > 0$. And if $N \geq 3$, X is precisely given by

$$X = \left\{ u \in L^{\frac{2N}{N-2}}(\mathbb{R}^N), \nabla u \in L^2(\mathbb{R}^N), G^-(u) \in L^1(\mathbb{R}^N) \right\}$$

while if $N = 2$, X is given by

$$X = \left\{ u \in L^1_{\mathrm{loc}}(\mathbb{R}^N), \nabla u \in L^2(\mathbb{R}^N), G^-(u) \in L^1(\mathbb{R}^N) \right\}.$$

THEOREM III.2. *We assume (40).*

(i) *If $N \geq 3$, we assume in addition (41). Then, any minimizing sequence of (38) is relatively compact in X up to a translation and there exists a minimum.*

(ii) *If $N = 2$, we assume in addition (42). Then, for any minimizing sequence (u_n) of (38), there exist $(y_n)_n$ in $\mathbb{R}^2$, $(\sigma_n)_n$ in $(0, \infty)$ such that the new minimizing sequence $u_n((\cdot + y_n)/\sigma_n)$ is relatively compact in X and there exists a minimum.*

Let us mention a remark on the following analogue of (38) where $\mathcal{E}, J$ are now given by

$$\mathcal{E}(u) = \tfrac{1}{2} \int_{\mathbb{R}^N} \sum_{i,j=1}^{N} a_{ij} \frac{\partial u}{\partial x_i} \frac{\partial u}{\partial x_j} \, dx$$

$$J(u) = \int_{\mathbb{R}^N} \frac{N-2}{2} \sum_{i,j=1}^{N} a_{ij} \frac{\partial u}{\partial x_i} \frac{\partial u}{\partial x_j} - NG(x, u) \, dx$$

Then, a minimum of (38) is a solution of (1)–(2), if g depends only on $\frac{x}{|x|}$. If this is the case, say $g \in C(S^{N-1} \times \mathbb{R})$, then we may consider, for instance, the case when $N \geq 3$ and we assume that (41) holds uniformly for $\omega \in S^{N-1}$ and that there exists $\varphi \in \mathcal{D}(\mathbb{R}^N)$ such that

$$\int_{\mathbb{R}^N} G\left(\frac{x}{|x|}, \varphi\right) dx > 0.$$

To define a problem at infinity we just observe that if $|x_n| \underset{n}{\to} \infty$, $\frac{x_n}{|x_n|} \underset{n}{\to} \omega \in S^{N-1}$ then $G\left(\frac{x + x_n}{|x + x_n|}, t\right) \to G(\omega, t)$ therefore we introduce

$$I_\omega = \inf\{\mathcal{E}(u) / u \in X, u \neq 0, \quad J^\omega(u) = 0\}$$

where $J^\omega(u) = \int_{\mathbb{R}^N} G(\omega, u(x)) \, dx$ and we set

$$I^\infty = \min_{\omega \in S^{N-1}} I_\omega.$$

Then, using again the concentration-compactness argument, we may show that any minimizing sequence is relatively compact in X up to a translation and any minimizing sequence is relatively compact in X if and only if $I < I^\infty$. In particular, if $I < I^\infty$, there exists a minimum. However, we have been unable to give an example where $I < I^\infty$ and all simple cases we considered in fact led to $I = I^\infty$. The existence of a minimum is thus an open question.

We now turn to a more general situation when $a_{ij} \equiv 0$ if $i \le 2, j \ge 3$ or if $j \le 2$, $i \ge 3$ and when $(a_{ij})_{i,j}, g$ do not depend on at least two variables, say x_1, x_2 where $x = (x_1, x_2, \dots, x_N)$ and we thus assume $N \ge 3$. The arguments which follow are adapted from O. Kavian [**35**]. The basic observation is that if u is a solution of (1)–(2) (say in $L^\infty_{\text{loc}}(\mathbb{R}^N)$, such that $\nabla u \in L^2(\mathbb{R}^N)$, $G(x,u) \in L^1(\mathbb{R}^N)$) then the invariance of $\mathbb{R}^N$ by dilations in (x_1, x_2) leads to the following identity:

$$\frac{1}{2}\int_{\mathbb{R}^N} |\nabla_{x'} u|^2\,dx = \int_{\mathbb{R}^N} G(x',u)\,dx \tag{43}$$

where $x = (x_1, x_2, x')$, $x' \in \mathbb{R}^{N-2}$, $\nabla_{x'} u = (\frac{\partial u}{\partial x_j})_{3 \le j \le N}$. We then denote by

$$\mathcal{E}(u) = \int_{\mathbb{R}^N} \tfrac{1}{2}|\nabla_1 u|^2\,dx,$$

$$J(u) = \int_{\mathbb{R}^N} \tfrac{1}{2}|\nabla_{x'} u|^2 - G(x',u)\,dx$$

where $\nabla_1 u = (\frac{\partial u}{\partial x_1}, \frac{\partial u}{\partial x_2})$. And following O. Kavian [**35**] we consider

$$I = \inf\{\mathcal{E}(u)/u \in H^1(\mathbb{R}^N), J(u) = 0, u \ne 0\} \tag{44}$$

and we assume for instance (many extensions and variants are possible) that g is given by (23), $(a_{ij})_{i,j}, a$ satisfy (6)–(7), f satisfies (25)–(27), (30) and $f(x,\cdot)$ is continuous on each interval $[-R,+R]$ $(\forall R < \infty)$ uniformly for $x \in \mathbb{R}^N$ (a.e.) and

$$\exists \varphi \in H^1(\mathbb{R}^N), \quad J(\varphi) \le 0. \tag{45}$$

Using (43), it is easy to check that a minimum of (44) is indeed a solution of (1). Then, we introduce limit systems $\mathcal{C}$ of $\mathcal{C}(y_n) = \{(a_{ij}(\cdot + y_n))_{i,j}, a(\cdot + y_n), f(\cdot + y_n, t)\}$ for arbitrary sequences $|y_n| \underset{n}{\to} \infty$ and we set

$$I = \inf\left\{\mathcal{E}(\tilde{u})/u \in H^1(\mathbb{R}^N), \quad \tilde{J}(u) = 0\right\}$$

where

$$\tilde{\mathcal{E}}(u) = \int_{\mathbb{R}^N} \tfrac{1}{2} \sum_{i,j=1}^{N} a_{ij}(x') \frac{\partial u}{\partial x_i} \frac{\partial u}{\partial x_j}\, dx,$$

$$\tilde{J}(u) = \int_{\mathbb{R}^N} \tfrac{1}{2} \sum_{i,j=1}^{N} \tilde{a}_{ij}(x') \frac{\partial u}{\partial x_i} \frac{\partial u}{\partial x_j}\, dx - \tilde{G}(x', u)\, dx.$$

Notice that is is enough to consider sequences y^n such that $y_1^n = y_2^n = 0$. Then, we define the problem at infinity by

$$I^\infty = \inf_{C} I(C) \tag{46}$$

where the infimum is taken over all limit systems C. And we have the

THEOREM III.3. *Under the above assumptions, we have*

$$I = \inf\{\mathcal{E}(u)/u \in H^1(\mathbb{R}^N), J(u) \geq 0, u \neq 0\} \leq I^\infty \tag{47}$$

and for any minimizing sequence $(u_n)_n$ of (44) there exist $(y'_n)_n$ in $\mathbb{R}^{N-2}$, $(\sigma_n)_n$ in $(0,\infty)$ such that $\left(u_n(\frac{x_1}{\sigma_n}, \frac{x_2}{\sigma_n}, x' + y'_n)\right)_n$ is relatively compact in $L^q(2 \leq q \leq \frac{2N}{N-2})$. Furthermore, if $I < I^\infty$, for any minimizing sequence $(u_n)_n$ of (44) there exists $(\sigma_n)_n$ in $(0,\infty)$ such that $\left(u_n(\frac{x_1}{\sigma_n}, \frac{x_2}{\sigma_n}, x')\right)_n$ is relatively compact in $H^1(\mathbb{R}^N)$ and there exists a minimum of (44).

REMARKS:

1. Exactly as in Theorem III.1, the problem at infinity is well-posed.
2. It is easy to give, exactly as in the preceding section, conditions which ensure the strict inequality $I < I^\infty$ and thus the existence of a minimum of (44).

III.2 The Mountain Pass Lemma approach.

In the previous sections, solutions of (1)–(2) were built using convenient minimization problems; we wish now to study the possibility of obtaining solutions of (1)–(2) directly by critical point arguments of the Mountain Pass Lemma type (see A. Ambrosetti and P. H. Rabinowitz [**2**]). Of course, as we explained in the Introduction and as we saw in the preceding sections, the usual compactness condition — known as Palais-Smale condition — does not hold and therefore the standard arguments do not apply.

Our analysis will be based upon one simple result which explains how Palais-Smale condition is possibly violated. Such results — for related losses of compactness — were obtained in different problems by P. Sacks and K. Uhlenbeck [**47**], P. L. Lions [**38**], H. Brezis and J. M. Coron [**13**], and M. Struwe [**50**]. In the context of semilinear elliptic problems in $\mathbb{R}^N$, variants of these results may be found in P. L. Lions [**36**], and V. Benci and G. Cerami [**8**].

We thus consider a sequence $(u^n)_n$ in $H^1(\mathbb{R}^N)$ satisfying

$$-\sum_{i,j=1}^{N} \frac{\partial}{\partial x_i}\left(a_{ij}\frac{\partial u^n}{\partial x_j}\right) - g(x,u^n) \underset{n}{\to} 0 \text{ in } H^{-1}(\mathbb{R}^N) \tag{48}$$

and we assume throughout this section that $a_{ij} \equiv a_{ji} \in L^\infty(\mathbb{R}^N)$ $(\forall 1 \leq i,j \leq N)$ and (3) holds, and that $g(x,0) \equiv 0$ a.e. in $\mathbb{R}^N$, $g \in L^\infty(\mathbb{R}^N \times (-R,R))$ $(\forall R < \infty)$ and $g(x,\cdot)$ is continuous on $(-R,R)$ $(\forall R < \infty)$ uniformly in $x \in \mathbb{R}^N$ (a.e.). There are many variants and extensions of the result which follows: to simplify the presentation, we give only one example of the type of results which may be obtained (we also indicate another one in the remarks which follow). We first recall that if $(y_n)_n \subset \mathbb{R}^N$ is a sequence such that $|y_n| \underset{n}{\to} \infty$, extracting a subsequence if necessary, we may assume that

$$\mathcal{C}(y_n) = \{(a_{ij}(\cdot + y_n))_{1\leq i,j\leq N}, g(\cdot + y_n, \cdot)\}$$

converges to some limit system $\mathcal{C} = \{(\tilde{a}_{ij})_{1\leq i,j\leq N}, \tilde{g}(\cdot,\cdot)\}$ (in the sense described in the preceding sections: G-convergence for the matrices, weak convergence for g). We next state the conditions of g we will use:

$$\exists \delta \in (0,1), -\frac{1}{\delta}t^2 \leq tg(x,t) \leq -\delta t^2 \text{ for } |t| \leq \delta, \text{ a.e. } x \in \mathbb{R}^N \tag{49}$$

$$\begin{cases} \lim\limits_{|t|\to\infty} |g(x,t)||t|^{-\frac{N+2}{N-2}} = 0 \text{ uniformly for } x \in \mathbb{R}^N \text{ a.e., if } N \geq 3 \\ \lim\limits_{|t|\to\infty} |g(x,t)|e^{|t|^\beta} = 0 \text{ uniformly for } x \in \mathbb{R}^2 \text{ a.e., where } \beta \in (0,2) \end{cases} \tag{50}$$

(no conditions are needed at infinity if $N = 1$). Then, we have the

THEOREM III.4. *We assume (49) and (50) and we consider a bounded sequence $(u^n)_n$ in $\mathbb{R}^N$ satisfying (48). Then, either u^n converges in H^1 to*

0 or there exist a subsequence of $(u^n)_n$ *that we still denote by* u^n, $m \geq 1$, *limit systems* $\mathcal{C}^k = \{(\tilde{a}^k_{ij})_{1\leq i,j\leq N}, \tilde{g}^k\}$ *for* $2 \leq k \leq m$ *(when* $m \geq 2$*), solutions* $(u_k)_{1\leq k\leq m}$ *of respectively*

$$(51)\quad \begin{cases} -\displaystyle\sum_{i,j=1}^{N} \frac{\partial}{\partial x_i}\left(a_{ij}\frac{\partial u_1}{\partial x_j}\right) = g(x,u_1) \text{ in } \mathbb{R}^N, u_1 \in C_0 \cap H^1, u_1 \not\equiv 0 \\ -\displaystyle\sum_{i,j=1}^{N} \frac{\partial}{\partial x_i}\left(\tilde{a}_{ij}\frac{\partial u_k}{\partial x_j}\right) = \tilde{g}^k(x,u_k) \text{ in } \mathbb{R}^N, u_k \in C_0 \cap H^1, u_k \not\equiv 0, \\ \qquad \text{for } 2 \leq k \leq m \end{cases}$$

such that we have for $2 \leq q \leq \frac{2N}{N-2}$ *(if* $N \geq 3$, $2 \leq q < \infty$ *if* $N = 2$*)*

$$(52)\qquad \left\| u^n - u_1 - \sum_2^m u_k(\cdot - y_n^k) \right\|_{L^q(\mathbb{R}^N)} \underset{n}{\to} 0$$

$$(53)\qquad \left| \int_{\mathbb{R}^N} \sum_{i,j=1}^{N} \left\{ a_{ij}\frac{\partial(u^n-u_1)}{\partial x_i}\frac{\partial(u^n-u_1)}{\partial x_j} + \right.\right. \\ \left.\left. - \sum_{k=2}^{m} \tilde{a}^k_{ij}(x-y^k_n)\frac{\partial u_k(x-y^k_n)}{\partial x_i}\frac{\partial u_k(x-y^k_n)}{\partial x_j} \right\} dx \right| \underset{n}{\to} 0$$

where $|y^k_n| \underset{n}{\to} \infty$, $\mathcal{C}(y^k_n) \underset{n}{\to} \mathcal{C}^k$, $|y^k_n - y^{k'}_n| \underset{n}{\to} \infty$ *if* $k \neq k'$ *for all* $2 \leq k, k' \leq m$ *(when* $m \geq 2$*). In addition, if* $a_{ij}(\cdot + y^k_n) \underset{n}{\to} \tilde{a}^k_{ij}$ a.e. *in* $\mathbb{R}^N (\forall 1 \leq i,j \leq N$, $\forall 2 \leq m \leq k)$ *then we also have*

$$(54)\qquad \left\| u^n - u_1 - \sum_2^m u_k(\cdot - y_n^k) \right\|_{H^1(\mathbb{R}^N)} \underset{n}{\to} 0.$$

Furthermore, if $u^n \geq 0$ a.e. *in* $\mathbb{R}^N$ *for all* $n > 1$, *then* $u^k > 0$ *in* $\mathbb{R}^N$ *for all* $1 \leq k \leq m$.

REMARKS:

1. If $m = 1$, we see that $u^n \underset{n}{\to} u_1$ in $H^1(\mathbb{R}^N)$.
2. A similar result holds for $N \geq 3$ if we replace (49) by

$$(55)\qquad \lim_{|t|\to 0} |g(x,t)||t|^{-\frac{N+2}{N-2}} = 0 \text{ uniformly for } x \in \mathbb{R}^N \text{ a.e.}$$

then the above result still holds except that $q = \frac{2N}{N-2}$ in (52).

3. Of course, we deduce from (52),(53) that if we denote by

$$\mathcal{E}(u) = \int_{\mathbb{R}^N} \tfrac{1}{2} \sum_{i,j=1}^{N} a_{ij} \frac{\partial u}{\partial x_i} \frac{\partial u}{\partial x_j} - G(x,u)\, dx, \quad \forall u \in H^1(\mathbb{R}^N)$$

$$\tilde{\mathcal{E}}^k(u) = \int_{\mathbb{R}^N} \tfrac{1}{2} \sum_{i,j=1}^{N} \tilde{a}^k_{ij} \frac{\partial u}{\partial x_i} \frac{\partial u}{\partial x_j} - \tilde{G}^k(x,u)\, dx, \quad \forall u \in H^1(\mathbb{R}^N)$$

where $G(x,t) = \int_0^t g(x,s)ds$, $\tilde{G}^k(x,t) = \int_0^t \tilde{g}^k(x,s)ds$.

(56) $$\mathcal{E}(u^n) \underset{n}{\to} \mathcal{E}(u_1) + \sum_{k=2}^{m} \tilde{\mathcal{E}}^k(u_k)$$

(as soon as $u^n \underset{n}{\not\to} 0$ in $H^1(\mathbb{R}^N)$ as it is the case if $\mathcal{E}(u^n) \underset{n}{\not\to} 0$).

In the remainder of this section, we will be interested only in positive solutions of (1). We then set

(57) $\bar{c}^\infty = \inf_{C} \inf\{\tilde{\mathcal{E}}(u)/u \text{ solution of (58)}\}$

(58) $$-\sum_{i,j=1}^{N} \frac{\partial}{\partial x_i}\left(\tilde{a}_{ij} \frac{\partial u}{\partial x_j}\right) = \tilde{g}(x,u) \text{ in } \mathbb{R}^N, u \in C_0 \cap H^1, u > 0 \text{ in } \mathbb{R}^N$$

where the infimum is taken over all limit systems $C = \{(\tilde{a}_{ij})_{1 \le i,j \le N}, \tilde{g}\}$ and $\tilde{\mathcal{E}}$ is the corresponding functional. As we will explain below, one often has $\bar{c}^\infty > 0$. Then, (56) immmediately implies the

COROLLARY III.1. *Under the assumptions of Theorem III.4, and if $\bar{c}^\infty > 0$, then either there exists a solution of (1)–(2) in $H^1 \cap C_0$ such that $\mathcal{E}(u) < 0$, or the following property holds for all $c < \bar{c}^\infty$*

(59) $$\begin{cases} \text{if } (u^n)_n \text{ is a bounded sequence in } H^1(\mathbb{R}^N), \text{ satisfying (48) and} \\ (u^n)^- \underset{n}{\to} 0 \text{ in } H^1, \mathcal{E}(u^n) \underset{n}{\to} c \text{ then } u^n \text{ is relatively compact.} \end{cases}$$

REMARKS:

1. If for all limit systems and all solutions of (52), $\tilde{\mathcal{E}}(u) = \bar{c}^\infty$ then the same "energy balance" yields the alternative: either there exists u solution of (1)–(2) in $H^1 \cap C_0$, or (59) holds for all $c \notin \{m\bar{c}^\infty / m \ge 1\}$. This is typically the case when $(a_{ij}(x))_{i,j}, g(x,\cdot)$ converge as $|x| \to \infty$

to $(\tilde{a}_{ij})_{i,j}, \tilde{g}(\cdot)$ and that, up to translation, there exists a unique positive solution of (58); we will come back on this point later on.
2. The above type of observations is closely related to those mentioned in H. Brezis and L. Nirenberg [**15**], and P. L. Lions [**38**]. And similar considerations appear in J. F. Yang [**51**].

We now explain why it is natural to assume that $\bar{c}^\infty > 0$. Indeed, if u is a solution of (58) for some limit system $\mathcal{C}$ we claim that $\tilde{\mathcal{E}}(u) > 0$ as soon as one of the conditions given below is satisfied for each $\mathcal{C}$

$$(60) \qquad \begin{cases} \sum_{j=1}^N \tilde{a}_{ij}\xi_j = \lambda(x)\xi_i \text{ a.e. } (\forall i) \text{ for some } \lambda \in L^\infty \\ \tilde{G}(\cdot,t), \tilde{a}_{ij}(\cdot) \text{ are independent of } x\cdot\xi \\ \qquad\qquad \text{for some } |\xi| = 1 (\forall t \geq 0, \forall 1 \leq i,j \leq N) \end{cases}$$

or

$$(61) \qquad \tfrac{1}{2}u(x)\tilde{g}(x,u(x) - \tilde{G}(x,u(x)) \geq 0 \text{ a.e. }, \not\equiv 0 \text{ a.e.}$$

Observe also that (61) holds immediately if, for instance, (6), (23), (25) and (30) hold and

$$(62) \qquad \sup_{t>0}(\theta t f(x,t) - F(x,t))^- \in L^1(\mathbb{R}^N), \text{ for some } \theta \in (0,\tfrac{1}{2})$$

(this condition is taken from W.Y. Ding and W.M. Ni [**27**]) since we then have

$$\tfrac{1}{2}u(x)\tilde{g}(x,u(x)) - \tilde{G}(x,u(x)) \geq (\tfrac{1}{2}-\theta)u(x)\tilde{f}(x,u(x)).$$

The fact that (60) implies that $\tilde{\mathcal{E}}(u) > 0$ is deduced from the partial dilations invariance identity (in the ξ direction, see O. Kavian [**35**] for related identities)

$$(63) \quad \frac{1}{2}\int_{\mathbb{R}^N}\left(\sum_{i,j=1}^N \tilde{a}_{ij}\xi_i\xi_j\right)\left|\frac{\partial u}{\partial xi}\right|^2 dx =$$
$$= \frac{1}{2}\int_{\mathbb{R}^N}\sum_{i,j=1}^N \tilde{a}_{ij}\left(\frac{\partial u}{\partial x_i} - \xi_i\frac{\partial u}{\partial \xi}\right)\left(\frac{\partial u}{\partial x_j} - \xi_j\frac{\partial u}{\partial \xi}\right)dx - \int_{\mathbb{R}^N}\tilde{G}(x,u)dx$$

hence we deduce

$$(64) \qquad \tilde{\mathcal{E}}(u) = \int_{\mathbb{R}^N}\left(\sum_{i,j=1}^N \tilde{a}_{ij}\xi_i\xi_j\right)\left|\frac{\partial u}{\partial \xi}\right|^2 dx > 0.$$

The fact that $\tilde{\mathcal{E}}(u) > 0$ does not quite imply in general that $\bar{c}^{\infty} > 0$ and we will see below that this is indeed the case when we complement the above conditions with some assumptions also needed for the existence.

We are now going to use the preceding results to study the possibility of obtaining solutions via the Mountain Pass Lemma (see [**2**]). This is why we consider (for instance)

(65) $$c = \inf_{h\in\Gamma} \max_{t\in[0,1]} \mathcal{E}(h(t))$$

where

$$\Gamma = \left\{h \in C([0,1]; H^1(\mathbb{R}^N))/h(0) = 0, \mathcal{E}(h(1)) < 0\right\}.$$

To simplify the presentation, we will assume all throughout the end of this section (6), (23), (25), (26), (27), (29) and (30) (even if similar results may be obtained in the situation corresponding to (31) when $N \geq 3$) and thus in particular Theorem III.4 and Corollary III.1 apply. Then, we deduce from [**2**] that there exists a sequence $(u_n)_n$ in $H^1(\mathbb{R}^N)$ satisfying (48) such that $\mathcal{E}(u^n) \underset{n}{\rightarrow} c$. Notice also that because of (6)

(66) $$0 < c.$$

Next, for each limit system $\mathcal{C} = \{\tilde{a}_{ij})_{1\leq i,j\leq N}, \tilde{g}\}$ we may consider

(67) $$c^{\infty}(\mathcal{C}) = \inf_{h\in\Gamma^{\infty}(\mathcal{C})} \max_{t\in[0,1]} \tilde{\mathcal{E}}(h(t))$$

where $\Gamma^{\infty}(\mathcal{C}) = \{h \in C([0,1]; H^1(\mathbb{R}^N))/h(0) = 0, \tilde{\mathcal{E}}(h(1)) < 0\}$. And we agree that $c^{\infty}(\mathcal{C}) = +\infty$ if $\Gamma^{\infty}(\mathcal{C}) = \emptyset$. Finally, we set

(68) $$c^{\infty} = \inf_{\mathcal{C}} c^{\infty}(\mathcal{C})$$

where the infimum is taken over all limit systems $\mathcal{C}$. As we will see below, one often has $c^{\infty} = \bar{c}^{\infty}$. Exactly as in [**37,38**] we find

(69) $$c \leq c^{\infty}.$$

In order to apply Theorem III.4 or Corollary III.1 we need the sequence $(u_n)_n$ mentioned above to be bounded in $H^1(\mathbb{R}^N)$. A simple condition that ensures automatically that any sequence in H^1 satisfying (48) and such that $\mathcal{E}(u_n) \underset{n}{\to} c$ (in fact $\mathcal{E}(u_n)$ bounded is enough) is the condition (62) taken from [**27**] that we recall

$$\sup_{t>0}(\theta t f(x,t) - F(x,t))^- \in L^1(\mathbb{R}^N), \text{ for some } \theta \in (0, \tfrac{1}{2}).$$

Indeed, multiplying (48) by u_n and integrating by parts we find

$$\left| \int_{\mathbb{R}^N} \frac{1}{2} \sum_{i,j=1}^{N} a_{ij} \frac{\partial u_n}{\partial x_i} \frac{\partial u_n}{\partial x_j} + \tfrac{1}{2} a |u_n|^2 \, dx - \int_{\mathbb{R}^N} \tfrac{1}{2} f(x, u_n) u_n \, dx \right| \leq \varepsilon_n \, \|u_n\|_{H^1}$$

where $\varepsilon_n \geq 0$, $\varepsilon_n \underset{n}{\to} 0$, while we have for some $C \geq 0$

$$\left| \int_{\mathbb{R}^N} \frac{1}{2} \sum_{i,j=1}^{N} a_{ij} \frac{\partial u_n}{\partial x_i} \frac{\partial u_n}{\partial x_j} + \tfrac{1}{2} a |u_n|^2 \, dx - \int_{\mathbb{R}^N} F(x, u_n) \, dx \right| \leq C.$$

These two inequalities combined with (62) immediately imply

$$(\tfrac{1}{2} - \theta) \int_{\mathbb{R}^N} \sum_{i,j=1}^{N} a_{ij} \frac{\partial u_n}{\partial x_i} \frac{\partial u_n}{\partial x_j} + a |u_n|^2 \, dx \leq C + \varepsilon_n \, \|u_n\|_{H^1}$$

and we conclude using (6).

To summarize, we deduce from the above arguments the

COROLLARY III.2. *Under the assumptions (6), (23), (25), (26), (27), (29), (30) and (62), and if in addition $\bar{c}^\infty > 0$, then either there exists a solution of (1)–(2) in $H^1 \cap C_0$ such that $\mathcal{E}(u) < 0$ or (P.S.-$\bar{c}$) holds for all $\bar{c} < \bar{c}^\infty$ where*

$$(\text{P.S.-}\bar{c}) \qquad \begin{cases} \text{if } (u^n)_n \subset H^1 \text{ satisfies (48) and } E(u^n) \underset{n}{\to} \bar{c} \text{ then} \\ (u^n)_n \text{ is relatively compact in } H^1 \end{cases}$$

In particular, if $c < \bar{c}^\infty$, then there exists a solution of (1)–(2) in $H^1 \cap C_0$.

Next, we want to check the condition $c < \bar{c}^\infty$ and this seems difficult to do without relating c^∞ and $\bar{c}^\infty$. Hence, we are going to explain why $c^\infty = \bar{c}^\infty$ in some general cases and at the same time we will show that if $c^\infty < +\infty$ there exist $\mathcal{C}$, a relatively explicit $h_0 \in \Gamma^\infty(\mathcal{C})$ and $t_0 \in (0,1)$ such that

$$c^\infty = c^\infty(\mathcal{C}) = \max_{t \in [0,1]} \tilde{\mathcal{E}}(h(t)) = \tilde{\mathcal{E}}(h(t_0)) \tag{70}$$

and $\tilde{u} = h(t_0)$ solves (58). At this stage, it will be then straightforward to compare c and c^∞ by imposing various conditions on the coefficients exactly as we did in Section II. In particular, we may then recover as particular cases all the examples treated in W. Y. Ding and W. M. Ni [**27**]. Since these considerations are routine adaptations of the arguments given in Section II, we will skip them.

We are able to prove the above claims if we assume that, for each limit system $C = \{(\tilde{a}_{ij})_{1\le i,j\le N}, \tilde{a}, \tilde{f}\}$ for which there exists a solution $\tilde{u}$ of (58), we have

$$\text{(71)}\qquad \begin{cases} \text{either (60) holds for } \xi = \xi_1, \xi_2, \xi_3 \text{ with } \xi_i \cdot \xi_j = \delta_{ij} \\ \qquad (\forall 1 \le i, j \le 3), \\ \text{or (60) holds and } \dfrac{\partial \tilde{f}}{\partial t} \text{ exists, is bounded, continuous} \\ \qquad \text{in } t \text{ uniformly on } \mathbf{R}^N \times [-R, +R] (\forall R), \\ \text{or } \dfrac{\tilde{f}(x,t)}{t} \text{ is nondecreasing in } t \ge 0, \text{ a.e. } x \in \mathbf{R}^N. \end{cases}$$

Indeed, in order to show that $c^\infty \le \bar{c}^\infty$, we just have to find for each solution $\tilde{u}$ of (58) a curve $h \in \Gamma^\infty(C)$ such that $\max\limits_{t\in[0,1]} \tilde{\mathcal{E}}(h(t)) = \tilde{\mathcal{E}}(\tilde{u})$.

In the third case, i.e. if $\frac{\tilde{f}(x,t)}{t}$ is nondecreasing in $t \ge 0$ (a.e. $x \in \mathbf{R}^N$) we observe that

$$\frac{d}{d\lambda}\{\tilde{\mathcal{E}}(\lambda\tilde{u})\} \le 0 \text{ for } 0 \le \lambda \le 1, \frac{d}{d\lambda}\{\tilde{\mathcal{E}}(\lambda\tilde{u})\} \ge 0 \text{ for } \lambda \ge 1$$

while $\tilde{\mathcal{E}}(\lambda\tilde{u}) \to -\infty$ as $\lambda \to +\infty$. Indeed, (62) yields

$$\int_{\mathbf{R}^N} \tilde{F}(x, \lambda\tilde{u})\, dx \ge \lambda^{1/\theta} \int_{\mathbf{R}^N} \tilde{F}(x, \tilde{u})\, dx$$

and since $\tilde{u} \not\equiv 0$, $\int_{\mathbf{R}^N} \tilde{F}(x, \tilde{u})\, dx > 0$. Notice that if we choose $\lambda_0 > 1$ such that $\tilde{\mathcal{E}}(\lambda_0\tilde{u}) < 0$, we may set

$$\text{(72)}\qquad h(t) = t\lambda_0\tilde{u}, \quad \forall t \in [0,1]$$

and our claim is proved.

In the first case, we may assume without loss of generality that $\xi_1 = (1,0,0,\ldots)$, $\xi_2 = (0,1,0,\ldots)$, $\xi_3 = (0,0,1,\ldots)$ and then (71) only means that $(\tilde{a}_{ij})_{1\le i,j}, \tilde{a}, \tilde{f}$ are independent of x_1, x_2, x_3 and $\tilde{a}_{ij} \equiv 0$ a.e. if i or $j \ge 4$, and if $1 \le i \ne j \le 3$. And (63) yields

$$\frac{1}{2}\int_{\mathbb{R}^N} \sum_{i=1}^{3} \tilde{a}_{ii} \left|\frac{\partial u}{\partial x_i}\right|^2 dx + \frac{3}{2}\int_{\mathbb{R}^N} \sum_{i,j\ge 4} \tilde{a}_{ij} \frac{\partial \tilde{u}}{\partial x_i}\frac{\partial \tilde{u}}{\partial x_j}\, dx = 3\int_{\mathbb{R}^N} \tilde{G}(x,\tilde{u})\, dx.$$

Next, if we consider for $\sigma > 0$, $\tilde{u}_\sigma(x) = \tilde{u}\left(\frac{x_1}{\sigma}, \frac{x_2}{\sigma}, \frac{x_3}{\sigma}, x_4, \dots\right)$, we deduce

$$\tilde{\mathcal{E}}(\tilde{u}_\sigma) = \frac{\sigma}{2}\int_{\mathbb{R}^N} \sum_{i=1}^{3} \tilde{a}_{ii} \left|\frac{\partial \tilde{u}}{\partial x_i}\right|^2 dx + \frac{\sigma^3}{2}\int_{\mathbb{R}^N} \sum_{i,j\ge 4} \tilde{a}_{ij} \frac{\partial \tilde{u}}{\partial x_i}\frac{\partial \tilde{u}}{\partial x_j}\, dx - \sigma^3 \int_{\mathbb{R}^N} \tilde{G}(x,\tilde{u})\, dx,$$

$$\frac{d}{d\sigma}\{\tilde{\mathcal{E}}(\tilde{u}_\sigma)\} = (1-\sigma^2)\frac{1}{2}\int_{\mathbb{R}^N} \sum_{i=1}^{3} \tilde{a}_{ii} \left|\frac{\partial \tilde{u}}{\partial x_i}\right|^2 dx.$$

And we conclude setting

$$h(t) = \tilde{u}_{t\sigma_0} \text{ for } t \in (0,1], \quad h(0) = 0 \tag{73}$$

(notice that $\tilde{u}_\sigma \to 0$ in H^1 as $\sigma \to 0_+$), where $\sigma_0 > 0$ is chosen such that $\tilde{\mathcal{E}}(\tilde{u}_{\sigma_0}) < 0$.

The last case is a bit more complicated and we will only sketch the argument. Again without loss of generality we may assume that $\xi = (1,0,0,\dots)$ and (71) then means that $(\tilde{a}_{ij})_{1\le i,j\le N}, \tilde{a}, \tilde{f}$ are independent of x_1. In particular this implies easily

$$-\sum_{i,j=1}^{N} \frac{\partial}{\partial x_i}\left(\tilde{a}_{ij}\frac{\partial}{\partial x_j}\left[\frac{\partial \tilde{u}}{\partial x_1}\right]\right) + \tilde{a}\frac{\partial \tilde{u}}{\partial x_1} = \left\{\frac{\partial}{\partial t}\tilde{f}(x,\tilde{u}(x))\right\}\frac{\partial \tilde{u}}{\partial x_1}$$

which in turn implies the existence of $\lambda_1 > 0$, $\varphi_1 \in H^1 \cap C_0$ such that

$$-\sum_{i,j=1}^{N} \frac{\partial}{\partial x_i}\left(\tilde{a}_{ij}\frac{\partial \varphi_1}{\partial x_j}\right) + \tilde{a}\varphi_1 - \left\{\frac{\partial}{\partial t}\tilde{f}(x,\tilde{u})\right\}\varphi_1 = \lambda_1\varphi_1$$

and $\varphi_1 > 0$ in $\mathbb{R}^N$. From this we deduce that there exists $\varepsilon_0 > 0$ such that for all $\varepsilon \in [0, \varepsilon_0]$

$$\begin{cases} -\sum_{i,j=1}^{N} \frac{\partial}{\partial x_i}\left(\tilde{a}_{ij}\frac{\partial}{\partial x_j}(\tilde{u} \pm \varepsilon\varphi_1)\right) \lesseqgtr \tilde{g}(x, \tilde{u} \pm \varepsilon\varphi_1) \\ \tilde{\mathcal{E}}(\tilde{u} \pm \varepsilon\varphi_1) \le \tilde{\mathcal{E}}(\tilde{u}). \end{cases} \tag{74}$$

We then set $\bar{u} = \tilde{u} - \varepsilon_0\varphi_1$, $\underline{u} = \tilde{u} - \varepsilon_0\varphi_1$ and we are going to find two continuous curves in the level set $\{v \in H^1 / \tilde{\mathcal{E}}(v) \leq \tilde{\mathcal{E}}(\tilde{u})\}$ connecting respectively $\underline{u}$ to 0 and $\bar{u}$ to "a point with negative energy". The resulting map h will then be obtained by following backwards the first curve from 0 to $\underline{u}$, then going from $\underline{u}$ to $\bar{u}$ by considering $\tilde{u} + \lambda\varphi_1$ for $-\varepsilon_0 \leq \lambda \leq \varepsilon_0$ and then following the second curve. To build these curves, we first observe that in view of the equations satisfied by $\tilde{u}$ and φ_1 we may assume without loss of generality that ε_0 has been chosen small enough so that $\underline{u} > 0$ in $\mathbb{R}^N$. Then, we use the heat flow associated with the equation, i.e. we consider the solution of the following semilinear parabolic equation

$$\frac{\partial w}{\partial t} - \sum_{i,j=1}^{N} \frac{\partial}{\partial x_i}\left(\tilde{a}_{ij}\frac{\partial w}{\partial x_j}\right) = \tilde{g}(x,w) \text{ in } \mathbb{R}^N \times (0,\infty)$$

and we observe that if we denote by $\underline{w}$ the maximal solution of this Cauchy problem corresponding to the initial condition $\underline{u}$, (74) implies that $\underline{w}(\cdot,t)$ is decreasing with $t \geq 0$, and satisfies (74) for all $t \geq 0$. Thus, the solution $\underline{w}(\cdot,t)$ exists for all $t \geq 0$ and we have

$$\begin{cases} 0 < \underline{w}(x,t') < \underline{w}(x,t) < \underline{u}(x) \quad \forall x \in \mathbb{R}^N \\ \tilde{\mathcal{E}}(\underline{w}(x,t)) \searrow \text{ as } t \nearrow +\infty. \end{cases}$$

It is then easy to conclude that $\underline{w}(\cdot,t) \to 0$ (in $H^1 \cap C_0$) as $t \to +\infty$ (it cannot converge to another positive solution since they are all unstable).

The curve starting from $\bar{u}$ is slightly more difficult to build for technical reasons. First of all, we claim that the arguments given above show that we may choose $\tilde{u}$ so that $\tilde{\mathcal{E}}(\tilde{u}) = \bar{c}^\infty$ and thus $\tilde{\mathcal{E}}(\bar{u}) < \bar{c}^\infty$. Indeed, if C^n is a sequence of limit systems such that $\tilde{u}^n$ is a sequence of corresponding positive solutions and $\tilde{\mathcal{E}}^n(\tilde{u}^n) \underset{n}{\to} \bar{c}^\infty$ (with obvious notations), then because of (71) we conclude easily provided we exclude the case when $u^n \underset{n}{\to} 0$ in $H^1(\mathbb{R}^N)$ (use Theorem III.4). This case is actually quite easily excluded since such a convergence would yield the existence of $\varepsilon_n > 0$, $\varepsilon_n \underset{n}{\to} 0$ such that

$$\|\tilde{u}_n\|_{H^1} \leq C \left\|\tilde{f}^n(x,\tilde{u}_n)\right\|_{H^{-1}} \leq C\varepsilon_n \|\tilde{u}_n\|_{H^1}$$

and for n large we reach a contradiction with the fact that $\tilde{u}_n > 0$.

Now, we consider the flow

$$\frac{\partial w}{\partial t} + w - \tilde{L}^{-1}(\tilde{f}(x,w) + Mw) = 0 \text{ in } \mathbb{R}^N, \quad w\Big|_{t=0} = \bar{u}$$

where $\tilde{L} = -\sum_{i,j=1}^{N} \frac{\partial}{\partial x_i}(\tilde{a}_{ij}\frac{\partial}{\partial x_j}) + \tilde{a} + M$, and $M \geq 0$ is chosen large enough such that $\tilde{f}(x,t) + Mt \geq \tilde{f}(x,\bar{u}(x)) + M\bar{u}(x)$ for all $t \geq \bar{u}(x)$. Then denoting by $|||\cdot|||$ the equivalent norm on $H^1(\mathbb{R}^N)$ given by

$$|||\varphi|||^2 = \int_{\mathbb{R}^N} \sum_{i,j=1}^{N} \tilde{a}_{ij} \frac{\partial w}{\partial x_i}\frac{\partial w}{\partial x_j} + \tilde{a}|w|^2\, dx$$

we deduce

$$\int_0^{\bar{T}} \left|\left|\left|\frac{\partial w}{\partial t}\right|\right|\right|^2 dt < \infty$$

which implies in particular that $\bar{T} = +\infty$. But, then, it is easy to build a sequence $t_n \underset{n}{\to} \infty$ such that $\frac{\partial w}{\partial t}(t_n) \underset{n}{\to} 0$ in H^1, therefore

$$\tilde{L}w_n - \tilde{f}(x,w_n) \underset{n}{\to} 0 \text{ in } H^{-1}(\mathbb{R}^N), \text{ where } w_n = w(t_n)$$

from which we deduce $w_n^- \underset{n}{\to} 0$ in H^1. Applying Theorem III.4, using (71) and the fact that

$$\bar{c}^\infty > \tilde{\mathcal{E}}(\bar{u}) \geq \tilde{\mathcal{E}}(w(t)) \searrow \text{ as } t \nearrow +\infty$$

we deduce that necessarily $w_n \underset{n}{\to} 0$ in H^1. One proves similarly that in fact $w(t) \to 0$ in H^1 as $t \to +\infty$. To conclude, we remark that $w(t) \geq \bar{u}$ for all $t \geq 0$; a contradiction which yields the desired curve.

We have thus proved that $0 < c^\infty \leq \bar{c}^\infty$, and there exists a limit system $\mathcal{C}$, an associated positive solution u and a curve $h \in \Gamma^\infty(\mathcal{C})$ (explicitly built above) such that

$$\bar{c}^\infty = \tilde{\mathcal{E}}(\tilde{u}) = \max_{t\in[0,1]} \tilde{\mathcal{E}}(h(t))$$

(in fact the existence of h holds for any limit system $\mathcal{C}$ and associated positive solution $\tilde{u}$ such that $\bar{c}^\infty = \tilde{\mathcal{E}}(\tilde{u})$). There only remains to show that $c^\infty \geq \bar{c}^\infty$. But by standard variational arguments we can find limit systems $\mathcal{C}_n = \{(\tilde{a}_{ij}^n)_{1\leq i,j\leq N}, \tilde{a}^n, \tilde{f}^n\}$, $u^n \in H^1$ such that with obvious notations

$$\begin{cases} -\displaystyle\sum_{i,j=1}^{N} \frac{\partial}{\partial x_i}\left(\tilde{a}_{ij}^n \frac{\partial u^n}{\partial x_j}\right) + \tilde{a}^n u^n - \tilde{f}^n(x,u^n) \underset{n}{\to} 0 \text{ in } H^{-1} \\ \tilde{\mathcal{E}}^n(u^n) \underset{n}{\to} c^\infty > 0 \end{cases}$$

and we may still apply Theorem III.4 (or a minor extension of this result) excluding the case when $u^n \underset{n}{\to} 0$ since $c^\infty > 0$, to deduce the existence of $m \geq 1$, limit systems $\mathcal{C}_1, \ldots, \mathcal{C}_m$, corresponding positive solutions $u_1, \ldots, u_m$ such that

$$c^\infty = \sum_{j=1}^{m} \mathcal{E}_1(u_i)$$

while by definition $\mathcal{E}_i(u_i) \geq \bar{c}^\infty > 0$.

This shows $m = 1$ and $c^\infty \geq \bar{c}^\infty$. To summarize, we conclude with the

COROLLARY III.3. *Under the assumptions (6), (23), (25), (26), (27), (29), (30), (62), and (71) we have*

(i) $0 < c^\infty = \bar{c}^\infty \leq +\infty$ *and if* $c^\infty < \infty$, *there exist a limit system* $\mathcal{C}$ *and a positive solution* $\tilde{u}$ *of the corresponding limit equation such that* $\bar{c}^\infty = \tilde{\mathcal{E}}(\tilde{u})$. *For any such* $(\mathcal{C}, \tilde{u})$ *there exists a curve* $h \in \Gamma^\infty(\mathcal{C})$ *containing* $\tilde{u}$ *(explicitly built above) such that* $c^\infty = \max_{t\in[0,1]} \tilde{\mathcal{E}}(h(t))$.

(ii) *If* $c < c^\infty$, *there exists a solution of (1)–(2) in* H^1.

REMARKS:

1. One can treat similarly the case corresponding to (55) (the so-called zero-mass case).
2. If we inspect the proofs above, we see that we only used (62) to get the boundedness in H^1 of "Palais-Smale sequences." Many variants are possible: let us only mention one:

$$(75) \qquad \begin{cases} \frac{1}{2}tf(x,t) - F(x,t) + \ell(x) \geq \nu_\varepsilon > 0 \text{ for } x \in \mathbb{R}^N, \\ \qquad\qquad t \in [\varepsilon, \frac{1}{\varepsilon}] \quad (\forall \varepsilon > 0) \\ \frac{1}{2}tf(x,t) - F(x,t) + \ell(x) \geq \delta |f(x,t)|^{\frac{2N}{N+2}} \text{ for } x \in \mathbb{R}^N, \quad t \geq \frac{1}{\delta} \end{cases}$$

where $\delta > 0$, $\ell \in L^1_+(\mathbb{R}^N)$.

But since we are only interested in building one critical point, we may in fact try to get the boundedness in H^1 of one "Palais-Smale sequence." To do so, we may, for example, approximate (1)–(2) by similar problems in balls B_R and let R go to $+\infty$. If we do so, we may replace in (75) the second inequality by

$$\frac{1}{2}tf(x,t) - F(x,t) + \ell(x) \geq \frac{1}{\delta} \left| \frac{f^+(x,t)}{t} \right|^{\frac{N}{2}} \quad \text{for } x \in \mathbb{R}^N, t \geq 1/\delta \quad (\forall \delta > 0).$$

IV. Unbounded domains and non-minimal solutions.

IV.1. Introduction and nonexistence results.

To simplify the presentation, we will consider in this section only the model problem

$$(76) \qquad -\Delta u + \lambda u = b(x)u^p \text{ in } \mathbb{R}^N, u \in H^1(\mathbb{R}^N) \cap C_0(\mathbb{R}^N), u > 0 \text{ in } \mathbb{R}^N$$

where $1 < p < \frac{N+2}{N-2}$, $\lambda > 0$, $b \in L^\infty(\mathbb{R}^N)$ and $\inf \operatorname{ess}_{\mathbb{R}^N} b > 0$. We will at the same time discuss the closely related problem

$$(77) \qquad -\Delta u + \lambda u = u^p \text{ in } \Omega, u \in H_0^1(\Omega) \cap C_0(\bar{\Omega}), u > 0 \text{ in } \Omega$$

where Ω is an unbounded domain such that $\operatorname{meas}(\Omega) = +\infty$.

In Section II, we saw that if $b(x) \to b^\infty$ as $|x| \to \infty$ then a "minimum-like" solution exists as soon as $b \geq b^\infty$ on $\mathbb{R}^N$ (or even if $b \geq b^\infty$ for x large and $b - b^\infty$ does not decay too fast) while such a "minimum-like" solution does not exist if $b \leq b^\infty$ on $\mathbb{R}^N$ and $b \not\equiv b^\infty$. We also saw that if $\sup \operatorname{ess}_{\mathbb{R}^N} b > b^\infty$ then a "minimum-like" solution exists for λ large. Finally, we also gave various conditions for the existence of "minimum-like" solutions in more general situations when b does not converge "at infinity."

Of course, one might expect to have solutions of (76) with a different "topology" (higher critical points of the functional). In general, this is not the case in view of the nonexistence results of M. J. Esteban and P. L. Lions [**28**]. For instance, if b satisfies

$$(78) \qquad \exists \chi \in \mathbb{R}^N, \quad |\chi| = 1, \quad \frac{\partial b}{\partial \chi} \geq 0 \text{ in } \mathbb{R}^N, \quad \frac{\partial b}{\partial \chi} \not\equiv 0$$

then there does not exist a solution of (76). This claim is very easy to check by multiplying the equation by $\frac{\partial u}{\partial \chi}$ and integrating by parts. Notice, however, that no nonexistence examples are known when $b(x) \to b^\infty$ as $|x| \to \infty$.

The picture is vaguely similar for (77). First of all, if Ω satisfies

$$(79) \qquad \forall R < \infty, \quad \exists y \in \Omega, \quad B(y, R) \subset \Omega,$$

then there does not exist a "minimum-like" solution (same arguments as in Section II). And if Ω is "star-shaped" with respect to infinity, i.e.

$$(80) \qquad \exists \chi \in \mathbb{R}^N, \quad |\chi| = 1, \quad n(x) \cdot \chi \geq 0 \text{ on } \partial\Omega, \quad n(x) \cdot \chi \not\equiv 0$$

where n denotes the unit outward normal to $\partial\Omega$ at x, then (see [**28**]) there does not exist a solution of (76). This is the case in particular if Ω is a half-space or a half-strip.

On the other hand, when (79) does not hold, "minimum-like" solutions may exist as it is the case for instance if $\Omega = \mathcal{O} \times \mathbb{R}^m$ where $m \geq 1$, $\mathcal{O}$ is a finite-measure domain in $\mathbb{R}^n (N = n + m)$ — see P. L. Lions [**37**]. And exactly as in the case when $b \to b^\infty$ as $|x| \to \infty$, we may discuss the existence of "minimum-like" solutions when Ω approaches asymptotically a strip $\Omega^\infty = \mathcal{O} \times \mathbb{R}^m$: if Ω contains Ω^∞ there is such a solution while if Ω^∞ contains Ω this is not so.

In the next sections, we want to explain a few additional existence results for (76) and (77); in general, these results yield nonminimum-like solutions.

IV.2. Symmetries.

If we use compactness lemmata due to W. Strauss [**48**] (radial case), P. L. Lions [**40**] (block radial case), we immediately see that if b is radial (resp. $\Omega = B_R^C$ for some $R > 0$) or if $b(x_1, \ldots, x_N)$ is spherically symmetric with respect to $(x_1, \ldots, x_{n_1})$, $(x_{n_1+1}, \ldots, x_{n_2+n_1})$, $\ldots$, $(x_{n_1+\cdots+n_{m-1}}, \ldots, x_N)$ with $n_j \geq 2 (1 \leq j \leq m)$ and $N = \sum_{j=1}^m n_j$ (resp. Ω has the corresponding symmetries) then there exists a solution of (76) (resp. (77)). In fact, the solution is obtained by minimizing the functional (with the constraint $\int b|u|^{p+1}\, dx = 1$) over the subspace of functions with the above symmetries. The way compactness is gained may be also proved directly (and very simply) by using the concentration-compactness lemma in [**37**]. Of course, one may also obtain other existence results by combining this argument with the preceding considerations (product situations with strips).

More generally, if $b(x)$ (resp. (Ω)) is invariant by a group G of orthogonal transformations on $\mathbb{R}^N$ then the "number of symmetries" plays an important role if we want to construct a solution of (76) (resp. (77)) by a restricted minimization problem. Following C. V. Coffman and M. M. Marcus [**21**], and P. L. Lions [38,41], we set

$$n_0 = \inf\left(\#\{g\cdot x/g \in G\}/x \in S^{N-1}\right) \tag{81}$$

(respectively

$$n_0 = \lim_{R\to\infty} \inf\left(\#\{g\cdot x/g \in G\}/x \in \bar{\Omega}, \quad |x| \geq R\right)). \tag{82}$$

Then, in the case of (76) assuming, for instance, that $b(x) \to b^\infty$ as $|x| \to \infty$ (meas $\{|b - b^\infty| > \delta\} < \infty$, $\forall \delta > 0$), we set

$$I^G = \inf\left\{\int_{\mathbb{R}^N} |\nabla u|^2 + \lambda u^2\, dx / \int_{\mathbb{R}^N} b|u|^{p+1}\, dx = 1; \right. \\ \left. u(x) = u(g\cdot x) \text{ on } \mathbb{R}^N, \quad \forall g \in G\right\} \tag{83}$$

$$I_t^\infty = \inf\left\{\int_{\mathbb{R}^N} |\nabla u|^2 + \lambda u^2\, dx / \int_{\mathbb{R}^N} b^\infty |u|^{p+1} = t\right\}, \text{ for } t \geq 0 \tag{84}$$

while in the case of (77) assuming for instance that Ω^c is bounded we set

$$I^G = \inf\left\{\int_\Omega |\nabla u|^2 + \lambda u^2\, dx / \int_{\mathbb{R}^N} |u|^{p+1}\, dx = 1; \quad u \in H_0^1(\Omega); \right. \\ \left. u(x) = u(g\cdot x) \text{ on } \mathbb{R}^N, \quad \forall g \in G\right\} \tag{85}$$

$$I_t^\infty = \inf\left\{\int_{\mathbb{R}^N} |\nabla u|^2 + \lambda u^2\, dx / \int_{\mathbb{R}^N} |u|^{p+1}\, dx = t\right\}. \tag{86}$$

With these assumptions and notations, one shows (cf. [38,41]) that all minimizing sequences of I^G are compact in $H^1(\mathbb{R}^N)$ if and only if

$$I^G < n_0 I^\infty_{1/n_0} = n_0^{\frac{p-1}{p+1}} I_1^\infty. \tag{87}$$

And by tedious computations (see A. Bahri and P. L. Lions [7]), it is possible to show that the strict inequality holds in the case of (77) as soon as $n_0 \geq 2$ while in the case of (76) it holds for $n_0 \geq 2$ if

$$(b^\infty - b)^+ = o\left(\exp(-2\sqrt{\lambda}|x|)|x|^{-\frac{N-1}{2}}\right) \text{ as } |x| \to \infty.$$

Let us conclude this section by mentioning the work by C. V. Coffman and M. M. Marcus [**22**] where it is shown that solutions of (77) still exist for small perturbations of Ω given by the complement of a ball (at least for "most radii").

IV.3. Critical points at infinity.

We saw in Theorem III.4 an explicit description of the loss of compactness of Palais-Smale sequences (which is still valid for (77)). In addition, if we argue by contradiction and assume that (76) and (77) does not have a positive solution, we see that (54) holds with $u_1 \equiv 0$ and each u_k is a positive solution of

$$(88) \qquad -\Delta u + \lambda u = u^p \text{ in } \mathbb{R}^N, \quad u \in H^1(\mathbb{R}^N) \cap C_0(\mathbb{R}^N), \quad u > 0 \text{ in } \mathbb{R}^N$$

(assuming that $b(x) \to 1$ as $|x| \to \infty$ for example).

By Gidas–Ni–Nirenberg results [**30,31**], one knows that, up to a translation, u is radial and decreasing. Now, if we know that there is a unique radial (decreasing) solution of (88), we would be able to conclude that each u_k may be taken equal to that solution, say ω. Unfortunately, such a uniqueness result is not known in full generality and the best known results are to be found in K. MacLeod and J. Serrin [**42**] (if $N = 2$, uniqueness is known for $1 < p \leq 3$). This detailed description of losses of compactness — modulo the preceding uniqueness statement — allows us to use various critical point methods.

In this direction, V. Benci and G. Cerami [**8**] first obtained in the case of (77) an existence result provided $\Omega = \bar{\mathcal{O}}^c$ and $\mathcal{O}$ is small. And, using the more powerful machinery due to A. Bahri [**5**] and A. Bahri and J. M. Coron [**6**] to solve various Yamabe-type equations, it is possible to show (see [**7**]) that (76), (77) admit a solution as soon as $b \geq b^\infty - C\exp(-\mu|x|) \quad (C, \mu > 0)$ for $|x|$ large or $\Omega = \bar{\mathcal{O}}^c$ and $\bar{\mathcal{O}}$ is bounded. Let us emphasize the fact that these statements hold modulo the uniqueness conjecture mentioned above and partially solved in K. MacLeod and J. Serrin [**42**] for some range of p's.

Ceremade, Université Paris-Dauphine, Place de Lattre de Tassigny, 75775 Paris Cedex 16, FRANCE

REFERENCES

1. A. Ambrosetti, *Existence and multiplicity results for some classes of nonlinear problems*, in "Proceedings of the International Congress of Mathematicians, 1983 – Warsaw," North-Holland, Amsterdam, 1984.
2. A. Ambrosetti and P. H. Rabinowitz, *Dual variational methods in critical point theory and applications*, J. Funct. Anal. **14** (1973), 369–381.
3. A. V. Atkinson and L. A. Peletier, *Ground states of* $-\Delta u = f(u)$ *and the Emden-Fowler equation*, Arch. Rat. Mech. Anal. **93** (1986), 103–128.
4. H. Attouch, "Variational convergence for functions and operators," Pitman, London, 1984.
5. A. Bahri, *The theory of critical points at infinity*, to appear.
6. A. Bahri and J. M. Coron, *On a nonlinear elliptic equation involving the Sobolev exponent.*
7. A. Bahri and P. L. Lions, in preparation.
8. V. Benci and G. Cerami, *Positive solutions of semilinear elliptic problems in exterior domains*, preprint.
9. H. Berestycki and P. L. Lions, *Nonlinear scalar field equations*, Arch. Rat. Mech. Anal. **82** (1983), 313–346; *II*, **82** (1983), 347–376; see also: C. R. Acad. Sci. Paris **287** (1978), 503–506; **288** (1979), 395–398; **297** (1983), 267–270.
10. H. Berestycki and P. L. Lions, *Existence of stationary states in nonlinear scalar field equations*, in "Bifurcation Phenomena in Mathematical Physics and Related Topics," Reidel, New York, 1980.
11. H. Berestycki and C. Taubes, personal communication, in preparation.
12. M. Berger, *On the existence and structure of stationary states for a nonlinear Klein-Gordon equation*, J. Funct. Anal. **9** (1972), 249–261.
13. H. Brézis and J. M. Coron, *Convergence de solutions de H-systèmes et applications aux surfaces à courbure moyenne constante*, C. R. Acad. Sci. Paris (1984).
14. H. Brézis and E. H. Lieb, *Minimum-action solutions of some vector-field equations*, Comm. Math. Phys. **96** (1984), 97–113.
15. H. Brézis and L. Nirenberg, *Positive solutions of nonlinear elliptic equations involving critical Sobolev exponents*, Comm. Pure Appl. Math **36** (1983), 437–477.
16. H. Brézis and R. E. L. Turner, *On a class of superlinear elliptic problems*, Comm. P. D. E. **2** (1977), 601–614.
17. E. Brüning, *On the variational approach to semilinear elliptic equations with scale covariance*, preprint.
18. C. V. Coffman, *A minimum-maximum principle for a class of nonlinear integral equations*, J. Anal. Math. **22** (1969), 391–419.
19. C. V. Coffman, *On a class of nonlinear elliptic boundary value problems*, J. Math. Mech. **19** (1970), 351–356.
20. C. V. Coffman, *Uniqueness of the ground state solution for* $\Delta u - u + u^3 = 0$ *and a variational characterization of other solutions*, Arch. Rat. Mech. Anal. **46** (1972), 81–95.
21. C. V. Coffman and M. M. Marcus, presonal communication.
22. C. V. Coffman and M. M. Marcus, *Existence theorems for superlinear elliptic Dirichlet problems in exterior domains*, preprint.
23. S. Coleman, V. Glazer and A. Martin, *Action minima among solutions to a class of Euclidian scalar field equations*, Comm. Math. Phys. **58** (1978), 211–221.
24. D. G. de Figueiredo, P. L. Lions and R. D. Nussbaum, *A priori estimates and existence of positive solutions of semilinear elliptic equations*, J. Maths. Pures Appl. **61** (1982), 41–63; see also: C. R. Acad. Sci. Paris **290** (1980), 217–220.
25. E. de Giorgi, *G-operators and* Γ*-convergence*, in "Proceedings of the International Congress of Mathematicians, 1983 – Warsaw," North-Holland, Amsterdam, 1984.
26. W. Y. Ding, *On a conformally invariant elliptic equation on* $\mathbb{R}^n$, preprint.

27. W. Y. Ding and W. N. Ni, *On the existence of positive entire solutions of a semilinear elliptic equation*, Arch. Rat. Mech. Anal. **91** (1986), 288–308.
28. M. J. Esteban and P. L. Lions, *Existence and non-existence results for semilinear elliptic problems in unbounded domains*, Proc. Roy. Soc. Edin. **93** (1982), 1–14; see also: C. R. Acad. Sci. Paris **290** (1980), 1083–1085.
29. M. J. Esteban and P. L. Lions, *Γ-convergence and the concentration-compactness method for some variational problems with lack of compactness*, to appear in Ric. Mat. Napoli.
30. B. Gidas, W. M. Ni and L. Nirenberg, *Symmetry of positive solutions of nonlinear elliptic equations in* $\mathbf{R}^n$, Advances in Math., Supplementary Studies **7** (1981), 369–402.
31. B. Gidas, W. M. Ni and L. Nirenberg, *Symmetry and related properties via the maximum principle*, Comm. Math. Phys. **68** (1979), 209–243.
32. B. Gidas and J. Spruck, *A priori bounds for positive solutions of nonlinear elliptic equations*, Comm. Part. Diff. Eq. **6** (1981), 883–901.
33. D. Gogny and P. L. Lions, *Hartree-Fock theory in nuclear physics*, RAIRO Modél. Math. et Anal. Num. **20** (1986), 571–638.
34. J. Hempel, *Multiple solutions for a class of nonlinear boundary value problems*, Ind. Univ. Math. J. **20** (1971), 983–996.
35. O. Kavian, *Minimum action solutions of nonlinear elliptic equations in unbounded domains containing a plane*, Proc. Roy. Soc. Edin. **102** (1986), 327–343.
36. P. L. Lions, *Solutions of Hartree-Fock equations for Coulomb systems*, Comm. Math. Phys. **109** (1987), 33–97.
37. P. L. Lions, *The concentration-compactness principle in the Calculus of Variations. The locally compact case*, Ann. Inst. H. Poincaré **1** (1984), 109–145; *II*, **1** (1984), 223–283; see also: C. R. Acad. Sci. Paris **294** (1982), 261–264; in "Contributions to nonlinear partial differential equations," Pitman, London, 1983.
38. P. L. Lions, *The concentration-compactness principle in the Calculus of Variations. The limit case*, Riv. Mat. Ibero-americana **1** (1985), 145–201; *II*, **1** (1985), 45–121; see also: Séminaire Goulaouic-Meyer-Schwartz 1982-83, exposé XIV, École Polytechnique, Palaiseau; C. R. Acad. Sci. Paris **302** (1986), 673–676.
39. P. L. Lions, *Solutions complexes d'équations elliptiques semilinéaires dans* $\mathbf{R}^N$, C. R. Acad. Sci. Paris **302** (1986), 673–676.
40. P. L. Lions, *Symétrie et compacité dans les espaces de Sobolev*, J. Funct. Anal. **49** (1982), 315–334.
41. P. L. Lions, *Symmetries and the concentration-compactness method*, in "Nonlinear Variational Problems," Pitman, London, 1985.
42. K. MacLeod and J. Serrin, *Uniqueness of solutions of semilinear Poisson equations*, Proc. Nat. Acad. Sci. USA **78** (1981), 6592–6595.
43. F. Murat and L. Tartar, *Calcul des variations et homogénéisation*, in "Les Méthodes de l'homogénéisation: théorie et applications en Physique," Eyrolles, Paris, 1985.
44. Z. Nehari, *On a nonlinear differential equation arising in nuclear physics*, Proc. Roy. Irish Acad. **62** (1963), 117–135.
45. S. I. Pohozaev, *Eigenfunctions of* $\Delta u + \lambda f(u) = 0$, Soviet Math. Dokl. **6** (1965), 1408–1411.
46. G. H. Ryder, *Boundary value problems for a class of nonlinear differential equations*, Pac. J. Math. **22** (1967), 477–503.
47. P. Sacks and K. Uhlenbeck, *The existence of minimal immersions of 2-spheres*, Ann. Math. **113** (1981), 1–24.
48. W. Strauss, *Existence of solitary waves in higher dimensions*, Comm. Math. Phys. **55** (1977), 149–162.
49. M. Struwe, personal communication.
50. M. Struwe, *A global compactness result for elliptic boundary value problems involving limiting nonlinearities*, Math. Z. **187** (1984), 511–517.
51. J. F. Yang, personal communication, in preparation.

The Behavior of Solutions of a Nonlinear Boundary Layer Equation

CHUNQING LU AND WILLIAM C. TROY

Abstract. We investigate the behavior of solutions of the equation $\phi''' + 1 - \phi + \lambda(\phi\phi'' - (\phi')^2) = 0$. This equation arises in modelling large scale ocean circulation with particular emphasis on the behavior of the Gulf Stream. There are two sets of physically interesting boundary conditions, namely $\phi(0) = \phi'(0) = 0$ and $\phi(\infty) = 1$ (no-slip conditions) or $\phi(0) = \phi'(0) = 0$ and $\phi(\infty) = 1$ (stress-free conditions). For each of these problems we prove that solutions exist if $|\lambda|$ is small. However, if $\lambda \leq -9$ then there is no solution of the no-slip problem. If $\lambda \leq -(2)^{1/3}$ then the stress-free problem has no solution.

Introduction.

We investigate a one layer model of large scale ocean circulation in the Gulf Stream. Under appropriate assumptions, the partial differential equations are transformed into a nonlinear third order differential equation with boundary conditions. Our goal is to analyze this equation and to rigorously prove the existence or non-existence of solutions over a wide range of parameter values. The paper is organized as follows: first, we present a brief development of the model followed by a summary of previous numerical studies. We then state our main results in Theorems 1 and 2. Sections 2 and 3 contain the proofs of our theorems.

Our development of the model follows that of Ierley and Ruehr [**1**] and more complete details are found in their paper.

The model stems from the so-called barotropic quasigeostrophic potential vorticity equation. In dimensionaless form, the steady state vorticity equation is given by

$$J(\psi, q) = w + K\nabla^2 q$$

where

$$q = y + \gamma\nabla^2\psi, \quad w = -\cos(\frac{\pi y}{2}), \quad J(\psi, q) = \frac{\partial(\psi, q)}{\partial(z, y)}.$$

Here q is the vorticity, ψ is the stream function, and the term w corresponds to forcing by the wind blowing steadily across a rectangular basin. The Jacobian represents nonlinear advection and the Laplacian the usual viscous drag. The relation between q and ψ is altered from the form one would expect by inclusion of the crucial effect of the Coriolis force due to the earth's variable rotation. It is assumed that the positive y-axis points north and the positive z-axis points east.

The boundary conditions are

$$\psi(\pm 1, y) = \psi(z, \pm 1) = 0,$$

and either

$$\frac{\partial \psi}{\partial x}(\pm 1, y) = \frac{\partial \psi}{\partial y}(z, \pm 1) = 0 \qquad \text{(no-slip)}$$

or

$$\frac{\partial^2 \psi}{\partial x^2}(\pm 1, y) = \frac{\partial^2 \psi}{\partial y^2}(z, \pm 1) = 0 \qquad \text{(stress-free)}.$$

They expand the solution in a Taylor series about $y = -1$ and assume that the stream function has the form

$$\psi \sim \phi(z)\pi(1 + y).$$

This leads to the fourth order equation (see [**1**] for details)

$$K\gamma\phi'''' = \pi\gamma(\phi'\phi'' - \phi\phi''') + \phi'.$$

Next, make the tranformation

$$x = (K\gamma)z^{-1/3}, \quad \lambda = \pi(\gamma K^{-2})^{1/3}$$

and consider the case $K \to 0$. With an appropriate matching process they find that $\phi \to 1$ as $x \to \infty$. One then may integrate the fourth order equation and arrive at the third order problem

(1) $$\phi''' = \lambda((\phi')^2 - \phi\phi'') + \phi - 1,$$

(2) $$\phi(\infty) = 1.$$

The appropriate initial conditions are either

(3) $$\phi(0) = \phi'(0) = 0 \qquad \text{(no-slip)}$$

or

(4) $$\phi(0) = \phi''(0) = 0 \qquad \text{(stress-free)}.$$

The numerical studies of Ierley and Ruehr [**1**] predict that solutions of the no-slip problem (1)–(2)–(3) behave as follows:

(i) for $\lambda \geq 0$ there is a unique solution,

(ii) for $-.7913 < \lambda < 0$ two solutions exist, and

(iii) for $\lambda < -.7913$ no solution exists.

Further, their computations show that solutions of the stress-free problem (1)–(2)–(4) satisfy

(i) if $\lambda > 0$ then exactly two solutions exist,

(ii) if $\lambda = 0$ one solution exists,

(iii) if $-.29657 < \lambda < 0$ there are two solutions, and

(iv) if $\lambda < -.29657$ then no solution exists.

Recently, Troy [**4**] has rigorously proved that existence of solutions for both (1)–(2)–(3) and (1)–(2)–(4) for all $\lambda \geq (27/4)^{1/3}$. In this parameter range solutions are eventually monotone and a topological shooting argument proves useful in showing that solutions exist.

We have two main goals in this paper. First, we investigate the existence of solutions of both (1)–(2)–(3) and (1)–(2)–(4) for small $|\lambda|$. Secondly, we address the non-existence of solutions for large negative λ. Before stating our main results we first need to discuss the special case $\lambda = 0$. Then equation (1) reduces to

$$\phi''' + 1 - \phi = 0.$$

Thus, for $\lambda = 0$ we find that the solution of the no-slip problem is

$$\phi_0 = 1 - \frac{e^{-x/2}}{\sqrt{3}} \sin(\frac{\sqrt{3}x}{2}) - e^{-x/2} \cos(\frac{\sqrt{3}x}{2}).$$

Likewise the solution of the stress-free problem (1)–(2)–(4) is

$$\phi_1 = 1 + e^{-x/2} \left(\frac{1}{\sqrt{3}} \sin(\frac{\sqrt{3}}{2}x) - \cos(\frac{\sqrt{3}}{2}x) \right).$$

For small $|\lambda|$ we expect solutions of (1)–(2)–(3) and (1)–(2)–(4) to behave like $\phi_0(x)$ and $\phi_1(x)$, respectively. We find that such behavior does occur and we summarize our results for small $|\lambda|$ in

THEOREM 1. *There is an $\epsilon_0 > 0$ such that if $|\lambda| < \epsilon_0$ then (1)–(2)–(3) has at least one solution and (1)–(2)–(4) has at least one solution. Furthermore these solutions have the following asymptotic behavior:*

(i) $(\phi(x) - 1)e^{x/8} \to 0$ as $x \to \infty$,

(ii) $|\phi'(x)e^{x/8}| \leq 3$ *and* $|\phi''(x)e^{x/8}| \leq 3$ *for all* $x \geq 0$,
(iii) *the solution of (1)–(2)–(3) satisfies* $\lim_{\lambda\to 0} |\phi(x) - \phi_0(x)| = 0$ *uniformly for* $x \in [0, \infty)$, *and*
(iv) *the solution of (1)–(2)–(4) satisfies* $\lim_{\lambda\to 0} |\phi(x) - \phi_1(x)| = 0$ *uniformly for* $0 \leq x < \infty$.

Our second main result concerns the non-existence of solutions of (1)–(2)–(3) and (1)–(2)–(4) for large negative λ. As described by Ierley (private communication) the non-existence of solutions for large negative λ may play a role in the explanation of why separation of the Gulf Stream occurs at a point considerably south of the observed "zero of wind stress curl."

THEOREM 2. *If* $\lambda \leq -9$ *then (1)–(2)–(3) has no solution. If* $\lambda \leq -(2)^{1/3}$ *then (1)–(2)–(4) has no solution.*

2. Proof of Theorem 1.

We find it convenient in the proof of both of our theorems to make the transformation $u = \phi - 1$. Then u satisfies the equation

$$u''' + \lambda u'' - u + \lambda(uu'' - (u')^2) = 0 \tag{5}$$

and either

$$u(0) = -1, \quad u'(0) = 0 \qquad \text{(no-slip)} \tag{6}$$

or

$$u(0) = -1, \quad u''(0) = 0. \qquad \text{(stress-free)} \tag{7}$$

In both cases the boundary condition at infinity is

$$u(\infty) = 0. \tag{8}$$

We prove Theorem 1 using the Schauder fixed point theorem (see, for example [**3**]). For mathematical simplicity we give the details only for the no-slip problem and remark that the proof of the stress-free problem is essentially the same. Thus we need to transform (5)–(6)–(8) into an integral equation. We then define an appropriate Banach space and an integral operator T on the space for which a fixed point corresponds to a solution of (5)–(6)–(8). We need to show that T satsifies the requirements of the Schauder theorem.

Our first step is to rewrite the problem (5)–(6)–(8) as an integral equation. We set $f_0(x) = u(x)$, $f_1 = u'$, $f_2 = u''$ and $f = (f_0, f_1, f_2)^T$. Then f satisfies

$$f' = Af + \lambda b \tag{9}$$

$$f_0(0) = -1, \quad f_1(0) = 0, \quad f_0(\infty) = 0 \tag{10}$$

where

$$A = \begin{bmatrix} 0 & 1 & 0 \\ 0 & 0 & 1 \\ 1 & 0 & -\lambda \end{bmatrix} \text{ and } b = \begin{pmatrix} 0 \\ 0 \\ f_1^2 - f_0 f_2 \end{pmatrix}.$$

We let $\Phi(t)$ denote a nonsingular matrix solution of

$$f' = Af. \tag{11}$$

Then the general solution of (9) is

$$f = f^0(x) + \lambda\Phi(x) \int_{x_0}^{x} \Phi^{-1}(t) b(t)\, dt \tag{12}$$

where $f^0(x)$ solves (11) and has the form $f^0 = \Phi(x)C$, $C = (C_1, C_2, C_3)^T$ is a constant column vector. The form for $\Phi(x)$ which we use is

$$\Phi(x) = \begin{bmatrix} e^{r_0 x} & e^{r_1 x} & e^{r_2 x} \\ r_0 e^{r_0 x} & r_1 e^{r_1 x} & r_2 e^{r_2 x} \\ r_0^2 e^{r_0 x} & r_1^2 e^{r_1 x} & r_2^2 e^{r_2 x} \end{bmatrix}$$

where r_0, r_1, r_2 are the eigenvalues of A and satisfy

$$r^3 + \lambda r^2 - 1 = 0. \tag{13}$$

For $|\lambda|$ sufficiently small, equation (13) has one positive solution $r_0 > 0$, $r_0 \to 1$ as $\lambda \to 0$, and two complex roots, r_1 and r_2 which tend to $-\frac{1}{2} \pm \frac{3}{2}i$ as $\lambda \to 0$. Furthermore the inverse of Φ for small λ is given by

$$\Phi^{-1}(x) =$$

$$\frac{1}{\Delta} \begin{bmatrix} r_1 r_2 (r_2 - r_1) e^{-r_0 x} & -(r_2 - r_1)(r_2 + r_1) e^{-r_0 x} & (r_2 - r_1) e^{-r_0 x} \\ r_0 r_2 (r_0 - r_2) e^{-r_1 x} & -(r_0 - r_2)(r_0 + r_2) e^{-r_1 x} & (r_0 - r_2) e^{-r_1 x} \\ r_0 r_1 (r_1 - r_0) e^{-r_2 x} & -(r_1 - r_0)(r_1 + r_0) e^{-r_2 x} & (r_1 - r_0) e^{-r_2 x} \end{bmatrix}$$

where $\Delta \equiv -(r_1 - r_0)(r_2 - r_1)(r_0 - r_2) \neq 0$.

From these observations it follows that (12) can be written in the form

$$f = \sum_{j=0}^{2} C_j \alpha_j e^{r_j x} + \lambda \sum_{j=0}^{2} K_j e^{r_j x} \int_{x_0}^{x} e^{-r_j t} b(t)\, dt \tag{14}$$

where $\alpha_j \equiv (1, r_j, r_j^2)^T$, C_j is an arbitrary constant,

$$K_j \equiv a_j \begin{pmatrix} 1/r_j & 1/r_j^2 & 1 \\ 1 & 1/r_j & r_j \\ r_j & 1 & r_j^2 \end{pmatrix} \quad \text{for } j = 0,1,2$$

and

$$\begin{aligned} a_0 &= -1/((r_1 - r_0)(r_0 - r_2)) \\ a_1 &= -1/((r_2 - r_1)(r_1 - r_0)) \\ a_2 &= -1/((r_0 - r_2)(r_2 - r_1)). \end{aligned}$$

Without loss of generality we may write equation (14) (i.e. adjust C_1 and C_2) as

(15)

$$f = \sum_{j=0}^{2} C_j \alpha_j e^{r_j x} + \lambda K_0 e^{r_0 x} \int_{x_0}^{x} e^{-r_0 t} b(t)\, dt + \lambda \sum_{j=1}^{2} K_j e^{r_j x} \int_{0}^{x} e^{-r_j t} b(t)\, dt.$$

We note that every solution of equation (5) can be transformed into a solution of (15) and vice-versa. Furthermore, the boundary conditions given by (6) and (8) correspond to those found in (10). Thus, our goal is to prove that for small $|\lambda|$ there exist constants x_0, C_0, C_1, C_2 and a vector function f satisfying (10) and (15).

We find that an appropriate Banach space for this problem is defined by

$$B = \left\{ f \in C[0,\infty) \mid f_i(x) e^{x/8} \to 0 \text{ as } x \to \infty, \quad i = 0,1,2 \right\}$$

with norm

$$\|f\| = \sup_{i,x} \left\{ \left| f_i(x) e^{x/8} \right| \right\}$$

(A similarly defined Banach space was used by McLeod [**2**] in his studies of the Von-Karman problem.). Next, let $\Omega \subset B$ be defined by

$$\Omega = \{ f \in B \mid f_0(0) = -1, \quad f_1(0) = 0, \quad \|f\| \leq 3 \}.$$

It is not difficult to show that Ω is closed and convex. Also, $\Omega \neq \phi$. For example, the vector with components given by $f_0 = e^{-2x} - 2e^{-x}$, $f_1 = -2(e^{-2x} - e^{-x})$, $f_2 = 4e^{-2x} - 2e^{-x}$ is in Ω. We define an operator on Ω by

$$(16)\quad Tf = f^* \equiv \sum_{j=1}^{2} C_j \alpha_j e^{r_j x} + \lambda K_0 e^{r_0 x} \int_{\infty}^{x} e^{-r_0 t} b(t)\, dt + \lambda \sum_{j=1}^{2} K_j e^{r_j x} \int_0^x e^{-r_j t} b(t)\, dt$$

where $b(t) = (0, 0, f_1^2 - f_0 f_2)^T$ and C_1, C_2 are determined by (10). That is,

$$(17)\quad \begin{aligned} C_1 + C_2 - \frac{\lambda}{(r_1 - r_0)(r_0 - r_2)} \int_{\infty}^{0} e^{-r_0 t}(f_1^2 - f_0 f_2)\, dt &= -1 \\ r_1 C_1 + r_2 C_2 - \frac{\lambda r_0}{(r_1 - r_0)(r_0 - r_2)} \int_{\infty}^{0} e^{-r_0 t}(f_1^2 - f_0 f_2)\, dt &= 0. \end{aligned}$$

It follows from (16) and (17) that $f_0^*(0) = -1$ and $f_1^*(0) = 0$. The improper integrals in (16)–(17) converge since $f \in \Omega$ and $r_0 > 0$. Furthermore, an algebraic manipulation of (17) shows that C_1 and C_2 are complex conjugates. It then follows that f^* is real since the input vector function $f \in \Omega$ is real. Next, we consider the individual components of f^* and show that they have the desired behavior as $x \to \infty$. For $j = 0, 1, 2$ we obtain

$$(18)\quad f_j^* = \sum_{i=1}^{2} C_i r_i^j e^{r_i x} + \lambda a_0 r_0^j e^{r_0 x} \int_{\infty}^{x} e^{-r_0 t}(f_1^2 - f_0 f_2)\, dt + \lambda \sum_{i=1}^{2} a_i r_i^j e^{r_i x} \int_0^x e^{-r_i t}(f_1^2 - f_0 f_2)\, dt.$$

If $\lambda \sim 0$ then $r_0 \sim 1$, $r_j \sim -\frac{1}{2} \pm \frac{\sqrt{3}}{2} i$. Thus, there exists $\epsilon_1 > 0$ such that if $|\lambda| < \epsilon_1$ then $f_j^*(x) e^{x/8} \to 0$ as $x \to \infty$ for $j = 0, 1, 2$. We have now shown that $f^* \epsilon B$. If there exists an $f \in \Omega$ such that $T(f) = f$ then f satisfies (9)–(10) and so f_0 satisfies (5)–(6). To complete our proof we use the Schauder fixed point theorem [**3**]. Thus it remains to prove the following properties of T:

(i) T maps Ω onto itself if $|\lambda| < \in$ and $\in$ is small,

(ii) T is continuous,

(iii) T is compact.

To complete the proof of (i) we need to show that $\|f^*\| \leq 3$. To do this we shall restrict $|\lambda|$ still further and re-estimate the size of $\|f^*\|$. It follows from (18) that

(19)

$$\|f^*\| \leq m\Big\{|C_1| + |C_2| + |\lambda| p\Big[\sup_x \Big|e^{(r_0+1/8)x}\int_\infty^x e^{-r_0 t}(f_1^2 - f_0 f_2)\,dt\Big| + \sum_{i=1}^{2} \sup_x \Big|e^{(r_1+1/8)x}\int_0^x e^{-r_i t}(f_1^2 - f_0 f_2)\,dt\Big|\Big]\Big\}$$

where $m \equiv \max_{i,j} |r_i^j|$ and $p \equiv \max_i |a_i|$. Furthermore, since $f \in \Omega$, we observe that

$$|f_i(x)| \leq 3e^{-x/8} (i = 0,1,2), \text{ and } |f_1^2 - f_0 f_2| \leq 18e^{-x/4}.$$

Thus it follows that

$$e^{(r_0+1/8)x}\int_x^\infty e^{-r_0 t}|f_1^2 - f_0 f_2|\,dt \leq \frac{72}{1+4r_0}e^{-x/8} \tag{20}$$

and

$$\Big|e^{(r_i+\frac{1}{8})x}\int_0^x e^{-r_i t}(f_1^2 - f_0 f_2)\,dt\Big| \leq \frac{18}{|\mathrm{Re}(r_i)+1/4|}\left[e^{-x/8} + e^{\mathrm{Re}(r_i)+1/8)x}\right]. \tag{21}$$

Besides these estimates, we also need to obtain upper bounds on $|C_1|$ and $|C_2|$. For this we refer to (17) and define

$$\ell(f) = \int_\infty^0 e^{-r_0 t}(f_1^2 - f_0 f_2)\,dt.$$

Solving (17) for C_1 and C_2, we obtain

$$\begin{aligned} C_1 &= -\frac{\lambda}{\Delta}(r_0 - r_2)\ell(f) - r_2/(r_2 - r_1) \\ C_2 &= -\frac{\lambda}{\Delta}(r_0 - r_1)\ell(f) - r_1/(r_2 - r_1). \end{aligned} \tag{22}$$

Thus, for small $|\lambda|$, we conclude that

$$\begin{aligned} |C_1| &\leq 18|\lambda|\left|\frac{r_2 - r_0}{\Delta}\right| + \left|\frac{r_2}{r_2 - r_1}\right| \\ |C_2| &\leq 18|\lambda|\left|\frac{r_2 - r_0}{\Delta}\right| + \left|\frac{r_1}{r_2 - r_1}\right|. \end{aligned} \tag{23}$$

From the definitions of m, p, r_i and Δ it follows that for each $\delta \in (0, 1/8)$ there exists $\in_2 > 0$, $0 < \in_2 < \in_1$, such that if $|\lambda| < \in_2$ then

$$(24) \qquad \begin{aligned} m < 1+\delta, \quad p < \tfrac{1}{3}+\delta, \quad 1-\delta < |r_i| < 1+\delta \\ \Delta > 3-\delta, \quad \sqrt{3}-\delta < |r_i - r_j| < \sqrt{3}+\delta \quad (i \neq j). \end{aligned}$$

It follows from (19)–(24) that if $|\lambda| < \in_2$ then

$$(25) \qquad \|f^*\| \leq \frac{2(1+\delta)^2}{\sqrt{3}-\delta} + |\lambda| M$$

where

$$(26) \qquad M = \frac{36(\sqrt{3}+\delta)(1+\delta)}{3-\delta} + \frac{18(1+\delta)(1/3+\delta)}{1-\delta} + \frac{72}{1/4-\delta}.$$

We conclude from (25)–(26) that there is an $\epsilon_0 \in (0, \epsilon_2)$ such that if $|\lambda| < \epsilon_0$ then $\|f^*\| \leq 3$. Thus, for $|\lambda| < \epsilon_0$, T maps Ω onto itself.

Next, we need to show that T is continuous. First, let $f, g \in \Omega$. Then $|f_1^2 - f_0 f_2 - g_1^2 + g_0 g_2| \leq 12 \|f-g\| e^{-x/8}$. Therefore it follows that

$$(27) \qquad e^{(r_0+1/8)x} \int_x^\infty e^{-r_0 t} |f_1^2 - f_0 f_2 - g_1^2 + g_0 g_2| \, dt \leq \frac{12}{r_0} \|f-g\|,$$

$$(28) \qquad e^{(r_i+1/8)x} \int_0^x e^{-r_i t} |f_1^2 - f_0 f_2 - g_1^2 + g_0 g_2| \, dt \leq \frac{12 \|f-g\|}{|\mathrm{Re}(r_i) + 1/4|}$$

and

$$(29) \qquad |\ell(f) - \ell(g)| \leq \frac{12}{r_0} \|f-g\|.$$

From (22) and (29) it follows that C_1 and C_2 are continuously dependent on f. In addition (27) and (28) imply that each integral term in (18) is continuously dependent upon f. Therefore we conclude that T is continuous.

Finally, we need to show that T is compact. Let $\{f^{(N)}\}_N$ denote an arbitrarily chosen sequence in Ω. We need to prove that $\{Tf^{(N)}\}_N$ has a subsequence convergent in the norm of B. First, we note that $\|f^{(N)}\| \leq 3$ for all N. Letting $f^{(N)*} \equiv Tf^{(N)}$ we note also that $\|f^{(N)(*)}\| \leq 3$ for all N so that $\{f^{(N)*}\}_N$ is a uniformly bounded sequence in Ω. That is $\{f_i^{(N)*} e^{x/8}\}_N$ is a uniformly bounded sequence for $i = 0, 1, 2$. Thus it follows that $f^{(N)*} \to 0$ uniformly in N as $x \to \infty$.

Next, we differentiate (18) and find that

$$\frac{df^{(N)*}}{dx} = Af^{(N)*} + \lambda b^{(N)}$$

where $b^{(N)} \equiv (0, 0, (f_1^{(N)})^2 - f_0^{(N)} f_2^{(N)})^T$. Therefore $\left\| \frac{df^{(N)*}}{dx} \right\|$ is uniformly bounded with respect to N so that $\frac{df^{(N)*}}{dx}$ is uniformly bounded in n and x, and $\frac{df^{(N)*}}{dx} \to 0$ uniformly in n as $x \to \infty$. Since $f^{(N)*}$ and its derivative approach 0 uniformly as $x \to \infty$ it follows from a simple extension of the Arzela-Ascoli Theorem that there is a subsequence, call it $\{f^{(N')*}\}$ of $\{f^{(N)*}\}$ which is uniformly convergent to a continuous vector function $u \in B$. This completes the proof of Theorem 1. □

3. Proof of Theorem 2.

We first consider the stress-free problem, that is

$$u''' + \lambda u'' - u + \lambda(uu'' - (u')^2) = 0 \tag{30}$$

$$u(0) = -1, \quad u''(0) = 0. \tag{31}$$

We let (E) denote the condition $u(\infty) = 0$, set $u'(0) = \beta$ and assume throughout that $\lambda < 0$. Our first goal is to eliminate the possibility that a solution exists for some $\beta < 0$. We do this in

LEMMA 1. *If $\beta \leq 0$ then the solution of (30)–(31) cannot satisfy (E).*

PROOF: From (30) we conclude that $u'''(0) = -1 + \lambda\beta^2$. Let $[0, x_0)$ denote the maximal interval of existence of the solution. It follows from (30) and (31) that $u < -1$, and $u' < 0$ and $u'' < 0 \quad \forall x \epsilon (0, x_0)$. Suppose, for the sake of contradiction, that there is a first $x_1 > 0$ for which $u''(x) = 0$. Then $u'''(x) \geq 0$. However, from (30) it follows that $u'''(x_1) = u(x_1) + \lambda(u'(x_1))^2 < 0$, a contradiction. Thus (E) cannot hold.

For the case $\beta > 0$ we shall use the following criterion to eliminate the existence of solutions:

LEMMA 2. *If there is a first $\hat{x} > 0$ with $u''(\hat{x}) = 1/\lambda$ then condition (E) cannot hold.*

PROOF: At $\hat{x}$ we have $u'''(\hat{x}) \leq -1$. If there is a next value $\bar{x} > \hat{x}$ where $u''(\bar{x}) = 1/\lambda$ then $u'''(\bar{x}) \geq 0$. However, from (30) it follows that $u'''(\bar{x}) \leq$

-1, a contradiction. Thus $u'' < 1/\lambda$ for all $x \geq \hat{x}$ and it follows that (E) cannot hold.

In the next lemma we complete the proof of Theorem 2 for the stress-free problem. For this we make use of the derivatives of equation (30), namely the equations

$$u^{(4)} = -\lambda(1+u)u''' + u' + \lambda u'u'' \tag{32}$$

and

$$u^{(5)} = -\lambda(1+u)u^{(4)} + u'' + \lambda(u'')^2. \tag{33}$$

LEMMA 3. *Let $\beta > 0$ and $\lambda \leq -2^{1/3}$. Then (E) cannot hold.*

PROOF: We assume for the sake of contradiction that condition (E) does hold for some $\beta > 0$ and $\lambda \leq -2^{1/3}$. Lemma 2 implies that $u'' > 1/\lambda \quad \forall x > 0$. Let $(0, a)$ denote the maximal interval over which $u' > 0$. We define $x_1 = (1 - \lambda\beta^2)/(2\beta)$ and consider two subcases:

(i) $a \geq x_1$: From this and (32) it follows that $u^{(4)} \leq 2u'$ on $0 < x < x_1$. Thus, since $u''(0) = 0$, $u'(0) = \beta$ and $u'''(0) < 0$, we conclude that $u'' < 0$, $u' < \beta$ and $u^{(4)} < 2\beta$ as long as $u''' < 0$ on $(0, x_1)$. An integration leads to $u''' < 2\beta x + \lambda\beta^2 - 1 < 0$ and $u'' \leq \beta x^2 + (\lambda\beta^2 - 1)x$ for all $x \in (0, x_1)$. In particular,

$$u''(x_1) \leq (-\lambda^2\beta^4 + 2\lambda\beta^2 - 1)/(4\beta) \leq 1/\lambda \tag{34}$$

if

$$2\lambda^2\beta^2 - 4\beta - \lambda \geq 0. \tag{35}$$

It easily follows that (35), and likewise (34), hold for all $\lambda \leq -2^{1/3}$. Thus, by Lemma 2, condition (E) cannot hold, a contradiction. Next, we consider

(ii) $a < x_1$: Here $u'(a) = 0$, $u''(a) < 0$, $u'''(a) < 0$ and $u(a) > -1$. Thus, from (32), $u^{(4)}(a) < 0$. It then follows from (33) that $u^{(4)}(x) < 0$ for all $x \geq a$. Integrating this inequality three times we conclude that $\lim_{x\to\infty} u(x) = -\infty$, again a contradiction. This completes the proof of the lemma.

We have now shown that the stress-free problem has no solution for $\lambda \leq -2^{1/3}$. To complete the proof of Theorem 2 we consider the no-slip problem, that is

$$u''' + \lambda u'' - u + \lambda(uu'' - (u')^2) = 0 \tag{36}$$

$$u(0) = -1, \quad u'(0) = 0. \tag{37}$$

Here we set $u''(0) = \beta$ and show that for each $\lambda < 0$ and $\beta \in R$ the solution of (36)–(37) cannot satisfy condition (E). We note that Lemma 2 holds for this problem and the application of this lemma plays a key role in aour analysis. As with the stress-free problem we first eliminate negative values of β in

LEMMA 4. *Let $\beta \leq 0$. Then the solution of (36)–(37) cannot satisfy condition (E).*

PROOF: Let $\beta < 0$ and let $[0, x_0)$ denote the maximal interval of existence of the solution. From (36), $u'''(0) = -1$. Therefore $u'' < 0$, $u' < 0$ and $u < -1$ on a maximal interval $(0, \delta)$. If there is a first $\hat{x} > 0$ for which $u''(\hat{x}) = 0$ then $u'''(\hat{x}) \geq 0$. However, since $u(\hat{x}) < -1$ it follows from (36) that $u'''(\hat{x}) < -1$, a contradiction. Therefore $u'' < 0$, $u' < 0$ and $u < -1$ for $x > 0$ as long as the solution exists, hence (E) cannot hold.

Next, we consider positive values of β. We show that for each $\lambda \leq -9$ and $\beta > 0$, the solution must satisfy $u'' = 1/\lambda$ at some $\hat{x} < 0$. It then follows from Lemma 2 that (E) cannot hold. We find it convenient to separate the range of β values into several cases. Our analysis rests upon the following two key lemmas:

LEMMA 5. *Let $\lambda \leq -9$. If there is a first $x_1 \geq 0$ with $u''(x_1) = -1/\lambda$, $u(x_1) \geq -1$, $u'(x_1) \geq 0$, $u'''(x_1) \leq -1$ and $u^{(4)}(x_1) \leq -1/4$ then condition (E) cannot hold.*

PROOF: Since (36) is autonomous we may assume without loss of generality that $x_1 = 0$, that is, $u(0) \geq -1$, $u'(0) \geq 0$, $u''(0) = -1/\lambda$, $u'''(0) \leq -1$ and $u^{(4)}(0) \leq -1/4$. Let $(0, a)$ denote the maximal interval over which $u' > 0$. There are two subcases to consider:

(i) $a \geq -\lambda/4$: We note that $u > -1$ over $(0, a)$. Thus, for $0 < x < a$, as long as $u^{(4)} < 0$ then $u''' < -1$, $u'' < -1/\lambda$ and it follows from (33) that $u^{(5)} < -1/\lambda$. Thus $u^{(4)} < -x/\lambda - 1/4 < 0$ for $0 \leq x \leq -\lambda/4$. Therefore $u''' < -x^2/(2\lambda) - x/4 - 1$ and

$u'' \leq -x^3/(6\lambda) - x^2/8 - x - 1/\lambda$ on $(0, -\lambda/4)$. We conclude that $u''(-\lambda/4) < 0$, $u'''(-\lambda/4) < -1$ and $u^{(4)}(-\lambda/4) < 0$. It then follows from (33) that $u^{(4)} < 0$, $u''' < -1$ and $u'' < 0$ for $x \geq -\lambda/4$ as long as the solution exists. Thus condition (E) cannot hold. We are left with the second possibility, namely

(ii) $0 < a < -\lambda/4$: Here $u'(a) = 0$, $u''(a) \leq 0$, $u'''(a) < 0$ and $u(a) > -1$. Again, from (32), $u^{(4)}(a) < 0$. If there were a first $b > a$ where $u^{(4)}(b) = 0$ then $u^{(5)}(b) \geq 0$. However, $u'' < 0$ and $u''' < 0$ on $[a, b]$ hence from (33), $u^{(5)}(b) < 0$, a contradiction. Therefore $u^{(5)} < 0$, $u''' < u'''(a) < 0$ for $x > a$ as long as the solution exists and it follows that condition (E) cannot hold.

LEMMA 6. *Let $\beta > -1/\lambda$. There is a first $x_1 > 0$ with $u''(x_1) = -1/\lambda$, and $u''' < 0$, $u^{(4)} < 0$ for $0 < x < x_1$.*

PROOF: From (36) and (32) we observe that $u'''(0) = -1$, $u^{(4)}(0) = 0$, and $u^{(5)}(0) = \beta + \lambda\beta^2 < 0$. It then follows from (32) and (33) that $u^{(4)} < 0$ and $u''' < -1$ for $x > 0$ as long as $u'' > -1/\lambda$. Thus the existence of x_1 is assured, completing the proof for the lemma.

We now proceed with the main arguments of our proof. First, we eliminate the possibility that $\beta \geq 1/2$ in the following:

LEMMA 7. *Let $\beta \geq 1/2$, and $\lambda \leq -9$. Then (E) cannot hold.*

PROOF: From Lemma 6 there is a first $x_0 > 0$ with $u''(x_0) = \frac{1}{2}$, $u'''(x_0) \leq -1$ and $u(x_0) \geq -1$. We consider two subcases:

(i) $u'(x_0) \geq 1/2$: From (32), $u^{(4)}(x_0) \leq -3/2$. Let $x_1 > x_0$ be the next value where $u'' = -1/\lambda$. Then (33) implies that $u^{(4)} < -3/2$, $u''' < -1$ and $u > -1$ on $(x_0, x_1]$. From this and Lemma 5 we see that (E) cannot hold.

(ii) $u'(x_0) \leq 1/2$: Let $x_1 > x_0$ be the next value where $u''(x_1) = -1/\lambda$. Since $u'' > -1/\lambda$ on (x_0, x_1), it follows from (32) that $u^{(4)} < 0$ and $u''' < -1$ on (x_0, x_1). Therefore $x_1 \in (x_0, x_0 + 1/2)$ and $u'(x_1) < 1$. Since our problem is autonomous we may assume for convenience that $x_1 = 0$; that is $u''(0) = -1/\lambda$, $u(0) > -1$, $0 < u'(0) < 1$ and $u'''(0) < -1$. We let $a > 0$ denote the next value for which $u' = 0$ and consider two subcases:

(iii) $a \geq 9/19$: It follows from (32) that for $0 \leq x \leq 9/19$, as long as $u''' < 0$ then $u'' < +\frac{1}{9}$, $u' < \frac{19}{18}$ hence $u^{(4)} \leq 19/9$. Integrating, we find that $u''' \leq (19x/9) - 1 = 0$ and $u'' \leq \frac{19x^2}{18} - x - \frac{1}{\lambda}$ for

$0 \leq x \leq 9/19$. In particular, at $x = 9/19$, $u''(9/19) \leq 1/\lambda$ if $\lambda \leq -9$. This and Lemma 2 show that (E) cannot hold. The second subcase we need to eliminate is:

(iv) $0 < a < 9/19$: Then $u(a) > -1$, $u'(a) = 0$ and $u''(a) \leq 0$. As above $u''' < 0$ on $(0, 9/19)$ so that $u'''(a) = 0$. Therefore, from (32), $u^{(4)}(a) < 0$. It follows from (33) that $u^{(4)} < 0$ and $u''' < u'''(a) < 0$ for $x > a$ as long as the solution exists. Thus (E) cannot hold and the lemma is proved.

The proof of Theorem 2 is completed in:

LEMMA 8. *Let $\lambda \leq -9$ and $0 < \beta < 1/2$. Then (E) cannot hold.*

PROOF: For the sake of contradiction we assume that (E) does hold for some $\lambda \leq -9$ and $0 < \beta < 1/2$. Again, we let $(0, a)$ denote the maximal interval over which $u' > 0$. For $0 < x < a$, as long as $u''' < -3/4$ then $u'' < -\frac{3x}{4} + \frac{1}{2}$ and $u' < -\frac{3x^2}{8} + \frac{x}{2}$. From these inequalities and (32) it follows that $u^{(4)} \leq 2u' \leq -\frac{3}{4}x^2 + x$. Integrating, we find that $u''' \leq -\frac{x^3}{4} + \frac{x^2}{2} - 1 \leq -\frac{3}{4}$ for $0 \leq x \leq 1$ as long as $u' > 0$. Suppose, first of all, that $a > 1$. Then $u''(1) \leq -1/4 \leq 1/\lambda$ for $\lambda \leq -4$. This and Lemma 2 show that (E) cannot hold. Therefore it must be the case that $0 < a < 1$. Then $u(a) \geq -1$, $u'(a) = 0$, $u''(a) \leq 0$ and $u'''(a) \leq -\frac{a^3}{4} + \frac{a^2}{2} - 1 \leq -\frac{3}{4}$. Therefore $u^{(4)}(a) < 0$. Once again (33) implies that $u^{(4)} < 0$ and $u''' < -\frac{3}{4}$ for $x > a$ as long as the solution exists. Therefore (E) cannot hold and we have reached a contradiction.

□

Department of Mathematics and Statistics
University of Pittsburgh, Pittsburgh PA 15260

The first author is on leave from Computing Center, Academic Sinica, Beijing, P.R. China

The second author is partially supported by NSF Grant DMS 8501531

Research at MSRI supported in part by NSF Grant DMS 812079-05.

REFERENCES

1. G.R. Ierley and O.G. Ruehr, *Analytic and numerical solutions of a nonlinear boundary-layer problem*, Studies in Applied Math. **75** (1986), 1–36.
2. J.B. McLeod, *The existence of axially symmetric flow above a rotating disk*, Proc. Roy. Soc. London **A 324** (1971), 391–414.
3. J. Smoller, "Shock waves and reaction-diffusion equations," Grundlehren der Math, Wissenschaften, Vol. 258, Springer-Verlag, New York, 1983.
4. W.C. Troy, *Solutions of a nonlinear boundary layer problem arising in physical oceanography*, SIAM J. Math. Anal. (submitted).

Asymptotic behavior of solutions of semilinear heat equations on S^1

HIROSHI MATANO

Abstract. We study the dynamical behavior of the initial value problem for the equation $u_t = u_{xx} + f(u, u_x)$, $x \in S^1 = \mathbf{R}/\mathbf{Z}$, $t > 0$. One of our main results states that any C^1-bounded solution approaches either a single periodic solution or a set of equilibria as $t \to \infty$. We also consider the case where the solution blows up in a finite time and prove that under certain conditions on f the blow-up set of any solution with nonconstant initial data is a finite set.

§1. Introduction.

Let $S^1 = \mathbf{R}/\mathbf{Z}$ be the circle of length 1 and consider the initial value problem

$$(1a) \qquad u_t = u_{xx} + f(u, u_x) \quad (x \in S^1, t > 0),$$

$$(1b) \qquad u(x,0) = u_0(x) \qquad (x \in S^1),$$

where $f : \mathbf{R} \times \mathbf{R} \to \mathbf{R}$ is a smooth function and $u_0 \in C^1(S^1)$. Problem (1) is of course equivalent to a problem on the interval $[0,1]$ with periodic boundary conditions, but for the convenience of the later argument we formulate the problem as in (1). It is well-known that problem (1) is well-posed and defines a local semiflow on the space $C^1(S^1)$ (see Mora [**20**]). Under some additional assumptions on f (see Remark 2.1) one can further show that (1) defines a local semiflow on $C(S^1)$, in which case u_0 can be an arbitrary continuous function on S^1.

The aim of this paper is to investigate the asymptotic behavior of solutions of (1) including those that blow up in a finite time. First we shall show that any C^1-bounded solution of (1) either tends to a time-periodic solution of a special form (which we call a *rotating wave* solution) or approaches a one-dimensional manifold of equilibrium solutions as $t \to \infty$ (Theorem A). Here, by a rotating wave solution we mean a function u of the form

$$(2a) \qquad u(x,t) = \varphi(x - ct),$$

where c is a nonzero constant and $\varphi(z)$ is a nonconstant solution of

$$(2b) \qquad c\frac{d\varphi}{dz} + \frac{d^2\varphi}{dz^2} + f\left(\varphi, \frac{d\varphi}{dz}\right) = 0 \qquad (z \in S^1).$$

Note that u in (2a) is an equilibrium solution if $c = 0$ or if φ is a constant solution of (2b). In deriving Theorem A we shall impose no condition on f other than smoothness. In particular, hyperbolicity of periodic solutions or equilibria is not assumed. The assumption of C^1-boundedness in Theorem A (as well as in the subsequent theorems) can be replaced by C^0-boundedness if $f(u,p)$ satisfies some growth condition with respect to p (Theorem C). Next we show that if the nonlinear term f has some kind of symmetry, say $f(u,-p) \equiv f(u,p)$, then any C^1-bounded solution of (1) converges to an equilibrium solution as $t \to \infty$ (Theorems B, B′). This result can be stated in terms of the uniqueness of the ω-limit point. A simple example to which Theorem B applies is the case where $f = f(u)$, with $f : \mathbf{R} \to \mathbf{R}$ being an arbitrary smooth function (Corollary D). Corollary D improves the result of Chen [**8**], who considered only a special class of functions f. Finally we consider the case where the solution u blows up in finite time. Under the assumption that $f(u,-p) \equiv f(u,p)$ and some growth conditions on f, we shall prove that the blow-up set of u is a finite subset of S^1 so far as the initial data $u_0(x)$ is not a constant function (Theorem E). Theorem E naturally extends to the problem on a compact interval in $\mathbf{R}$ under the Dirichlet or the Neumann boundary conditions.

As regards the convergence of solutions as $t \to \infty$, more extensive studies have been made in earlier works including [**24**], [**18**] and [**14**] in the case where the spatial domain is a compact interval in $\mathbf{R}$. More precisely, let $L > 0$ and consider the initial boundary value problem

$$
\begin{aligned}
&(3a) \qquad && u_t = u_{xx} + f(x,u,u_x) && (0 < x < L, t > 0),\\
&(3b) && u(x,0) = u_0(x) && (0 < x < L),\\
&(3c) && (1-\alpha)u(0,t) - \alpha u_x(0,t) = 0 && (t > 0),\\
& && (1-\beta)u(L,t) + \beta u_x(L,t) = 0 && (t > 0),
\end{aligned}
$$

where $f : [0,L] \times \mathbf{R} \times \mathbf{R} \to \mathbf{R}$ is smooth and α, β are constants satisfying $0 \le \alpha \le 1$, $0 \le \beta \le 1$. It is essentially the result of [**24**], [**18**] and [**14**] that any solution of (3) that remains bounded in $C^1(S^1)$ converges to an equilibrium solution as $t \to \infty$. Since problem (3) has a nice Liapunov function, one easily sees that the solution approaches a set of equilibria as $t \to \infty$. So the question discussed in the above-mentioned articles concerns whether the solution picks up precisely one element from the set of equilibria as its ω-limit point. The method used in [**24**] is based on the fact that any one-dimensional manifold of equilibria — if it exists — is "hyperbolic", or in other words, the zero eigenvalue of the linearized eigenvalue problem at each point of the manifold of equilibria is always a simple eigenvalue, which follows from the uniqueness theorem for ordinary differential equations. A similar idea has been used in [**14**] to study higher dimensional problems as well. On the other hand, the method in [**18**] is based on the maximum principle and Jordan curve theorem — in a fashion similar to what is now classical in deriving the variation decreasing property of one-dimensional diffusion equations (cf. Nickel [**21**] and Matano [**19**]).

In the present problem in which the spatial domain is S^1, the behavior of solutions can be more complicated, and there are several difficulties in extending the methods of [**24**], [**18**] and [**14**] to the present problem. The first difficulty is the lack of a nice Liapunov function for problem (1). Those Liapunov functions used in the arguments of [**24**], [**18**] or [**14**] to investigate problem (3) do not work for problem (1) except in very special cases such as the case where $f = f(u)$. The second difficulty arises from the fact that S^1 has no boundary. Since the method in [**18**] depends on the investigation of the behavior of solutions on the boundary, the same approach does not apply to (1). Note also that the method in [**24**] or [**14**] does not work either, since the linearized problem on S^1 may have a degenerate zero eigenvalue, therefore periodic solutions or manifolds of equilibria are not always hyperbolic. Such difficulties remain a bottleneck even in the simple case where $f = f(u)$ despite the existence of a nice Liapunov function for this case.

In the present paper we shall make a quite different approach to problem (1). More precisely, we take advantage of the fact that equation (1a) is equivariant with respect to the shift $(x,t) \mapsto (x+a, t+\tau)$ and study carefully how the "oscillation number" of the function $u(x,t) - u(x+a,t+\tau)$ changes as $t \to \infty$ and as (a,τ) varies in $S^1 \times \mathbb{R}^+$. Roughly speaking, for each (a,τ) this oscillation number plays the role of a Liapunov function whose value is a nonnegative integer. The "invariance principle" for this Liapunov function will be established for each parameter (a,τ). For each fixed (a,τ) this invariance principle provides too little information to be useful, but collecting that information for every $(a,\tau) \in S^1 \times \mathbb{R}^+$ provides substantial knowledge that is sufficient for proving Theorem A. Angenent and Fiedler [**4**] use a different method to obtain a result similar to — but much weaker than — our Theorem A. They also consider orbital connections for problem (1).

In the special case where $f(u,p) \equiv f(u,-p)$, equation (1a) is equivariant also with respect to the reflection $(x,t) \mapsto (2a-x,t)$ for every $a \in S^1$. Comparing solution $u(x,t)$ and its reflection $u(2a-x,t)$ for every $a \in S^1$, and adding this information to Theorem A, we obtain Theorem B. A similar reflection method can be found in [**8,9**] and also in [**10,11**], in which the same idea is used for the study of isolated singularities of a semilinear elliptic equation in $\mathbb{R}^2$.

Concerning the blow-up problem, our main result Theorem E, as mentioned earlier, applies also to problems on an interval in $\mathbb{R}$:

$$u_t = u_{xx} + f(u,u_x) \qquad (0 < x < L, t > 0),$$

with either the Dirichlet or the Neumann boundary conditions at $x = 0$, $x = L$, where f is, typically, such a function as e^u, $|u|^{q-1}u \quad (q > 1)$ or $u(\log(|u|+1))^r \quad (r > 2)$. Since we make no assumption on u_0 except that u_0 be a nonconstant function and $u_0 \geq 0$, Theorem E is a considerable generalization of the results of Caffarelli and Friedman [**6**], Friedman and McLeod [**13**] and Weissler [**23**]. As in Theorem B, the proof of Theorem E uses extensively the reflectional symmetry of the equation. See the forthcoming paper [**9**] for a more extensive treatment of this problem.

This paper is organized as follows: in §2 we state the main results (Theorems A, B, B′, C, E and Corollary D). We prove Theorem A in §3, Theorems B, B′, C and Corollary D in §4 and Theorem E in §5. Finally, in the Appendix, we show how to construct a Liapunov function for problem (3), or more precisely, for problem (32) which is a much more generalized version of problem (3) involving possibly nonlinear boundary conditions. The existence of such a Liapunov function, though already proved in [**24**], does not seem to be widely known except for the simpler case where f is independent of u_x. Our construction of the Liapunov function is different from that of [**24**] and requires weaker smoothness assumptions on f and other coefficients.

The author wishes to express his gratitude to the Mathematical Sciences Research Institute in Berkeley for their hospitality and support. Thanks are extended to Prof. Bernold Fiedler for stimulating discussions and to Mr. Xu-Yan Chen, who has made a substantial contribution to the development of Theorem E. He also would like to point out that a recent result of Angenent [**3**], which was obtained after the author's lecture at Berkeley in 1986, helped improve the earlier version of Theorem A and reduce the size of the proof of other theorems substantially. After this work was completed, the author was informed that P. Massatt had recently obtained independently a result similar to Theorem A under the additional assumption that f be analytic [**17**].

§2. Main results.

Throughout this paper, unless otherwise stated, we assume that $f : \mathbf{R} \times \mathbf{R} \to \mathbf{R}$ is of class C^2. For each $c \in \mathbf{R}$ we set

$$\Lambda_c = \{\varphi \in C^2(S^1) \mid \varphi \text{ satisfies } (2b)\}, \tag{4a}$$

$$\Lambda = \bigcup_{c \in \mathbf{R}} \Lambda_c. \tag{4b}$$

Note that Λ_0 coincides with the set of all equilibrium solutions of (1). Let $\Phi = \{\Phi_t\}_{t \geq 0}$ be the local semiflow on $C^1(S^1)$ generated by (1). More precisely, for each $\psi \in C^1(S^1)$ we let $u(x, t; \psi)$ be the solution of (1) with initial data $u_0 = \psi$ and

$$s(\psi) = \sup\{T > 0 \mid u(\cdot, t; \psi) \text{ exists for } 0 \leq t < T\}. \tag{5a}$$

We then define

$$\Phi_t(\psi) = u(\cdot, t; \psi) \tag{5b}$$

for $\psi \in C^1(S^1)$ and $0 \leq t < s(\psi)$. From the standard local existence theorem, it is easily seen that

$$\lim_{t \to s(\psi)} \|\Phi_t(\psi)\|_{C^1(S^1)} = \infty$$

if $s(\psi) < \infty$, where

$$\|w\|_{C^1(S^1)} = \max_{x \in S^1} |w(x)| + \max_{x \in S^1} |w'(x)|.$$

REMARK 2.1: Suppose that the function $f(u, p)$ satisfies the following: there exists $\varepsilon > 0$ such that for each $M > 0$ there exists a constant $C_M > 0$ such that

$$|f_u(u, p)| \leq C_M(1 + |p|^{2-\varepsilon}),$$
$$|f_p(u, p)| \leq C_M(1 + |p|^{1-\varepsilon}),$$

for any $(u,p) \in [-M,M] \times \mathbf{R}$. Then (1) defines a local semiflow on the space $C(S^1)$, the space of continuous fucntions on S^1. This follows from a general theory for abstract parabolic equations (see [**2**]).

DEFINITION 2.2: For each $u_0 \in C^1(S^1)$ with $s(u_0) = \infty$, we define its *ω-limit set* by

$$\omega(u_0) = \bigcap_{t>0} \text{closure } \{\Phi_\tau(u_0) \mid \tau \geq t\}, \tag{6}$$

where the closure is taken in the topology of $C^1(S^1)$.

It is well-known that $\varphi \in \omega(u_0)$ if and only if there exists a sequence $0 \leq t_1 < t_2 < t_3 < \cdots \to \infty$ such that $u(\cdot, t_k) \to \varphi$ as $k \to \infty$.

THEOREM A. *Let u be a solution of (1) such that $\|u(\cdot,t)\|_{C^1(S^1)}$ remains bounded as $t \to \infty$. Then $\omega(u_0)$ is a nonempty compact connected set in $C^2(S^1)$ and there exist $c \in \mathbf{R}$ and $\varphi \in \Lambda_c$ such that $\omega(u_0) \subset \{\varphi(\cdot + a) \mid a \in S^1\}$.*

THEOREM B. *Let the assumption in Theorem A hold. Assume further that $f(u,-p) = f(u,p)$ for $u \in \mathbf{R}$, $p \in \mathbf{R}$. Then $\omega(u_0)$ contains exactly one element. In other words, $u(\cdot,t)$ converges to an equilibrium solution as $t \to \infty$.*

THEOREM B′. *Let the assumption in Theorem A hold. Assume further that $f(-u,p) = -f(u,p)$ for $u \in \mathbf{R}$, $p \in \mathbf{R}$ and that there exists $M > 0$ (possibly $M = \infty$) such that $f(u,0) \geq 0$ for $0 \leq u < M$ and $f(u,0) \leq 0$ for $M < u < \infty$. Then the same conclusion as in Theorem B holds.*

THEOREM C. *Assume that for each $M > 0$ there exists a constant $C_M > 0$ such that*

$$|f(u,p)| \leq C_M(1+p^2) \tag{7}$$

for $(u,p) \in [-M,M] \times \mathbf{R}$. Then Theorems A, B and B′ remain true if the boundedness of $\|u(\cdot,t)\|_{C^1(S^1)}$ is replaced by that of $\|u(\cdot,t)\|_{L^\infty(S^1)}$.

Combining Theorems B and C, we obtain the following:

COROLLARY D. *Let $u(x,t)$ be a solution of*

$$u_t = u_{xx} + f(u) \quad (x \in S^1, t > 0), \tag{8a}$$

$$u(x,0) = u_0(x) \quad (x \in S^1) \tag{8b}$$

where $f : \mathbb{R} \to \mathbb{R}$ is of class C^2. Suppose that $\|u(\cdot,t)\|_{L^\infty(S^1)}$ remains bounded as $t \to \infty$. Then $u(x,t)$ converges as $t \to \infty$ to an equilibrium solution.

Next we consider the blow-up problem. In addition to the C^2-smoothness of f, we assume the following:

(9a) $\quad f(u,-p) \equiv f(u,p)$;

(9b) $\quad f(0,0) = 0$;

(9c) $\quad$ there exists $M_0 \geq 0$ such that $f \geq 0$ for any $u \geq M_0, p \in \mathbb{R}$;

(9d) $\quad f_p$ is bounded in $\mathbb{R}^+ \times \mathbb{R}^+$.

Furthermore, we assume that there exist $M_0 \geq 0$, $\tilde{\varepsilon} > 0$ and a smooth function $F : [M_0, \infty) \to \mathbb{R}$ such that

(10a) $\quad F(u) > 0, \quad F'(u) \geq 0, \quad F''(u) \geq 0 \quad$ for $u \geq M$;

(10b) $\quad F f_u - F' f - \tilde{\varepsilon} F F' \geq 0 \quad$ for $u \geq M_0, p \in \mathbb{R}$;

(10c) $\quad \displaystyle\int_{M_0}^{\infty} \frac{du}{F(u)} < \infty.$

Conditions of the form (10) are due to [**6**] and [**13**], though the present ones are slightly more simplified than those of [**6**,**13**].

DEFINITION 2.3: Let u be a solution of (1) such that $s(u_0) < \infty$ (hence $\|u(\cdot,t)\|_{C^1(S^1)} \to \infty$ as $t \to s(u_0)$). The *blow-up set* of u is defined by

$$B(u) = \big\{x \in S^1 \mid \text{there exist } x_1, x_2, x_3, \cdots \to x \text{ and } 0 < t_1 < t_2 < \cdots \to s(u_0) \text{ such that } |u(x_k,t_k)| \to \infty \text{ as } k \to \infty\big\}.$$

Note that $B(u)$ may be empty even if $s(u_0) < \infty$ since $\|u(\cdot,t)\|_{L^\infty}$ may remain bounded. However, under condition (7), $B(u)$ is always a nonempty set.

THEOREM E. *Assume (9), (10) and let u be a solution of (1) with $u_0 \geq 0$. Suppose that $s(u_0) < \infty$. Then $u(x,t) \geq 0$ for $x \in S^1$, $0 \leq t < s(u_0)$ and*

$$\max_{x \in S^1} u(x,t) \to \infty \text{ as } t \to s(u_0).$$

Moreover, if the initial data is not a constant function, then the blow-up set $B(u)$ is a finite subset of S^1.

REMARK 2.4: One cannot drop conditions (9), (10) in Theorem E, as the counterexample of Lacey [**16**] shows.

REMARK 2.5: Theorem E remains true if problem (1) is replaced by a problem on an interval:

$$\begin{aligned} u_t &= u_{xx} + f(u, u_x) \quad && (0 < x < L, t > 0), \\ u(x, 0) &= u_0(x) && (0 < x < L) \end{aligned}$$

under the Dirichlet boundary conditions

$$u(0, t) = u(L, t) = 0 \quad (t > 0)$$

or the Neumann boundary conditions

$$u_x(0, t) = u_x(L, t) = 0 \quad (t > 0).$$

In fact, these problems can be converted into those with periodic boundary conditions (which are equivalent to (1)) by symmetrically extending the solution u as

$$\tilde{u}(x, t) = u(|x|, t) \quad (-L \leq x \leq L, t \geq 0)$$

in the case of the Neumann boundary conditions or as

$$\tilde{u}(x, t) = \begin{cases} u(x, t) & (0 \leq x \leq L, t > 0) \\ -u(-x, t) & (-L \leq x < 0, t > 0) \end{cases}$$

together with

$$\tilde{f}(u, p) = \begin{cases} f(u, p) & (\text{if } u \geq 0) \\ -f(-u, p) & (\text{if } u < 0) \end{cases}$$

in the case of the Dirichlet boundary conditions. In the latter case, the assumption $u_0 \geq 0$ in Theorem E is not satisfied after symmetric extension, but this case can be treated by using the condition $f(-u, p) \equiv -f(u, p)$. For details and a more extensive treatment of the blow-up problem, see the forthcoming article [**9**].

REMARK 2.6: Conditions (9) and (10) are satisfied if, for example

$$\begin{aligned} f(u, p) &= e^u + g(u, p), \\ f(u, p) &= |u|^{q-1}u + g(u, p) \end{aligned}$$

or

$$f(u,p) = u|\log(1+u)|^r + g(u,p),$$

where $q > 1$, $r > 2$ and $g(u,p)$ is a smooth function such that $g(u,-p) \equiv g(u,p)$ and such that $g_p(u,p)$ is bounded in $\mathbf{R}^+ \times \mathbf{R}^+$ and $e^{-u} g_u(u,p) \to 0$ as $u \to \infty$ uniformly in $p \in \mathbf{R}$ (respectively $u^{-q+1} g_u(u,p) \to 0$, $(\log u)^{1-r} g_u(u,p) \to 0$ in the second and third cases). In fact, conditions (9) are easily seen to hold and conditions (10) can be verified by letting $F(u) = e^{\alpha u}$ $(0 < \alpha < 1)$, $F(u) = u^{\alpha q}$ $(1/q < \alpha < 1)$ and $F(u) = u|\log u|^{\alpha r}$ $(1/r < \alpha \leq 1 - 1/r)$.

REMARK 2.7: Rotating wave solutions of (1) — if they exist — are unstable by virtue of the result of Hirsch [**15**], and, roughly speaking, "almost all" bounded solutions tend to equilibria. Moreover, by extending the result of Chafee [**7**], one can show that an equilibrium solution of (1) is stable if and only if it is a constant function.

§3. Proof of Theorem A.

For each $a \in S^1$, we define a shift operator $\sigma_a : C^1(S^1) \to C^1(S^1)$ and a reflection operator $\rho_a : C^1(S^1) \to C^1(S^1)$ by

$$\sigma_a w = w(\cdot + a), \tag{11a}$$

$$\rho_a w = w(2a - \cdot). \tag{11b}$$

For each $a \in S^1$ and $\tau \geq 0$, a space-time shift $\sigma_{a,\tau} : C^1(S^1) \to C^1(S^1)$ is defined by

$$\sigma_{a,\tau} = \Phi_\tau \sigma_a \quad (= \sigma_a \Phi_\tau), \tag{11c}$$

where Φ is the local semiflow associated with problem (1) (as defined in (5)). It is implied in (11c) that $\sigma_{a,\tau}$ operates on $\psi \in C^1(S^1)$ if and only if $\Phi_\tau \sigma_a$ does as well, that is, if and only if $\sigma(\psi) > \tau$.

NOTATION 3.1: Given $w \in C^1(S^1)$, we define $\nu(w)$ to be the number of connected components of the set $\{x \in S^1 \mid w(x) \neq 0\}$. We understand that $\nu(w) = 0$ if and only if $w \equiv 0$ on S^1.

DEFINITION 3.2: We say that a function $w \in C^1(S^1)$ *has only simple zeroes* if $w'(x) \neq 0$ for any $x \in S^1$ satisfying $w(x) = 0$. We denote by Σ the set of all C^1-functions on S^1 that have only simple zeroes.

Note that Φ_t commutes with σ_a, ρ_a, and $\sigma_{a,\tau}$. ν defines a function $C^1(S^1) \to \mathbb{N} \cup \{0, \infty\}$. It is clear that $\nu(w) < \infty$ if $w \in \Sigma$.

LEMMA 3.3. *Let $w(x,t)$ be a continuous function on $S^1 \times (T_1, T_2)$ for some $-\infty \le T_1 < T_2 \le +\infty$. Suppose that w is C^2 in x and C^1 in t and that it satisfies the linear differential equation*

$$w_t = w_{xx} + q(x,t)w_x + r(x,t)w \quad (x \in S^1, T_1 < t < T_2), \tag{12}$$

where q, q_x, q_t and r are bounded on any compact subset of $S^1 \times (T_1, T_2)$. Then

(i) *$\nu(w(\cdot,t)) < \infty$ for any $t \in (T_1, T_2)$;*
(ii) *$\nu(w(\cdot,t))$ is monotone nonincreasing and right-continuous in t;*
(iii) *given $t_0 \in (T_1, T_2)$, $\lim_{t \uparrow t_0} \nu(w(\cdot,t)) = \nu(w(\cdot,t_0))$ if and only if either $w(\cdot,t_0) \in \Sigma$ or w is identically equal to zero on $S^1 \times (T_1, T_2)$;*
(iv) *if $w \not\equiv 0$ then for any $\delta \in (T_1, T_2)$ there exists a (possibly empty) finite set $A \subset [\delta, T_2)$ such that $w(\cdot,t) \in \Sigma$ for every $t \in [\delta, T_2) \backslash A$.*

Statement (ii) of the above lemma is well-known (see for example [**21**,**19**]). Statements (i) and (iii) are due to Angenent [**3**]. Statement (iv) is an immediate consequence of (i), (ii) and (iii). Historically, a prototype of (iv) was first obtained by Matano [**19**, Lemma 5.4], in which it was shown that $w(\cdot,t) \in \Sigma$ for every t in an open dense subset of (T_1, T_2). The same result was later obtained in [**5**] by using exactly the same method as in the above-mentioned lemma of [**19**]. The present lemma is a much improved version of those results.

LEMMA 3.4. *Let u be a solution of (1) that exists globally in $S^1 \times [0, \infty)$. Given $a \in S^1$, $\tau \ge 0$, let $w(x,t) = u(x,t) - u(x+a, t+\tau)$. Then w satisfies an equation of the form (12) with q, q_x, q_t, r bounded and $T_1 = 0$, $T_2 = \infty$.*

The above lemma follows immediately from the fact that both $u(x,t)$ and $u(x+a, t+\tau)$ are solutions of equation (1a). The following two lemmas are also obvious:

LEMMA 3.5. *Let $\{\varphi_k\}_{k \in \mathbb{N}}$ be a sequence of C^1 functions on S^1 converging to φ as $k \to \infty$ in the topology of $C^1(S^1)$. Then*

$$\liminf_{k \to \infty} \nu(\varphi_k) \ge \nu(\varphi).$$

If, in addition, $\varphi \in \Sigma$, then $\nu(\varphi_k) = \nu(\varphi)$ for all large k.

LEMMA 3.6. *Let $h : \mathbb{R}^+ \to \mathbb{R}$ be a bounded uniformly continuous function and assume that there exist $\alpha > 0$ and $t^* \geq 0$ such that $h(t) - h(t+\alpha) \neq 0$ for any $t \geq t^*$. Then there exists a function $\tilde{h} : \mathbb{R} \to \mathbb{R}$ such that $\tilde{h}(t+\alpha) = \tilde{h}(t)$ for $t \in \mathbb{R}$ and that $|h(t) - \tilde{h}(t)| \to 0$ as $t \to \infty$.*

In what follows we let $u(x,t)$ be an arbitrary solution of (1) satisfying the assumption of Theorem A, and we fix this solution u throughout the remainder of this section. Schauder's a priori estimate for (1) and a standard argument in dynamical systems theory show that the ω-limit set $\omega(u_0)$ is a nonempty compact connected and invariant subset of $C^{2+\alpha}(S^1)$. By virtue of Lemma 3.4 and Lemma 3.3 (i), (ii), one can define a function $\xi : S^1 \times \mathbb{R}^+ \to \mathbb{N} \cup \{0\}$ by

$$\xi(a,\tau) = \lim_{t\to\infty} \nu(u(\cdot,t) - \sigma_{a,\tau}u(\cdot,t)) \quad (a \in S^1, \tau \geq 0). \tag{13}$$

It is clear that $\xi(a,\tau) \equiv 0$ if and only if u is a constant function. Now let

$$\xi_0 = \min\{\xi(a,\tau) \mid a \in S^1, \quad \tau \geq 0, \quad \xi(a,\tau) > 0\} \tag{14}$$

if $u \not\equiv$ constant and set $\xi_0 = 0$ if $u \equiv$ constant. The quantity ξ_0 will be important in the later argument.

LEMMA 3.7. *Let $\varphi \in \omega(u_0)$ and suppose that $\varphi - \sigma_{a,\tau}\varphi \notin \Sigma$ for some $a \in S^1$, $\tau \in \mathbb{R}$. Then $\varphi = \sigma_{a,\tau}\varphi$.*

PROOF: Since $\omega(u_0)$ is an invariant set, there exists a bounded function $v(x,t)$ on $S^1 \times \mathbb{R}$ which satisfies

$$v_t = v_{xx} + f(v, v_x) \quad (x \in S^1, t \in \mathbb{R}), \tag{15a}$$

$$v(x,0) = \varphi(x) \qquad (x \in S^1). \tag{15b}$$

In fact, if we choose a sequence $0 \leq t_1 < t_2 < t_2 < \cdots \to \infty$ appropriately, then we can show that the limit

$$\lim_{k\to\infty} u(\cdot, t_k + t)$$

exists for every $t \in \mathbb{R}$ and that the limit function $v(\cdot,t)$ satisfies (15a) and (15b). Let $w(x,t) = v(x,t) - v(x+a, t+\tau)$. By Lemma 3.4 w is a solution of a linear parabolic equation of the form (12) with $T_1 = -\infty$, $T_2 = +\infty$. By Lemma 3.5

$$\liminf_{k\to\infty} \nu\big(u(\cdot,t_k+t) - \sigma_{a,\tau}u(\cdot,t_k+t)\big) \geq \nu\big(v(\cdot,t) - \sigma_{a,\tau}v(\cdot,t)\big).$$

It follows that

$$\xi(a,\tau) \geq \nu(w(\cdot,t)) \tag{16}$$

for any $t \in \mathbf{R}$. Moreover equality in (16) holds for any $t \in \mathbf{R}$ such that $w(\cdot,t) \in \Sigma$. Now suppose that $w \not\equiv 0$. Then by Lemma 3.3 (iv) there exists a discrete set $A \subset \mathbf{R}$ such that $w(\cdot,t) \in \Sigma$ for any $t \in \mathbf{R}\backslash A$. Considering this and the fact that $\nu(w(\cdot,t))$ is monotone nonincreasing in t, we see that $\xi(a,\tau) = \nu(w(\cdot,t))$ for every $t \in \mathbf{R}$. It then follows from Lemma 3.3 (iii) that $w(\cdot,t) \in \Sigma$ for every $t \in \mathbf{R}$, contradicting the supposition that $w(\cdot,0) = \varphi - \sigma_{a,\tau}\varphi \notin \Sigma$. This shows $w \equiv 0$, completing the proof of Lemma 3.7.

LEMMA 3.8. *Let $\varphi \in \omega(u_0)$. Then*

$$\nu(\varphi - \sigma_{a,\tau}\varphi) \in \{\xi(a,\tau), 0\} \tag{17}$$

for any $a \in S^1$, $\tau \geq 0$. Furthermore, if $\nu(\varphi - \sigma_{a_0,\tau_0}\varphi) \neq 0$ for some $a_0 \in S^1$, $\tau_0 \geq 0$, then $\xi(a,\tau) = \xi(a_0,\tau_0)$ for any $(a,\tau) \in S^1 \times \mathbf{R}^+$ sufficiently close to (a_0,τ_0).

PROOF: It follows from (16) that $\xi(a,\tau) \geq \nu(\varphi - \sigma_{a,\tau}\varphi)$. Moreover it is clear from the latter part of the proof of Lemma 3.7 that $\xi(a,\tau) = \nu(\varphi - \sigma_{a,\tau}\varphi)$ if $\varphi - \sigma_{a,\tau}\varphi \neq 0$, or, equivalently, if $\nu(\varphi - \sigma_{a,\tau}\varphi) \neq 0$. This proves (17). Next suppose that $\nu(\varphi - \sigma_{a_0,\tau_0}\varphi) \neq 0$. Then by Lemma 3.7 $\varphi - \sigma_{a_0,\tau_0}\varphi \in \Sigma$. It follows from Lemma 3.5 that $\nu(\varphi - \sigma_{a,\tau}\varphi) = \nu(\varphi - \sigma_{a_0,\tau_0}\varphi)$ for any (a,τ) sufficiently close to (a_0,τ_0). This, together with (17), proves Lemma 3.8.

NOTATION 3.9: For each $w \in C(S^1)$, we let

$$m(w) = \max_{x\in S^1} w(x)$$

LEMMA 3.10. *Let $\varphi \in \omega(u_0)$. Then $m(\sigma_{a,\tau}\varphi) = m(\varphi)$ for any $a \in S^1$, $\tau \geq 0$.*

PROOF: The assertion is trivial when $\tau = 0$. Suppose that there exists $\tau_0 > 0$ such that $m(\sigma_{a,\tau_0}\varphi) \neq m(\varphi)$ for some $a \in S^1$. Then this holds for any $a \in S^1$ since the left-hand side is independent of a. It follows from Lemmas 3.7 and 3.5 that

$$\varphi - \sigma_{a,\tau_0}\varphi \in \Sigma \tag{18a}$$

$$\nu(\varphi - \sigma_{a,\tau_0}\varphi) = \nu(\varphi - \sigma_{a_0,\tau_0}\varphi) \tag{18b}$$

for any $a \in S^1$, where a_0 is some fixed point of S^1. Let $0 \le t_1 < t_2 < t_3 < \cdots \to \infty$ be such that $u(\cdot, t_k) \to \varphi$ in $C^1(S^1)$ as $k \to \infty$. By (18a), (18b), Lemma 3.5 and the compactness of the set $\{\sigma_{a,\tau_0}\varphi \mid a \in S^1\}$, there exists an integer $k_0 > 0$ such that

$$\nu\big(u(\cdot, t_k) - \sigma_{a,\tau_0}u(\cdot, t_k)\big) = \nu(\varphi - \sigma_{a_0,\tau_0}\varphi)$$

for any $k \ge k_0$, $a \in S^1$. In view of this and the fact that $\nu\big(u(\cdot,t) - \sigma_{a,\tau_0}u(\cdot,t)\big)$ is monotone nonincreasing, we see that

$$\nu\big(u(\cdot,t) - \sigma_{a,\tau_0}u(\cdot,t)\big) = \nu(\varphi - \sigma_{a_0,\tau_0}\varphi)$$

for any $t \ge t_{k_0}$ and $a \in S^1$. It follows from Lemma 3.3 (iii) that $u(\cdot,t) - \sigma_{a,\tau_0}u(\cdot,t) \in \Sigma$ for any $t \ge t_{k_0}$ and $a \in S^1$, hence

$$m(u(\cdot,t)) \ne m(u(\cdot, t+\tau_0))$$

for any $t \ge t_{k_0}$. Note that $m(u(\cdot,t))$ is a bounded uniformly continuous function of t by virtue of the boundedness of u and a standard a priori estimate from problem (1). Applying Lemma 3.6, we see that $m(u(\cdot,t))$ tends to a τ_0-periodic function of t as $t \to \infty$. It follows that $m(\Phi_{\tau_0}(\varphi)) = m(\varphi)$, hence $m(\sigma_{a,\tau_0}\varphi) = m(\varphi)$ for $a \in S^1$, contradicting the supposition. This contradiction proves the lemma.

LEMMA 3.11. $\omega(u_0) \subset \Lambda$.

PROOF: By Lemma 3.10, for each $\tau \ge 0$ there exists $a(\tau) \in S^1$ such that $\varphi - \sigma_{a(\tau),\tau}\varphi \notin \Sigma$. It follows from Lemma 3.7 that $\varphi = \sigma_{a(\tau),\tau}\varphi$ for any $\tau \ge 0$. This implies

$$v(x + a(\tau), \tau) = \varphi(x) \quad (x \in S^1, \tau \ge 0),$$

where v is the solution of (15). Let $\tau_1 > \tau_2 > \tau_3 > \cdots \to 0$ be a sequence such that $a(\tau_k)$ converges as $k \to \infty$, say to a_0. Then $\varphi(x + a_0) \equiv \varphi(x)$. Letting $b_k = a(\tau_k) - a_0$, we have

$$\begin{aligned} &v(x + b_k, \tau_k) = \varphi(x) \quad (x \in S^1, k \in \mathbb{N}) \\ &\tau_k \to 0, \quad b_k \to 0 \qquad \text{as } k \to \infty. \end{aligned} \tag{19}$$

Replacing $\{\tau_k\}$, $\{b_k\}$ by their subsequences if necessary, we may assume without loss of generality that either

$$|b_k|/\tau_k \to \infty \text{ as } k \to \infty \tag{20a}$$

or there exists $c \in \mathbb{R}$ such that

$$b_k/\tau_k \to c \text{ as } k \to \infty. \tag{20b}$$

Here we have regarded b_k as a number in R; such identification is possible without confusion since $b_k \to 0$. By (19),

$$\frac{\tau_k v_t(x,0) + b_k v_x(x,0)}{|\tau_k| + |b_k|} \to 0 \text{ as } k \to \infty.$$

Therefore, if (20a) holds, we have $v_x(x,0) \equiv 0$ in S^1, which means that $\varphi(x) = v(x,0)$ is a constant function, hence $\varphi \in \Lambda$. On the other hand, if (20b) holds, then $v_t(x,0) = -cv_x(x,0)$, hence

$$cv_x(x,0) + v_{xx}(x,0) + f\big(v(x,0), v_x(x,0)\big) = 0,$$

which implies $\varphi = v(\cdot,0) \in \Lambda_c$. In either case we have $\varphi \in \Lambda$, as claimed.

LEMMA 3.12. *Let φ, $\psi \in \Lambda$ and suppose that $\varphi - \sigma_{a,\tau}\psi \notin \Sigma$ for some $a \in S^1$, $\tau \in \mathbb{R}$. Then $\varphi = \sigma_{a,\tau}\psi$.*

PROOF: Let c, $\tilde{c}$ be such that $\varphi \in \Lambda_c$, $\psi \in \Lambda_{\tilde{c}}$. Then $w(\cdot,t) = \Phi_t(\varphi) - \Phi_t(\sigma_{a,\tau}\psi)$ is a solution of an equation of the form (12). And it is easily seen that $w(\cdot,t) \notin \Sigma$ for any $t \geq 0$ with $(c - \tilde{c})t \equiv 0 \pmod 1$. In particular $w(\cdot,t) \notin \Sigma$ for infinitely many values of t in $[1,\infty)$, which implies $w \equiv 0$ by Lemma 3.3 (iv). The lemma is proved.

LEMMA 3.13. *Let $\varphi \in \omega(u_0)$ and suppose that $\varphi(x)$ is not a constant function. Then there exist finitely many points $a_1, a_2, \ldots, a_m \in S^1$ such that $\nu(\varphi - \sigma_a\varphi) = \xi_0$ for any $a \in S^1 \backslash \{a_1, a_2, \ldots, a_m\}$, where ξ_0 is as in (14).*

PROOF: By Lemma 3.11 there exists $c \in \mathbb{R}$ such that $\varphi \in \Lambda_c$. Let $A = \{a \in S^1 \mid \varphi = \sigma_a\varphi\}$. A is a closed subgroup of S^1. Since φ is not a constant function, A is a finite set, say $A = \{a_1, a_2, \ldots, a_m\}$. By Lemma 3.7, $\varphi - \sigma_a\varphi \in \Sigma$ for any $a \in S^1 \backslash A$. Consequently, $\varphi - \sigma_{a,\tau}\varphi \in \Sigma$ for any $(a,\tau) \notin D$, where $D = \{(a_i + ct, t) \mid t \geq 0, i = 1, 2, \ldots, m\}$. It follows from Lemma 3.5

that $(a,\tau) \mapsto \nu(\varphi - \sigma_{a,\tau}\varphi)$ is constant (and nonzero) on each connected component of $S^1 \times \mathbb{R}^+ \backslash D$. Observe that

$$\begin{aligned}\nu(\varphi - \sigma_{a_i-\varepsilon}\varphi) &= \nu(\varphi - \sigma_{-\varepsilon}\varphi) = \nu(\sigma_\varepsilon\varphi - \sigma_\varepsilon\sigma_{-\varepsilon}\varphi)\\ &= \nu(\sigma_\varepsilon\varphi - \varphi) = \nu(\varphi - \sigma_{a_i+\varepsilon}\varphi)\end{aligned}$$

for any sufficiently small $\varepsilon > 0$. In view of this, we easily find that $\nu(\varphi - \sigma_{a,\tau}\varphi)$ is constant (and nonzero) on $S^1 \times \mathbb{R}^+ \backslash D$. Hence, by Lemma 3.8, there exists $\xi_1 \in \mathbb{N}$ such that

$$\xi(a,\tau) = \nu(\varphi - \sigma_{a,\tau}\varphi) = \xi_1$$

for any $(a,\tau) \in S^1 \times \mathbb{R}^+ \backslash D$. And from the last part of Lemma 3.8 it is easily seen that $\xi(a,\tau) \in \{\xi_1, 0\}$ for any $(a,\tau) \in D$. Consequently $\xi_1 = \xi_0$, which proves the lemma.

PROOF OF THEOREM A: The assertion that $\omega(u_0)$ is a nonempty compact connected set in $C^2(S^1)$ follows from Schauder's estimate and a standard argument in dynamical systems theory, so we only prove the latter part. In the case where $\omega(u_0)$ consists only of constant functions on S^1, the uniqueness of the ω-limit point follows easily from the comparison theorem since in this case $\omega(u_0)$ is a totally ordered set of equilibria. We therefore assume that $\omega(u_0)$ contains a nonconstant function, say φ. Let

$$\Gamma = \{\sigma_a\varphi \mid a \in S^1\}.$$

Supposing that $\omega(u_0) \not\subset \Gamma$, we shall derive a contradiction. Since $\omega(u_0)$ is compact and connected, there exists $\tilde{\varphi} \in \Gamma \cap \omega(u_0)$ such that $\tilde{\varphi}$ is an accumulation point of $\omega(u_0)\backslash\Gamma$. Without loss of generality we may assume that $\varphi = \tilde{\varphi}$, since otherwise we can simply replace φ by $\tilde{\varphi}$ without changing Γ. Since φ is nonconstant, by Lemma 3.13 and Lemma 3.7 there exists $a_0 \in S^1$ such that $\nu(\varphi - \sigma_{a_0}\varphi) = \xi_0$, $\varphi - \sigma_{a_0}\varphi \in \Sigma$. Using Lemma 3.5 and the fact that φ is an accumulation point of $\omega(u_0)\backslash\Gamma$, we see that there exists $\psi \in \omega(u_0)\backslash\Gamma$ such that $\nu(\varphi - \sigma_{a_0}\psi) = \xi_0$. Since $\psi \notin \Gamma$, we have, by Lemma 3.12, that

$$\varphi - \sigma_a\psi \in \Sigma, \text{ for every } a \in S^1, \tag{21a}$$

hence by Lemma 3.5,

$$\nu(\varphi - \sigma_a\psi) = \xi_0 \text{ for every } a \in S^1. \tag{21b}$$

In view of (21a) and (21b), and considering the fact that $\varphi, \psi \in \omega(u_0)$ and that S^1 is compact, we see that there exist $t_1, t_2 \geq 0$ such that

$$\nu\big(u(\cdot,t_1) - \sigma_a u(\cdot,t_2)\big) = \xi_0 \tag{22}$$

for any $a \in S^1$. Without loss of generality we may assume $t_2 > t_1$. Let $\tau_0 = t_2 - t_1$. By (22) and Lemma 3.3 (ii) we have

$$\nu\big(u(\cdot,t) - \sigma_{a,\tau_0} u(\cdot,t)\big) \leq \xi_0 \tag{23}$$

for any $t \geq t_1$ and $a \in S^1$. Letting $t \to \infty$, we get $\xi(a,\tau_0) \leq \xi_0$, hence, by the minimality of ξ_0, either $\xi(a,\tau_0) = \xi_0$ or $\xi(a,\tau_0) = 0$. The latter is impossible, for otherwise the left-hand side of (23) would become zero in a finite time while positive at $t = t_1$, which contradicts the uniqueness theorem for backward parabolic equations (see [**12**]; this also follows from Lemma 3.3 (iii) of the present paper). It follows that the equality in (23) holds for any $t \geq t_1$. Consequently, by Lemma 3.3 (iii),

$$u(\cdot,t) - \sigma_{a,\tau_0} u(\cdot,t) \in \Sigma$$

for any $t > t_1$, $a \in S^1$. In particular, we have

$$m(u(\cdot,t)) \neq m(u(\cdot,t+\tau_0)) \text{ for } t > t_1.$$

By Lemma 3.6 there exists a function $\tilde{h}(t)$, with $\tilde{h}(t+\tau_0) \equiv \tilde{h}(t)$, such that $|m(u(\cdot,t)) - \tilde{h}(t)| \to 0$ as $t \to \infty$. Since $\varphi \in \omega(u_0)$, there exists $\tau_1 \in [0,\tau_0)$ such that $m(\Phi_t(\varphi)) = \tilde{h}(t+\tau_1)$ for any $t \geq 0$. On the other hand, by Lemma 3.10 $m(\Phi_t(\varphi))$ is independent of t, hence so is $\tilde{h}(t)$. It follows that $m(u(\cdot,t))$ converges to a constant as $t \to \infty$. Consequently $m(\psi) = m(\varphi)$. But this, together with Lemma 3.12, contradicts the supposition that $\psi \notin \Gamma = \{\sigma_a \varphi \mid a \in S^1\}$. This contradiction proves $\omega(u_0) \subset \Gamma$, completing the proof of Theorem A.

§4. Proof of Theorems B, B′ and C.

In the previous section the argument relied on the fact that equation (1a) is equivariant with respect to the space-time shift. In this section, assuming that f is symmetric in some sense, we add one more equivariance relation to equation (1a). For example, if $f(u,-p) \equiv f(u,p)$, then (1a) is equivariant not only with respect to the shift but also with respect to the reflection ρ_a as defined in (11b). We begin with the following lemma, which is essentially due to Chen [**8**], though the proof below is a little different from that of [**8**]. An interesting application of a similar lemma can be found in [**11**].

LEMMA 4.1. *Assume that $f(u,-p) \equiv f(u,p)$. Let u be a solution of (1). Then*

$$\lim_{t \to s(u_0)} \operatorname{sgn}(u_x(x,t))$$

exists, where $s(u_0)$, $0 < s(u_0) \leq \infty$, is as in (5a) and $\operatorname{sgn}(w) = -1, 0, 1$ *depending on whether $w < 0$, $w = 0$, or $w > 0$.*

PROOF: Given $a \in S^1$, let

$$w(x,t) = u(x,t) - u(2a - x, t).$$

By the assumption $f(u,-p) \equiv f(u,p)$, both $u(x,t)$ and $u(2a - x, t)$ are solutions of (1a), therefore w satisfies a linear parabolic equation of the form (12) with $T_1 = 0$, $T_2 = s(u_0)$, and q, q_x, q_t, r locally bounded. Note also that

$$\text{(24a)} \qquad w(a,t) = 0 \qquad 0 \leq t < s(u_0),$$

$$\text{(24b)} \qquad w_x(a,t) = 2u_x(a,t), \quad 0 \leq t < s(u_0).$$

Suppose that $\operatorname{sgn}(u_x(a,t))$ does not converge as $t \to s(u_0)$. Then by (24b), $\operatorname{sgn}(w_x(a,t))$ does not converge as $t \to s(u_0)$. It follows that there exists a sequence $0 < t_1 < t_2 < t_3 < \cdots \to s(u_0)$ such that $w_x(a,t_k) = 0$ for $k = 1,2,3,\ldots$. This, together with (24a), implies that $w(\cdot, t_k) \notin \Sigma$ for $k = 1,2,3,\ldots$, hence by Lemma 3.3 (iv) we have $w \equiv 0$. Therefore $\operatorname{sgn}(u_x(a,t)) = 0$ for any $t \in [0, s(u_0))$, contradicting the supposition that this quantity does not converge as $t \to s(u_0)$. This contradiction proves the lemma.

PROOF OF THEOREM B: Let $\varphi \in \omega(u_0)$. Then by Theorem A, $\varphi \in \Lambda_c$ for some $c \in \mathsf{R}$ and $\omega(u_0) \subset \Gamma = \{\sigma_a \varphi \mid a \in S^1\}$. If φ is a constant function, then $\Gamma = \{\varphi\}$ and the conclusion of the theorem follows immediately. Assume therefore that φ is not a constant function and let $x_0 \in S^1$ be such that $\varphi(x)$ attains its maximum at $x = x_0$. Then $\varphi'(x_0) = 0$ and

$\varphi''(x_0) \neq 0$, the latter being the consequence of the uniqueness theorem for the ordinary differential equation (2b) and the fact that φ is not a constant function. Now suppose that $\omega(u_0)$ contains an element other than φ. Then, by the connectedness of $\omega(u_0)$, there exists a non-degenerate arc segment $\gamma \subset S^1$ such that

$$\{\sigma_a \varphi \mid a \in \gamma\} \subset \omega(u_0).$$

Let b be an interior point of γ. Then $(\sigma_b \varphi)'(x_0 - b) = 0$, $(\sigma_b \varphi)''(x_0 - b) \neq 0$. Therefore $(\sigma_b \varphi)'(x_0 - b)$ changes sign when a varies in γ. Consequently there exist $\psi_1, \psi_2 \in \omega(u_0)$ such that $\psi_1'(x_0 - b) > 0$ and $\psi_2'(x_0 - b) < 0$. On the other hand, by Lemma 4.1, $\psi_1'(x_0 - b) > 0$ implies $\psi_2'(x_0 - b) \geq 0$, a contradiction. This contradiction proves Theorem B.

PROOF OF THEOREM B′: Under the assumption $f(-u, p) \equiv -f(u, p)$, equation (1a) is equivariant with respect to the transform $u(\cdot, t) \to -\rho_a u(\cdot, t)$. In view of this, we consider the function

$$w(x, t) = u(x, t) + u(2a - x, t).$$

Arguing as in Lemma 4.1, we see that

$$\lim_{t \to \infty} \operatorname{sgn}(u(x, t)) \tag{25}$$

exists for each $x \in S^1$. Let $\varphi \in \Lambda$ be such that $\omega(u_0) \subset \{\sigma_a \varphi \mid a \in S^1\}$. We assume that φ is a nonconstant function since otherwise the assertion $\omega(u_0) = \{\varphi\}$ is trivial. Then from the sign condition on $f(u, 0)$ one easily finds that $\varphi(x)$ changes sign on S^1. Let $x_0 \in S^1$ be such that $\varphi(x_0) = 0$. Then we have $\varphi'(x_0) \neq 0$ by virtue of the uniqueness theorem for equation (2b). Arguing as in the proof of Theorem B, we see that there exist $b \in S^1$ and $\psi_1, \psi_2 \in \omega(u_0)$ such that $\psi_1(x_0 - b) > 0$, $\psi_2(x_0 - b) < 0$. But this contradicts the existence of the limit (25). The theorem is proved.

PROOF OF THEOREM C: Under the hypothesis (7), the boundedness of $\|u(\cdot, t)\|_{C(S^1)}$ implies that of $\|u(\cdot, t)\|_{C^1(S^1)}$ (see Amann [**1**]), from which the assertion of the theorem follows immediately.

We omit the proof of Corollary D since it is an obvious consequence of Theorems B and C.

§5. Proof of Theorem E.

As in the proof of Theorem B, Lemma 4.1 in the previous section plays an important role. Once we have this lemma, the rest of the proof of Theorem E is similar to that of [**6**] and [**13**], though our argument is slightly modified. See the forthcoming paper [**9**] for a more extensive treatment of the problem.

LEMMA 5.1. *Let $f(u,-p) \equiv f(u,p)$ and let u be a solution of (1) with u_0 a nonconstant function and $s(u_0) < \infty$. Then there exist $t^* \in [0, s(u_0))$ and C^1-curves $\eta_1, \eta_2, \ldots, \eta_{2m} : [t^*, s(u_0)) \to S^1$ such that*

(i) *for each $t \in [t^*, s(u_0))$ the points $\eta_1(t), \eta_2(t), \ldots, \eta_{2m}(t)$ lie on S^1 in this order (counterclockwise) and $\eta_i(t) \neq \eta_j(t) \quad (i \neq j)$;*

(ii) *for each $t \in [t^*, s(u_0))$,*

$$\{x \in S^1 \mid u_x(x,t) = 0\} = \{\eta_1(t), \eta_2(t), \ldots, \eta_{2m}(t)\};$$

(iii) *the limit*

$$y_j = \lim_{t \to s(u_0)} \eta_j(t) \tag{26}$$

exists for each $j = 1, 2, \ldots, 2m$.

Note that the above lemma does not exclude the possibility of $y_j = y_{j+1}$ for some j. The proof is immediate from Lemma 4.1 and Lemma 3.3 (i), (iv), so we omit it. Our goal is to show that the blow-up set $B(u)$ is contained in $\{y_1, y_2, \ldots, y_{2m}\}$.

In what follows we let γ be an arbitrary (non-degenerate) closed arc segment contained in $S^1 \backslash \{y_1, y_2, \ldots, y_{2m}\}$. In order to prove Theorem E, it suffices to show that $B(u) \cap \gamma = \emptyset$. Since γ is diffeomorphic to a closed interval $[a,b] \subset \mathbb{R}$ through the local diffeomorphism $\mathbb{R} \to \mathbb{R}/\mathbb{Z} = S^1$, hereafter we shall identify γ with $[a,b]$.

LEMMA 5.2. *Let the assumptions of Lemma 5.1 hold, and let $\gamma = [a,b]$ be as above. Then there exists $t_0 \in [t^*, s(u_0))$ such that $u_x(x,t)$ does not change sign in the rectangular region $[a,b] \times [t_0, s(u_0))$.*

The proof is immediate from Lemma 5.1.

LEMMA 5.3. *Let the assumptions of Lemma 5.2 hold, with* $u_x(x,t) > 0$ *in* $D = [a,b]\times[t_0, s(u_0))$. *Assume further that* $u(x,t) \geq 0$, $f(u(x,t), u_x(x,t)) \geq 0$ *in* D *and that* $\bar{a} \in B(u)$ *for some* $\bar{a} \in (a,b)$. *Then* $\lim_{t\to s(u_0)} u(x,t) = \infty$ *for any* $x \in (\bar{a}, b]$.

PROOF: Choose $a_1 \in (\bar{a}, b)$ arbitrarily. By the assumption of the lemma, there exist sequences $x_1, x_2, x_3, \ldots, \to \bar{a}$ and $t_1 < t_2 < t_3 < \cdots \to s(u_0)$ such that $u(x_k, t_k) \to \infty$ as $k \to \infty$. Clearly $(x_k, t_k) \in [a, a_1] \times [t_0, s(u_0))$ for every sufficiently large k, say $k \geq k_0$. Since $u_x \geq 0$ on D, we have $u(x, t_k) \to \infty$ as $k \to \infty$ uniformly in $x \in [a_1, b]$. In view of this, and seeing that $f \geq 0$, $u \geq 0$ and the fact that $s(u_0)$ is finite, we easily find by using a comparison argument that $u(x,t) \to \infty$ as $t \to s(u_0)$ for every $x \in (a_1, b]$. Since a_1 can be chosen arbitrarily close to $\bar{a}$, the lemma is proved.

PROOF OF THEOREM E: By condition $f(0,0) = 0$, $u_0 \geq 0$ and the comparison theorem, $u(x,t) \geq 0$ for $x \in S^1$, $0 \leq t < s(u_0)$. Note also that condition (9d) implies that (7) holds for any $M > 0$, therefore, by Theorem C and $u \geq 0$, we have $\lim_{t\to s(u_0)} \max_{x\in S^1} u(x,t) = \infty$. In what follows, following the notation of Lemma 5.2, we shall prove $B(u) \cap (a,b) = \emptyset$.

Suppose the contrary and let $\bar{a} \in B(u) \cap (a,b)$. Define

$$c(x) = \left(\sin \frac{x-\bar{a}}{b-\bar{a}}\pi\right)^2 \quad (\bar{a} \leq x \leq b) \tag{27}$$

and

$$w(x,t) = u_x(x,t) - \varepsilon c(x) F(u(x,t)), \tag{28}$$

where F is as in (10) and ε is a positive constant to be chosen later. (Here, as already mentioned, the arc segment $[a,b]$ is identified with a closed interval in R through the local diffeomorphism R $\to$ R/Z $= S^1$, so the expression (27) makes sense.) A simple calculation shows

$$\begin{aligned} &w_t - w_{xx} - q(x,t)w_x - r(x,t)w \\ &= \varepsilon c\left\{F f_u - F' f + \frac{c'' + c' f_p}{c} F + \varepsilon(c f_p + 2c')FF'\right\} + \varepsilon c F''(u_x)^2, \end{aligned} \tag{29}$$

where

$$\begin{aligned} q(x,t) &= f_p(u(x,t), u_x(x,t)), \\ r(x,t) &= \varepsilon c(x) F'(u) f_p(u, u_x) + f_u(u, u_x) + 2\varepsilon c'(x) F'(u). \end{aligned}$$

Considering that $F'(u) \to \infty$ as $u \to \infty$ by virtue of (10a) and (10c), and that $(c''+c'f_p)/c$ is bounded from below in $(\bar{a}, b) \times [t_0, s(u_0))$, then choosing $\varepsilon > 0$ sufficiently small and using (9d), (10b) and $F'' \geq 0$, we find that

$$(30a) \qquad w_t - w_{xx} - qw_x - rw \geq 0 \quad (\bar{a} < x < b, t_0 < t < s(u_0)).$$

Also, by the positivity of u_x (and assuming that ε is chosen sufficiently small), we have

$$(30b) \qquad w(\bar{a}, t) > 0, \quad w(\bar{b}, t) > 0 \quad (t_0 < t < s(u_0)),$$

$$(30c) \qquad w(x, t_0) > 0 \quad (\bar{a} < x < b).$$

Applying the maximum principle to (30), we obtain $w \geq 0$, or

$$\frac{u_x(x,t)}{F(u(x,t))} \geq \varepsilon c(x) \quad (\bar{a} \leq x \leq b, t_0 \leq t < s(u_0)).$$

Integrating this inequality over $\bar{a} \leq x \leq b$ yields

$$(31) \qquad \int_{u(\bar{a},t)}^{u(b,t)} \frac{du}{F(u)} \geq \varepsilon \int_{\bar{a}}^{b} c(x)\, dx \quad (t_0 \leq t < s(u_0)).$$

The right-hand side of (31) is a positive constant, while the left-hand side tends to zero as $t \to \infty$ by virtue of condition (10c), a contradiction. This contradiction shows that $(a, b) \cap B(u) = \emptyset$. Since $[a, b]$ is an arbitrary closed arc segment of $S^1 \backslash \{y_1, y_2, \ldots, y_{2m}\}$, the theorem is proved.

Appendix: Construction of Liapunov Function.

We construct a Liapunov function for the problem

$$(32a) \qquad u_t = a(x, u, u_x)u_{xx} + f(u, u, u_x), \quad 0 < x < L, t > 0,$$

$$(32b) \qquad u(x, 0) = u_0(x) \qquad 0 < x < L,$$

under the boundary conditions either of the form

$$(32c) \qquad u(0, t) = u(L, t) = 0, \quad t > 0,$$

or of the form

$$\text{(32}d\text{)} \qquad \begin{aligned} u_x(0,t) &= g_0(u(0,t)), \quad t > 0, \\ u_x(L,t) &= g_1(u(L,t)), \quad t > 0, \end{aligned}$$

where $a(x,u,p)$ and $f(x,u,p)$ are of class C^1 with $a > 0$ in $[0,L]\times\mathbf{R}\times\mathbf{R}$, and g_0, g_1 are continuous functions. The construction we give here is different from that of [**24**] and requires fewer regularity conditions on a and f.

We shall find a Liapunov function $J(u)$ of the form

$$\text{(33)} \quad J(u(\cdot,t)) = \int_0^L G(x,u(x,t),u_x(x,t))\,dx + H_0(u(0,t)) + H_1(u(L,t)),$$

where $G(x,u,p)$, $H_0(u)$ and $H_1(u)$ are smooth functions defined on $[0,L]\times\mathbf{R}\times\mathbf{R}$, $\mathbf{R}$ and $\mathbf{R}$ respectively. We only consider the case where the boundary condition is of the form (32d), since the case (32c) can be treated quite similarly and more simply. (In fact, the very same function $G(x,u,p)$ can be put into (33) for both (32c) and (32d). $H_0 \equiv H_1 \equiv 0$ in the case of (32c).)

Suppose that $G(x,u,p)$ satisfies the equation

$$\text{(34}a\text{)} \qquad \frac{f}{a}G_{pp} = pG_{pu} + G_{px} - G_u \quad (x \in [0,L], u \in \mathbf{R}, p \in \mathbf{R})$$

and that H_0, H_1 are given by

$$\text{(34}b\text{)} \qquad \begin{aligned} H_0(u) &= \int_0^u G_p(0,v,g_0(v))\,dv, \\ H_1(u) &= -\int_0^u G_p(L,v,g_1(v))\,dv \end{aligned}$$

for $u \in \mathbf{R}$. Then if we define $J(u)$ as in (33) and if u is a solution of (32) with u, u_x, u_{xx}, u_t, u_{tx} continuous, then a simple calculation shows

$$\text{(35)} \qquad \frac{d}{dt}J(u(\cdot,t)) = -\int_0^L \frac{G_{pp}(x,u(x,t),u_x(x,t))}{a(x,u(x,t),u_x(x,t))}\{u_t(x,t)\}^2\,dx.$$

In order to find a solution G of (34a) with $G_{pp} > 0$, let $\varphi(x,u,p)$ be a solution of the following first-order differential equation

$$\text{(36)} \qquad \varphi_x + p\varphi_u - \frac{f}{a}\varphi_p = \left(\frac{f}{a}\right)_p \quad (x \in [0,L], u, p \in \mathbf{R}).$$

A solution to equation (36) can always be found by looking at the characteristic curves. Next define

$$\tilde{G}(x,u,p) = \int_0^p \int_0^q \exp(\varphi(x,u,r))\,dr\,dq. \tag{37}$$

Then one easily sees that there exists a smooth function $k(x,u)$ such that

$$\frac{f}{a}\tilde{G}_{pp} = p\tilde{G}_{pu} + \tilde{G}_{px} - \tilde{G}_u + k(x,u). \tag{38}$$

Finally, let

$$G(x,u,p) = \tilde{G}(x,u,p) + \int_0^u k(x,v)\,dv. \tag{39}$$

Then G satisfies (34a). Summarizing, we have

PROPOSITION. *Let $G(x,u,p)$ be as in (39), where $\tilde{G}$ and k are given by (36), (37) and (38). Let H_0, H_1 be as in (34b) and define J as in (33). Then J is a Liapunov function for problem (32a), (32b), and (32d) in the sense that*

$$\frac{d}{dt}J(u(\cdot,t)) = -\int_0^L \frac{1}{a}e^{\varphi}(u_t)^2\,dx \le 0 \tag{40}$$

for any solution u with u, u_x, u_{xx}, u_t continuous on $[0,L]\times \mathbb{R}^+$ and u_{xt} continuous in $(0,L)\times\mathbb{R}^+$. In the case of the Dirichlet boundary conditions (32c), we define J by (33) with $H_0 \equiv H_1 \equiv 0$. Then (40) holds for any solution of (32a), (32b), (32c) with u, u_x, u_{xx}, u_t continuous on $[0,L]\times\mathbb{R}^+$ and u_{xt} continuous in $(0,L)\times\mathbb{R}^+$.

Department of Mathematics, Hiroshima University, Hiroshima 730, JAPAN

REFERENCES

1. H. Amann, *Existence and multiplicity theorems for semilinear elliptic boundary value problems*, Math. Z. **150** (1976), 281–295.
2. H. Amann, *Semigroups and nonlinear evolution equations*, Lin. Alg. Appl. **84** (1986), 3–32.
3. S. Angenent, *The zero set of a solution of a parabolic equation*, to appear in J. Reine Angew. Math.
4. S. Angenent and B. Fiedler, *The dynamics of rotating waves in scalar reaction diffusion equations*, to appear in Trans. Amer. Math. Soc.
5. P. Brunovský and B. Fiedler, *Simplicity of zeroes in scalar parabolic equations*, J. Diff. Equ. **62** (1986), 237–241.
6. L. A. Caffarelli and A. Friedman, *Blow-up of solutions of nonlinear heat equations*, to appear in J. Math. Anal. Appl.
7. N. Chafee, *Asymptotic behavior for solutions of a one-dimensional parabolic equation with homogeneous Neumann boundary conditions*, J. Diff. Equ. **18** (1975), 111–134.
8. X.-Y. Chen, *Uniqueness of the ω-limit point of solutions of a semilinear heat equation on the circle*, Proc. Japan Acad., Ser A. **62** (1986), 335–337.
9. X.-Y. Chen and H. Matano, *Convergence, asymptotic periodicity and finite-point blow-up in one-dimensional semilinear heat equations*, in preparation.
10. X.-Y. Chen, H. Matano and L. Véron, *Singularités anisotropes d'équations elliptiques semilinéaires dans le plan*, C. R. Acad. Sci. Paris, Sér. I **303** (1986), 963–966.
11. X.-Y. Chen, H. Matano and L. Véron, *Anisotropic singularities of solutions of nonlinear elliptic equations in $\mathbf{R}^2$*, submitted to J. Funct. Anal..
12. A. Friedman, "Partial Differential Equations of Parabolic Type," Prentice-Hall, Englewood Cliffs, N. J., 1964.
13. A. Friedman and B. McLeod, *Blow-up of positive solutions of semilinear heat equations*, Indiana Univ. Math. J. **34** (1985), 425–447.
14. J. K. Hale and P. Massatt, *Asymptotic behavior of gradient-like systems*, in "Univ. of Florida Symposium on Dyn. Systems II," Academic Press, 1982.
15. M. W. Hirsch, *Systems of differential equations that are competitive or cooperative II: convergence almost everywhere*, SIAM J. Math. Anal. **16** (1985), 423–439.
16. A. A. Lacey, *Global blow-up of a nonlinear heat equation*, Proc. Royal Soc. Edinburgh, Sec. A **104** (1986), 161–167.
17. P. Massatt, private communication.
18. H. Matano, *Convergence of solutions of one-dimensional semilinear parabolic equations*, J. Math. Kyoto Univ. **18** (1978), 221–227.
19. H. Matano, *Nonincrease of the lap-number of a solution for a one-dimensional semilinear parabolic equation*, J. Fac. Sic. Univ. Tokyo, Sec. IA **29** (1982), 401–441.
20. X. Mora, *Semilinear parabolic problems define semiflows in C^k spaces*, Trans. Amer. Math. Soc. **28** (1983), 21–55.
21. K. Nickle, *Gestaltsaussagen über Lösungen parabolischen Differentialgeichungen*, J. Reine Angew. Math. **211** (1962), 78–94.
22. M. H. Protter and H. F. Weinberger, "Maximum principles in differential equations," Prentice-Hall, Englewood Cliffs, N. J., 1967.
23. F. B. Weissler, *Single point blow-up of semilinear initial value problems*, J. Diff. Equ. **55** (1984), 202–224.
24. T. J. Zelenyak, *Stabilization of solutions of boundary value problems for a second-order parabolic equation with one space variable*, Diff. Equ. **4** (1968), 17–22. (translated from Differential'nye Uravneniya).

Some Uniqueness Theorems for Exterior Boundary Value Problems

KEVIN MCLEOD

1. Introduction.

The existence of non-trivial solutions of

$$(1) \qquad \begin{aligned} &\Delta u + f(u) = 0 \text{ in } \mathbb{R}^n \\ &u(x) \to 0 \text{ as } |x| \to \infty \end{aligned}$$

has been studied by many authors [**1,2,3,7,10,13**]. In particular, Berestycki and Lions [**2**] prove the existence of a positive radial solution and of infinitely many (radial) solutions, while Jones and Küpper [**7**] prove the existence of a radial solution with any prescribed number of zeros. The conditions they impose on f are stronger than those of [**2**], but their result does cover the model case of $f(u) = -u + u^p$, $1 < p < \frac{n}{n-2}$.

A natural conjecture would be that, for "nice" functions f, there is a unique radial solution of (1) with any given number of zeros. For the model case, the uniqueness of the positive solution has been studied in [**4**] and [**9**]. Uniqueness — again for the positive solution — in case f has sublinear growth has been studied in [**8,11,12**]. In attempting to extend these uniqueness results to solutions of (1) which change sign, we were led to the exterior boundary value problem

$$(2) \qquad \begin{aligned} &u'' + \frac{n-1}{r}u' + f(u) = 0 \text{ in } (R, \infty) \\ &u(R) = u_0 \geq 0 \\ &u > 0 \text{ in } (R, \infty), \quad u \to 0 \text{ as } r \to \infty. \end{aligned}$$

Our aim in this note is to point out that the techniques in [**11,12**] can be applied to this problem, although the results are not as complete as in those papers.

We will assume throughout that f satisfies the following conditions:

(i) f is defined and locally Lipschitz on $[0, \infty)$.

(ii) $\lim_{u \to 0^+} \frac{f(u)}{u} = -m < 0$.

(iii) There are constants α, β, γ with $0 < \alpha < \beta < \gamma \leq \infty$ such that $f < 0$ in $(0, \alpha)$, $f > 0$ in (α, γ) and $F(\beta) = 0$, where $F(u) \equiv \int_0^u f(s)ds$. We note, however, that these conditions can be weakened by applying ideas from [**8**]. Note also that (ii) requires $f(0) = 0$.

2. Monotone Solutions.

We first note from the differential equation in (2) that any critical point of a solution where $u > \alpha$ must be a maximum; if $0 < u < \alpha$, it must be a minimum. (By the uniqueness theorem for initial value problems, no solution of (2) can have a critical point on the line $u = \alpha$.) Now, multiplying the differential equation by u', we obtain

$$\left(\frac{1}{2}u'^2 + F(u)\right)' = -\frac{n-1}{r}u'^2$$

and since the critical points of a solution are isolated, this tells us that the "energy"

$$E(r) \equiv \frac{1}{2}u'(r)^2 + F(u(r))$$

is strictly decreasing. Since $E(r) \to 0$ as $r \to \infty$ for any solution of (2), we see also that $E(r) > 0$ in $[R, \infty)$. In particular, at any critical point of a solution,

$$0 < E(r) = F(u(r))$$

so that $u > \beta$, and the critical point is a maximum. Thus, any non-monotone solution of (2) can have only a single critical point, which occurs above the line $u = \beta$. In this section, we will study the monotone solutions of (2), for which the arguments of [**11,12**] carry through without change; non-monotone solutions will be studied in §3. We quote the following results from [**11**]; the proofs are exactly the same as for entire solutions.

(A) Two monotone solutions of (2) cannot intersect below or on the line $u = \beta$.

(B) Let u and v be two solutions of (2) (not necessarily monotone). For large r, u and v are monotone; let $r(u)$ and $s(u)$ be their inverse functions. Then ultimately (as $u \to 0^+$) the quantity $r(u) - s(u)$ has one sign and, if $r(u) - s(u) > 0$ for small u, then $r(u) - s(u)$ is a decreasing function of u for small u.

Furthermore, on any interval on which it is positive, $r(u) - s(u)$ has a single critical point, which is a maximum.

Our first theorem on monotone solutions generalizes a corollary in [**5**].

THEOREM 1. *For any $R > 0$ and $u_0 \in (0, \beta]$, problem (2) has a unique solution such that $u(r) \leq \beta$ on $[R, \infty)$. Furthermore, this solution is monotone.*

PROOF: We have seen that any critical point of a non-monotone solution must occur above the line $u = \beta$. It follows that any solution with $u(r) \leq \beta$ must be monotone. The monotonicity of the solution having been established, uniqueness follows from result (A), since any two solutions of (2) automatically intersect at (R, u_0) and $u_0 \leq \beta$ by hypothesis.

It remains only to prove existence of the solution, which we do by a simple shooting argument. We consider the initial value problem

$$v'' + \frac{n-1}{r} v' + f(v) = 0$$
$$v(R) = u_0, \quad v'(R) = a,$$

and we let $v_a(r)$ be the solution of this problem and denote its energy by $E_a(r)$. It suffices to find one value of a for which v_a is bounded away from zero, and one value for which v_a has a zero at some finite value $r_0 > R$. (For details of this type of argument, see [9].) It is easy to see on energetic grounds that if $a = 0$ the corresponding solution $v_0(r)$ is bounded away from 0. (Recall that $u_0 \leq \beta$, so $E(r_0) = F(u_0) \leq 0$.) We finish the proof by showing that for large, negative a the solution $v_a(r)$ has a zero.

Suppose, for contradiction, that $v_a > 0$ in $[R, \infty)$ for all large, negative a. We let $M = \sup_{u \in [0,\beta]} |f(u)|$ and we let $r_1 > R$ be the first point at which $v_a' = \frac{1}{2}a$ or $v_a' = \frac{3}{2}a$ (whichever occurs first). Then $\frac{3}{2}a \leq v_a' \leq \frac{1}{2}a$ in $[R, r_1]$. (Clearly, if r_1 did not exist, we would have our contradiction.) Now, for sufficiently large a we have, in $[R, r_1]$,

$$\begin{aligned} |v_a''| &\leq -\frac{n-1}{R} v_a' + M \\ &\leq -2\frac{n-1}{R} v_a' \\ &\leq -3\frac{n-1}{R} a. \end{aligned}$$

By the mean-value theorem

$$|v_a'(r_1) - v_a'(R)| \leq \frac{3|a|(n-1)}{R}(r_1 - R)$$

from which we obtain, using the definition of r_1,

$$r_1 - R \geq \frac{1}{2}|a| \Big/ \frac{3|a|(n-1)}{R} = \frac{R}{6(n-1)}.$$

We have shown that $v_a' < \frac{1}{2}a$ on an interval whose length $r_1 - R$ is bounded independently of a for large, negative a. This contradicts the assumption that $v_a > 0$ in $[R, \infty)$, and finishes the proof of the theorem.

If we allow $u > \beta$ somewhere (in particular if $u_0 > \beta$) we require additional hypotheses on f to get uniqueness of the monotone solution. One possibility is the following star-shaped condition, which is a slight generalization of condition (S) in [**11**]. We will say that f satisfies condition (S_λ) if

$$(S_\lambda) \qquad \frac{d}{du}\left(\frac{f(u)}{u-\lambda}\right) \leq 0 \text{ on } (\lambda, \gamma).$$

(By our standing assumptions (i)–(iii), this forces $\alpha \leq \lambda \leq \gamma$.) The arguments of [**11**] then give

(C) If f satisfies condition (S_λ) and $u_0 \geq \lambda$, then two solutions of (2) cannot intersect above the line $u = \lambda$.

We remark that the proof of (C) (essentially [**11**,Lemma 10]) does not use the fact that u or v is monotone, only that certain scaled solutions intersect tangentially above the line $u = \lambda$. To ensure this in the non-monotone case, we must assume $u_0 \geq \lambda$. Since we know $\lambda \geq \alpha > 0$, this unfortunately prevents us from handling the Dirichlet problem, $u_0 = 0$.

We now give a theorem on uniqueness of the monotone solution for large values of u_0.

THEOREM 2. *Suppose f satisfies (S_λ) for some $\lambda \in [\alpha, \beta]$. Then, for any $R, u_0 > 0$, problem (2) has at most one monotone solution.*

PROOF: If $u_0 \leq \beta$, the result follows from Theorem 1, while if $u_0 > \beta$ and $\lambda \leq \beta$, two monotone solutions intersecting at (R, u_0) would contradict (C).

We do not know whether a monotone solution always exists for $u_0 > \beta$.

3. Non-monotone solutions.

We have the following uniqueness result for the case of large initial values.

THEOREM 3. *Suppose f satisfies (S_λ) for some $\lambda \in [\alpha, \beta]$. Then, for any $R > 0$, $u_0 \geq \lambda$, problem (2) has at most one non-monotone solution.*

PROOF: Although it is not essential for the proof, we will show first that any two non-monotone solutions of (2) must intersect at some value $R_1 >$

R. (This does not require $u_0 \geq \lambda$.) Indeed, if not, we would have the following situation (see Figure 1): the horizontal separation $r(u) - s(u)$ is an increasing function of u at $u = \bar{v}$, and a decreasing function of u near $u = 0$, by (B). Thus, it must have a minimum, which contradicts the last part of (B).

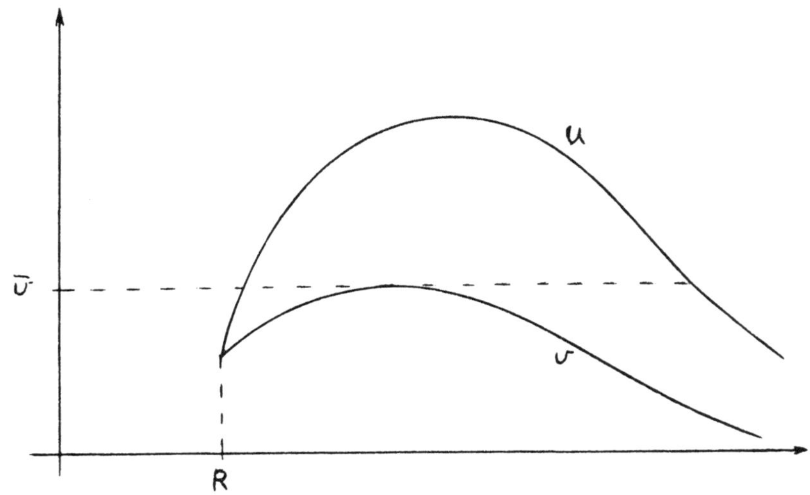

Figure 1

Where then is the last intersection of the two solutions? It cannot be where both solutions are decreasing, since this would contradict Theorem 2 (with R equal to the value of r at the final intersection). Hence it must be where (at least) one solution is increasing, and so above $u = \lambda$. But this contradicts (C). Thus, two solutions cannot intersect after all, and so the solution must in fact be unique.

As in the monotone case, we do not know whether there is a solution at all if u_0 is large. However, there is the following existence result, due to Garaizar [**6**]. (I would like to thank Garaizar for pointing out that the methods of his thesis, where $u_0 = 0$, can be applied to the case $u_0 > 0$ also.)

THEOREM 4 (Garaizar). *Suppose that there is a constant $k > -1$ and positive constants b, c_1, c_2 such that*

$$(G) \qquad c_1 u^k \leq f(u) \leq c_2 u^k \text{ for } u > b.$$

Then, for any $R > 0$, $u \in [0, \beta]$, problem (2) has a non-monotone solution.

It is interesting that there is no Pohozaev growth condition in this theorem. (Of course, the exterior of a ball is not a star-shaped domain.) Similar results have been obtained in $\mathbb{R}^3$ by Coffman and Marcus [5].

The conditions of Garaizar's theorem force $\gamma = \infty$. Note that conditions (G) and (S_λ) are compatible if $k \leq 1$, and so we obtain our final theorem.

THEOREM 5. *Let f satisfy conditions (G) and* (S_λ) *for some* $\lambda \in [\alpha, \beta]$ *and suppose* $\lambda \leq u_0 \leq \beta$, $R > 0$. *Then problem (2) has exactly two solutions: one monotone and one non-monotone.*

Department of Mathematics, Indiana University, Bloomington IN 47405

REFERENCES

1. F. V. Atkinson and L. A. Peletier, *Ground state and Dirichlet problems for* $-\Delta u = f(u)$ *in* $\mathbb{R}^2$, Arch. Rat. Mech. Anal. **96** (1986), 147–165.
2. H. Berestycki and P.-L. Lions, *Nonlinear scalar field equations I: existence of a ground state; II: existence of infinitely many solutions*, Arch. Rat. Mech. Anal. **82** (1983), 313–375.
3. H. Berestycki, P.-L. Lions and L. A. Peletier, *An ODE approach to the existence of positive solutions for semilinear problems in* $\mathbb{R}^n$, Indiana Univ. Math. J. **30** (1981), 141–167.
4. C. V. Coffman, *Uniqueness of the ground state solution for* $\Delta u - u + u^3 = 0$ *and a variational characterization of other solutions*, Arch. Rat. Mech. Anal. **46** (1972), 12–95.
5. C. V. Coffman and M. Marcus, *Superlinear elliptic Dirichlet problems in almost spherical domains*, Arch. Rat. Mech. Anal. **96** (1986), 167–197.
6. X. Garaizar, *Existence of positive radial solutions for semilinear elliptic equations in the annulus*, J. Diff. Eqs. **70** (1987), 69–92.
7. C. Jones and T. Küpper, *On the infinitely many solutions of a semilinear elliptic equation*, SIAM J. Math. Anal. **17** (1986), 803–835.
8. H. G. Kaper and M. K. Kwong, *Uniqueness of nonnegative solutions of a class of semilinear elliptic equations*, in "Proceedings of the Microprogram on Nonlinear Diffusion Equations and their Equilibrium States," J. Serrin, ed., Springer-Verlag, 1988.
9. K. McLeod and J. Serrin, *Uniqueness of positive radial solutions of* $\Delta u + f(u) = 0$ *in* $\mathbb{R}^n$, Arch. Rat. Mech. Anal. **99** (1987), 115–145.
10. A. Nehari, *On a nonlinear differential equation arising in nuclear physics*, Proc. R. Ir. Acad. Sect. A **62** (1963), 117–135.
11. L. A. Peletier and J. Serrin, *Uniqueness of positive solutions of semilinear equations in* $\mathbb{R}^n$, Arch. Rat. Mech. Anal. **81** (1983), 181–197.
12. L. A. Peletier and J. Serrin, *Uniqueness of nonnegative solutions of semilinear equations in* $\mathbb{R}^n$, J. Diff. Eqs. **61** (1986), 380–397.
13. W. Strauss, *Existence of solitary waves in higher dimensions*, Comm. Math. Phys. **55** (1977), 149–162.

Some Aspects of Semilinear Elliptic Equations on $\mathbb{R}^n$

WEI-MING NI

Introduction.

In this expository paper, I intend to give a *partial* survey of semilinear elliptic equations on $\mathbb{R}^n$.

Mathematically, such problems are of interest due to the lack of compact embedding theorems of Sobolev type, therefore new techniques and methods have to be devised. On the other hand, many such equations arose in geometry and physics and they have provided interesting model problems for mathematicians to attack. In recent years, some of those equations have received extensive investigations. The following is a list of problems I shall cover in this paper:

$$\frac{4(n-1)}{n-2}\Delta_g u - ku + Ku^{n^*} = 0 \text{ and } u > 0 \text{ on } \mathbb{R}^n, n \geq 3, \tag{1}$$

$$\Delta u + Ku^{n^*} = 0 \text{ and } u > 0 \text{ on } \mathbb{R}^n, n \geq 3, \tag{1'}$$

$$\Delta_g u - k + Ke^{2u} = 0 \text{ on } \mathbb{R}^2, \tag{2}$$

$$\Delta u + Ke^{2u} = 0 \text{ on } \mathbb{R}^2, \tag{2'}$$

$$\begin{cases} \Delta u - u + Q(x)u^p = 0, \quad 1 < p < n^*, \\ u > 0 \text{ on } \mathbb{R}^n, \\ u \to 0 \text{ at } \infty, \end{cases} \tag{3}$$

$$\begin{cases} \Delta u + \dfrac{|x|^{\lambda-2}}{(1+|x|^2)^{\lambda-2}} u^p = 0, \quad p > 1, \quad \lambda > 0, \\ u > 0 \text{ on } \mathbb{R}^n, \quad n \geq 3 \end{cases} \tag{4}$$

$$\begin{cases} \Delta u + \dfrac{1}{1+|x|^2} e^{2u} = 0, \\ u > 0 \text{ on } \mathbb{R}^n, \quad n \geq 3, \end{cases} \tag{5}$$

where Δ_g in (1) and (2) is the Laplace-Beltrami operator of the metric g on $\mathbb{R}^n$ and $\mathbb{R}^2$, respectively, and $n^* = (n+2)/(n-2)$. Equation (1) (and the special case (1′)) is known as the *conformal scalar curvature equation* and equation (2) (and the special case (2')) is the *conformal Gaussian curvature equation* in Riemannian geometry. Such equations arise when one tries to conformally deform an existing metric g on $\mathbf{R}^n$ so that the scalar (or Gaussian) curvature of the new metric is the prescribed function K. When $Q(x) \equiv 1$, equation (3) is the *scalar field equation.* (It also arises in phase transitions.) However, with $Q \not\equiv 1$, equation (3) does not seem to have much physical meaning although it does provide a simple mathematical model for us to understand equations on $\mathbf{R}^n$. Equation (5) was proposed by Eddington around 1915 to model globular clusters of stars. When $\lambda = 2$ in (4), the equation was proposed by Matukuma in 1930 to improve Eddington's model. The equation (4) (with $\lambda > 0$) recently appeared in a paper of Batt, Faltenbacher and Horst [BFH]. (For more detailed background and motivation of the equations (1)–(5) above, we refer the readers to §1–§4 below.) Besides the above equations, I shall also discuss the equation

$$\Delta u + f(u) = 0 \tag{6}$$

in $\mathbf{R}^n$, $n \geq 2$, to some extent. All the equations above ((1)–(6)) have been studied by various authors and some rather *striking* results and examples have been obtained. I shall only mention a few such results here (and I hope the readers will also find them interesting) and leave the rest in §1–§4 below.

First, if $Q \equiv 1$, then all solutions of (3) are radially symmetric [**GNN2**]. Whether this statement holds for other equations such as (1'), (4), (5) or (6) (with the extra condition "$u \to 0$ at ∞" added) is, however, not known (although partial results have been obtained, see [**GNN1,GNN2,LN**].

Our second example concerns equation (1'). In case $K \equiv 1$ in (1'), all the radial solutions are known ($r = |x|$):

$$u(r) = \frac{\left[n(n-2)\lambda^2\right]^{(n-2)/4}}{(\lambda^2 + r^2)^{(n-2)/2}}, \quad r \geq 0,$$

where $\lambda > 0$ serves as a parameter. It is easy to see that they all behave like $C|x|^{2-n}$ at ∞ and they give rise to the standard metric on $S^n\backslash\{\text{a point}\}$ through the stereographic projection (which is conformal). Note that this metric is *incomplete.* In [**DN2**] it was proved that (see the figure below)

(i) if $K \equiv K(r) = 1 + \eta(r)$, then for every $\alpha > 0$, there exists a radial entire solution u of (1') with $u(0) = \alpha$. Furthermore, all radial solutions of (1′) behave like $C|x|^{(2-n)/2}$ at ∞ (which implies that they all give rise to *complete non-compact* metrics on $\mathbb{R}^n$);

(ii) if $K \equiv K(r) = 1 - \eta(r)$, then equation (1′) has no radial entire solutions;

where $\eta(r)$ is a non-negative non-increasing function on $[0, 1)$ with $\eta(0) < 1$ (and can otherwise be arbitrarily small and with compact support).

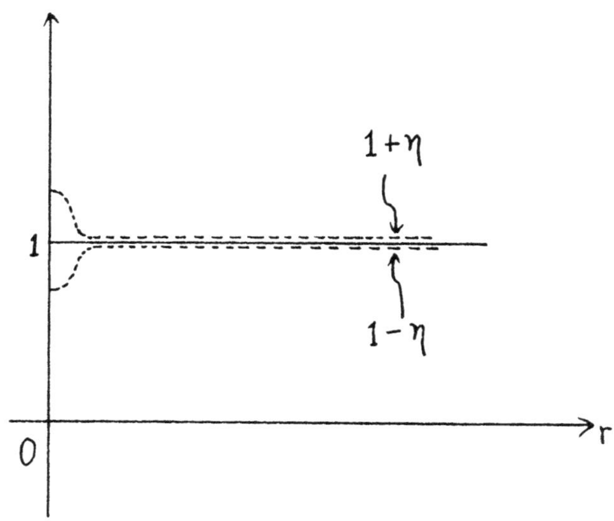

This example shows that equation (1′) is quite delicate and sensitive even to (arbitrarily) small perturbations. However, one open question arises immediately: is it true that equation (1′) does not possess any entire solutions (not just radial ones) for $K(r) = 1 - \eta(r)$?

The last example has to do with a basic question in partial differential equations. In [**LSW**], it was established that if $L \equiv \partial_i(a_{ij}(x)\partial_j)$ where (a_{ij}) is uniformly elliptic in, $\mathbb{R}^n$, $n \geq 3$, then L has a (global) fundamental solution on $\mathbb{R}^n$ with the asymptotic behaviour $C|x|^{2-n}$ both at ∞ and at the pole. A natural question arises: does a second order uniformly elliptic operator of divergence form on $\mathbb{R}^2$ possess a (global) fundamental solution with the asymptotic behaviour $C \log |x|$ both at infinity and at the pole? This seems to be a question of fundamental importance. However, it seems that there had been no answer (although such a fundamental solution may have been assumed to exist by many mathematicians) until the recent paper [**KN2**]. An affirmative answer was obtained by applying the method developed for equation (2) in [**KN2**].

In general, lots of important questions still remain open in this primitive area waiting to be explored.

By now I have roughly described most of the equations that I shall discuss in this expository paper. As one can see that almost all of them are special equations. It seems to me that the most interesting mathematics in nonlinear partial differential equations usually lies in the study of special equations, especially those arising in geometry, physics and mathematical biology. (It also seems to be a fact of life that we simply do not have one unifying approach which works for all such equations.) The choices of topics here only reflect my personal taste and should not be taken as a value judgement on the omitted material.

This paper is organized as follows. In §1, we discuss the conformal scalar curvature equation (1) and (1′). In §2, we discuss the conformal Gaussian curvature equation (2) and (2′). The Eddington equation (5) and Matukuma equation (4) are included in §3, and various aspects of the generalized scalar field equation (3) are described in §4. Finally, we end our paper by collecting some concluding remarks (including a number of open problems) in §5. Almost no proofs are given (due to the length of this paper), although the approaches or procedures of the proofs are generally discussed. The paper (again, due to its length) is written in such a way that *each section may be read independently.* For instance, if a reader is only interested in equation (3), then he (or she) can go to §4 directly.

Part of this paper was presented in a series of lectures at National Tsing Hua University at Hsinchu, Taiwan, Republic of China. I wish to thank Professor Hwai-chiuan Wang for the invitation and the staff of the Mathematics Department of National Tsing Hua University for their warm hospitality.

§1. Conformal scalar curvature equations.

Geometric background. Let (M, g) be a Riemannian manifold of dimension n. In various circumstances, one is led to the problem of finding new metrics on M with desired properties. The "prescribing curvature problem" was posed and studied by several mathematicians in recent years, notably, by Nirenberg, Moser, Berger, Kazdan and Warner, and others, while the equations appeared in this problem go back at least to H. Poincaré and is also related to the Yamabe problem.

At each point P on M, there is a quantity k called the "scalar curvature" of M at P (in the metric g) which is essentially the sum (or, the average) of all sectional curvatures of M at P (in case $n \geq 3$). k measures the deviation of the volume of geodesic balls from that of Euclidean balls:

$$\text{Vol}\, B_g(P;r) - \omega_n r^n = -Ckr^{n+2} + o(r^{n+2}) \tag{1.1}$$

for $r > 0$ small, where ω_n is the volume of the unit ball in $\mathbb{R}^n$ and C is a universal constant.

The "prescribing curvature problem" may be stated as follows:

Let (M, g) be a Riemannian manifold of dimension $n \geq 3$, and K be a given smooth function on M. *Question*: can one find a new metric g_1 on M such that

(i) K is the scalar curvature of (M, g_1), and
(ii) g_1 is conformal to g (i.e. $g_1 = \varphi g$ for some function $\varphi > 0$)?

(The condition (ii) gives some "control" on g_1 in terms of the original metric g.)

If one computes the curvature of $g_1 = \varphi g$ and equate that to K, one obtains a nonlinear elliptic equation (in φ) on M. To simplify the equation (i.e. to eliminate the first order term of φ in this equation), we set $\varphi = u^{\frac{4}{n-2}}$, $u > 0$, the equation then becomes

$$\frac{4(n-1)}{(n-2)}\Delta_g u - ku + Ku^{\frac{n+2}{n-2}} = 0 \tag{1.2}$$

where k is the scalar curvature of (M, g) and Δ_g is the Laplace-Beltrami operator on (M, g)

$$\Delta_g = \frac{1}{\sqrt{|g|}} \sum_{i,j=1}^{n} \frac{\partial}{\partial x_i}\left(\sqrt{|g|}g^{ij}\frac{\partial}{\partial x_j}\right),$$

with

$$|g| = \det(g_{ij}) \text{ and } (g^{ij}) = (g_{ij})^{-1}.$$

In case M is compact, the equation (1.2) above has been studied rather extensively by Berger, Moser, Kazdan and Warner, Chen and Ding, Bahri, and others, and is related to the Yamabe problem (which was recently solved by R. Schoen [**S1**]). We refer the reader to [**K**] for a survey.

In case M is complete, noncompact, little was known until recently. The first natural case is of course that $M = \mathbb{R}^n$ and g is the usual metric. We shall mostly concentrate on this special case in which the equation (1.2) reduces to

$$\frac{4(n-1)}{n-2}\Delta u + K u^{\frac{n+2}{n-2}} = 0 \tag{1.3}$$

in $\mathbb{R}^n$, where $\Delta = \sum_{i=1}^{n} \frac{\partial^2}{\partial {x_i}^2}$. In this case lots of work has been done (see our discussion below) and by now we have fairly good understanding of this equation. Other cases have also been considered. For instance, when $M = H^n(-1)$ (Poincaré disk), i.e. $M = \{x \in \mathbb{R}^n \mid |x| < 1\}$ and $g_{ij} = \delta_{ij}\left(\frac{2}{1-r^2}\right)^2$, $r = |x|$, the equation (1.2) reduces to

$$\frac{4(n-1)}{(n-2)}\Delta_g u + n(n-1)u + K u^{\frac{n+2}{n-2}} = 0 \tag{1.4}$$

in M, where

$$\Delta_g = \left(\frac{1-r^2}{2}\right)^2 \Delta + (n-2)\left(\frac{1-r^2}{2}\right)\sum_{i=1}^{n} x_i \frac{\partial}{\partial x_i}.$$

Equation (1.4) has been studied by [**AM**] and by [**CCY**], and will not be considered here.

Returning to (1.3), we remark that it is equivalent to the following equation

$$\Delta u + K u^{n^*} = 0 \tag{1.5}$$

in $\mathbb{R}^n$, $n \geq 3$ and $n^* = (n+2)/(n-2)$. From now on we shall assume that K is locally Hölder continuous and we are only interested in positive solutions of (1.5). Almost all the results described below concerning (1.5) do admit generalizations of various degrees (i.e. they also hold for more general equations than (1.5) above); however, we shall not even try to include the most general forms of those theorems.

Equation (1.5), although looks very simple, was not well-understood until the first general work on this by Ni [**N1**] appeared in early 1980's. On the surface, several difficulties are encountered immediately once one is interested in solving (1.5). First of all, elliptic equations (including linear ones) on $\mathbb{R}^n$ were not, and are still not, well-understood. All the Sobolev compact embedding theorems do not hold for $\mathbb{R}^n$. Moreover, the role of the critical exponent n^* in (1.5) which lies on the borderline of compact and continuous embeddings in case of bounded domains is also not clear in the entire space case. One could go into more concrete approaches, e.g. the variational approach, the barrier method, the finite domain approximations, etc.; however, difficulties are associated with each individual case. We shall see how to overcome those difficulties under various special circumstances later.

The fast decay case: existence.

The first general existence and non-existence results concerning positive solutions of (1.5) are due to Ni [**N1**] in 1980. Since then there have been lots of research on this equation. The main existence result of [**N1**] may be described as follows (here we write $x = (x_1, x_2) \in \mathbb{R}^3 \times \mathbb{R}^{n-3} \equiv \mathbb{R}^n$).

THEOREM 1.6. *If $|K(x)| \leq C/|x_1|^\ell$ for some $\ell > 2$ at $x_1 = \infty$ and uniformly in x_2, then equation (1.5) has infinitely many bounded positive solutions which are also bounded away from zero.*

We shall see that the same conclusion holds if we replace the exponent n^* in (1.5) by an arbitrary number $\tau > 1$; i.e. the equation

$$\Delta u + K u^\tau = 0 \tag{1.5'}$$

in $\mathbb{R}^n$, where $\tau > 1$. Note that K is only required to be bounded in $\mathbb{R}^n$ and to decay in *three* directions in the above theorem. Also notice that the solutions in the above theorem give rise to complete, non-compact metrics in the original geometric problem. Various extensions of the above theorem have been obtained by many authors: Kawano [**Ka**], Naito [**Na**], and others. The following result is due to Naito.

THEOREM 1.7. *Suppose that there exists a locally Hölder continuous function $k^*(r) \geq 0$ on $[0, \infty)$ such that $|K(x)| \leq k^*(|x|)$ for all $x \in \mathbb{R}^n$ and*

$$\int_0^\infty r k^*(r) dr < \infty. \tag{1.8}$$

Then there exists a $\delta > 0$ such that for every constant $c \in (0, \delta)$, there is a positive solution u of (1.5) on $\mathbb{R}^n$ with $u(x) \to c$ as $|x| \to \infty$.

(This theorem also holds for the more general equation (1.5′) with the same proof.)

There was an attempt in solving (1.5) by P.-L. Lions [**L**] who wished to apply the so-called "concentration-compactness" principle. This attempt seemed unsuccessful (see [**L**]). However, this does bring up a natural question: *Does (1.5) possess positive solutions which tend to zero under conditions similar to the ones above?* This question has also been studied independently and rather extensively by Japanese mathematicians earlier, such as Kusano and his colleagues. Only very recently, the following result was obtained by Li and Ni [**LN**] which, in some sense, gives a negative answer to the above question.

THEOREM 1.9. *Suppose that $K \equiv K(r)$ is strictly decreasing in $[0, \infty)$ and that $K(r) = 0(r^{-\ell})$ at ∞ for some $\ell > 2$. Then equation (1.5) does not possess any positive solution (radial or not) which tends to 0 at ∞.*

REMARK: In case $K(r)$ is not strictly decreasing, there may exist positive solutions of (1.5) which tends to 0 at ∞ under suitable conditions, see e.g. [**LN,LL**]. Moreover, Theorem 1.9 also admits extensions to non-radial K's and (1.5′).

From the three theorems above and various examples constructed in [**N1,LN**] and [**NY1,2**], our understanding of the structure of solutions of (1.5) in case K decays fast enough seems fairly complete. The point here seems to be that in case K decays fast enough, the situation is basically similar to that of the Euclidean space.

Before we go to other types of K, we first describe the proofs of those theorems above. The proof of Theorem 1.9 consists of two steps:

(i) any positive solution u of (1.5) which tends to 0 at ∞ must have the asymptotic behavior $\sim C|x|^{2-n}$ at ∞;

(ii) equation (1.5) does not have any positive solutions which tend to 0 at ∞. (This is achieved by using a version of Rellich-Pohozaev identity together with the asymptotic behaviour of u and ∇u at ∞.) We omit the details here and refer the interested readers to [**LN**].

The basic approach used in proving Theorems 1.6 and 1.7 is the "barrier method", and it turns out that the kind of equations we can treat using this approach is quite general; in particular, it includes the equation (1.2) for uniformly elliptic metrics g. (Of course the general case (1.2) requires more work.) We first describe the general method.

BARRIER METHOD. *Let $u_1 \geq u_2$ in $\mathbb{R}^n$ be such that*

$$\begin{cases} Lu_1 + f(x, u_1) \leq 0 & \text{(i.e. } u \text{ is a super-solution)}, \\ Lu_2 + f(x, u_2) \geq 0 & \text{(i.e. } u \text{ is a sub-solution)} \end{cases}$$

where f is locally Hölder continuous in (x, u) and locally Lipschitz in u, L is an elliptic operator of second order. Then there exists a solution u of $Lu + f(x, u) = 0$ with $u_1 \geq u \geq u_2$.

REMARKS:

(i) Since this theorem is very general, the most difficult step in this approach is to find those suitable barriers. This may often be achieved by solving some linear equations on $\mathbb{R}^n$ with some additional care on the asymptotic behaviour of solutions. In the proof of Theorem 1.7 this may be done fairly easily. However, for the more general equation (1.2), we need to first solve an exterior Dirichlet problem and a (interior) Dirichlet problem, then, we match their solutions on a ring-shaped domain to obtain an entire solution. This technique was used in [**F**], and is systematically studied in combination with the barrier method above by Kenig and Ni [**KN1**] in order to extend various results in [**N1**].

(ii) The above barrier method, in its present form, was due to Ni [**N1**]. A special case was proved earlier by Ako and Kusano [**AK**]. The proof is standard. We first solve the equation $Lu + f(x, u) = 0$ on B_R. Then, letting $R \to \infty$, we obtain a solution on $\mathbb{R}^n$ by a diagonal process.

EXTENSIONS: As we mentioned earlier, the barrier method does apply to the more general equation (1.2) in certain cases. The main technical point here is to construct suitable barriers for the equation (1.2) which, in turn, reduces to constructing bounded entire solutions with suitable asymptotic behavior at ∞ for the *linear* inhomogeneous equation

$$\Delta_g u - ku = f(x).$$

This approach is systematically studied in [**KN1**] for equation (1.2). Here we state the result and outline its proof.

THEOREM 1.10. *Suppose that the Riemannian metric (g_{ij}) has all its eigenvalues of the order of magnitude $|x|^{-b}$ at ∞ for some constant $b \in (-\infty, 2)$, and that the scalar curvature k of $(\mathbb{R}^n, g)$ satisfies*

$$0 \leq k(x) \leq C_1|x|^{-(2+\varepsilon_1-b)} \text{ at } \infty,$$

where C_1, ε_1 are positive constants. Then, for any K with

$$|K(x)| \leq C_2|x|^{-(2+\varepsilon_2-b)} \text{ at } \infty,$$

where C_2, ε_2 are some positive constants, there exist infinitely many bounded positive solutions of (1.2) which are also bounded below by positive constants in $\mathbf{R}^n$. That is, there exist infinitely many metrics g_1 such that

(a) *g_1 is conformal to g and K is the scalar curvature of g_1,*

(b) *the eigenvalues of g_1 have the same order of magnitude as those of g at ∞.*

REMARK: First, note that this theorem allows certain degeneracy in the ellipticity of Δ_g. Furthermore, the conclusion of this theorem actually holds for more general equations than (1.2) (e.g. (1.11) below).

To prove this theorem, it is best to free ourselves from geometry and consider

$$Lu - ku + Ku^{\tau} = 0 \tag{1.11}$$

in $\mathbf{R}^n$, $n \geq 3$, where $\tau > 1$, $L \equiv \sum_{i,j=1}^n \frac{\partial}{\partial x_i}\left(a_{ij}(x)\frac{\partial}{\partial x_j}\right)$ and (a_{ij}) is a symmetric, positive definite matrix with its eigenvalues of the order of magnitude $C|x|^{-a(n-2)}$ at ∞, where $a \in (-\infty, 1)$. (In case $a_{ij} = \sqrt{|g|}g^{ij}$, we have $a = b/2$.) It is straightforward to verify that the following statement implies our theorem: *if there exists an $\varepsilon > 0$ such that the following conditions hold for $|x|$ large*

$$|K(x)| \leq \frac{C_1}{|x|^{2+\varepsilon+a(n-2)}}, \quad \textit{and}$$

$$0 \leq k(x) \leq \frac{C_2}{|x|^{2+\varepsilon+a(n-2)}},$$

for some positive constants C_1, C_2, then equation (1.11) possesses infinitely many bounded positive solutions with the property that each of these solutions is also bounded below by a positive constant. It is not hard to see that this follows from the barrier method and the following theorem for linear equations (see [**KN1**] for the details).

THEOREM 1.12. *Under the above hypotheses on L, if $0 \leq c(x) \in L^r_{\text{loc}}(\mathbf{R}^n)$ for some $r > \frac{n}{2}$, and $f(x)$ is locally bounded with*

$$|f(x)| \leq \frac{C}{|x|^{2+\varepsilon+a(n-2)}} \quad \text{at } |x| = \infty, \tag{1.13}$$

where $0 < \varepsilon < (1-a)(n-2)$ then there exists a bounded entire solution v of the linear equation

$$Lv - cv = f$$

in $\mathbb{R}^n$. *Moreover,* $|v(x)| \le C/|x|^{\varepsilon}$ *at* $|x| = \infty$.

Nonexistence.

Before we go any further, we wish to describe some nonexistence results so that we have a general picture of the situation before we enter the more interesting but complicated range of K (in terms of its asymptotics).

The first nonexistence result concerning positive entire solutions of (1.5) goes back to Haviland around 1944 and was later extended to more general equations or to manifolds (instead of $\mathbb{R}^n$) by many mathematicians including Osserman, Keller, Calabi, Kazdan and Warner, Cheng and Yau, to name a few. It may be stated as follows.

PROPOSITION 1.14. *If* $K \le -\varepsilon < 0$ *on* $\mathbb{R}^n$, *then equation (1.5) has no positive entire solutions in* $\mathbb{R}^n$.

The proof of this result was based on a comparison argument which enables us to only consider the simpler case $K \equiv -1$ in (1.5). In this connection, we should mention the work of Loewner and Nirenberg [**LoN**] on this case in an arbitrary bounded smooth domain in $\mathbb{R}^n$.

In the case $K \le 0$ in $\mathbb{R}^n$, the following nonexistence result seems to be best possible in view of our existence results, e.g. Theorem 1.6 or 1.7.

THEOREM 1.15. *If* $K \le 0$ *on* $\mathbb{R}^n$ *and* $K(x) \le -C|x|^{-\ell}$ *at* ∞ *for some* $\ell \le 2$, $C > 0$, *then equation (1.5′) possesses no positive entire solutions in* $\mathbb{R}^n$, *for any* $\tau > 1$.

The case $\ell < 2$ was established by Ni [**N1**] in 1982 and was improved to include the case $\ell = 2$ by F.-H. Lin [**Ln**] in 1985. In fact, one does not need the "pointwise" estimate on K; an averaged version suffices. The proof for the case $\ell < 2$ consists of several steps:

(i) to obtain an ordinary differential inequality for the spherical mean $\bar{u}$ of a solution u of (1.5′), namely,

$$\bar{u}'' + \frac{n-1}{r}\bar{u}' + \tilde{K}(r)\bar{u}^{\tau} \ge 0 \tag{1.16}$$

where $\tilde{K}(r)$ is another kind of average of K on $|x| = r$;

(ii) to show that $\bar{u} \ge Cr^{\sigma}$ at ∞ for some $\sigma > 0$;

(iii) to show that $\bar{u} \ge Cr^{\rho}$ at ∞ for any $\rho > 0$;

(iv) to show that $\bar{u}$ must become ∞ at some *finite* r.

The case $\ell = 2$ was left open in [**N1**] because step (ii) above was only obtained in [**N1**] with r^σ replaced by a power of $\log r$. Using Harnack inequality, F.-H. Lin [**Ln**] was able to improve Ni's estimate and obtain step (ii) for this critical decay of K. (See [**Ln**] for the details.) Recently, a simple, elementary proof was obtained in [**CL**].

Following Calabi's argument [**C**], Kenig and Ni [**KN1**] extended Theorem 1.15 to complete manifolds with nonnegative Ricci curvatures.

For positive curvature K, the first nonexistence result was also due to Ni [**N1**].

THEOREM 1.17. *If $K \geq C|x|^2$ at ∞ (K may change sign at finite part), then equation (1.5) possesses no positive entire solutions in $\mathbb{R}^n$.*

REMARK: Again the pointwise condition "$K \geq C|x|^2$ at ∞" may be replaced by

$$\int^{\infty} t^{-\frac{n}{n-2}} \tilde{K}\left(t^{\frac{1}{n-2}}\right) dt = \infty \tag{1.18}$$

where $\tilde{K}(r)$ is some kind of average of K on the sphere S_r. (Naturally, $\tilde{K} \equiv K$ if K is already radial.) Intuitively speaking, this theorem says that in order to solve (1.5) the "total curvature" (weighted in some sense) cannot be infinite.

To prove this result, we first average as before to get an ordinary differential inequality for the spherical mean $\bar{u}$ of u

$$\bar{u}'' + \frac{n-1}{r}\bar{u}' + \tilde{K}(r)\bar{u}^{n^*} \leq 0. \tag{1.19}$$

(Observe that the inequality signs in (1.16) and (1.19) are opposite to each other.) Next we make the following change of variables

$$s = r^{n-2}, \varphi(s) = s\bar{u}(r),$$

and we arrive at

$$\varphi_{ss} + \frac{1}{(n-2)^2} \frac{\tilde{K}\left(s^{\frac{1}{n-2}}\right)}{s^{1-\frac{2}{n-2}+n^*}} \varphi^{n^*} \leq 0.$$

Then we work out an "oscillation theorem" for this nonlinear differential inequality to conclude that if (1.18) is satisfied, then φ cannot be positive at ∞. We leave it to the reader to fill in the details.

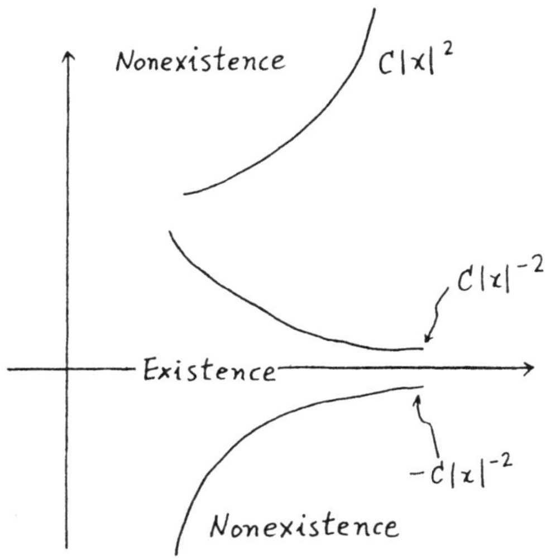

Other nonexistence results have been obtained by Cheng and Lin [**CL**].

The more interesting range of K.

The schematic picture above sums up what we have done so far.

Thus the range of K which is to be discussed here is basically between $C|x|^{-2}$ and $C|x|^2$ for $|x|$ large. As we have seen in the Introduction, there are interesting and striking results if K lies in this range. However, our results below also seem to indicate that there is no "clean" theorem in this interesting range of K (while Theorems 1.6, 1.7, 1.9, 1.15, and 1.17 certainly look very clean and simple).

The first set of results in this direction is due to Ding and Ni [**DN2**] in 1985. The following is a special case.

PROPOSITION 1.20. *Suppose that $K \geq 0$ and $x \cdot \nabla K(x) \leq 0$ for all $x \in \mathbb{R}^n$. Suppose further that K is symmetric (i.e. even) with respect to each variable $x_1, \ldots, x_n$. Then for every $\alpha > 0$, (1.5′) possesses a positive solution u in $\mathbb{R}^n$ with $|u|_{L^\infty} = \alpha$, for every $\tau \geq n^*$.*

REMARK: As far as the existence of a positive solution of (1.5′) with $\tau \geq n^*$ is concerned, the condition that "K is symmetric w.r.t. x_j, $j = 1, 2, \ldots, n$" may be relaxed to that "K is almost symmetric with respect to n independent directions".

For the properties of the solutions guaranteed by the above proposition, we have

PROPOSITION 1.21. *Let $K \geq \alpha > 0$ be C^1 and $u > 0$ be a solution* *of (1.5). If $x \cdot \nabla K(x) \leq 0$ and $\not\equiv 0$ on $\mathbf{R}^n$, then*

$$\int_{\mathbf{R}^n} |Du|^2 = \infty, \text{ and } \int_{\mathbf{R}^n} u^{2n/(n-2)} = \infty. \tag{1.22}$$

REMARK: When $K \equiv 1$, then all the known solutions of (1.5) are radial and can be explicitly written down as

$$u_\lambda(x) = \frac{\left[n(n-1)\lambda^2\right]^{(n-2)/4}}{(\lambda^2 + |x|^2)^{(n-2)/2}}.$$

It is easy to check that for every λ it holds that

$$\int_{\mathbf{R}^n} |Du_\lambda|^2 < \infty \text{ and } \int_{\mathbf{R}^n} u_\lambda^{2n/(n-2)} < \infty$$

On the other hand, we may choose K as close to the constant 1 as we like and satisfying all the hypotheses in Propositions 1.20 and 1.21. Then there are positive solutions of (1.5) and all positive solutions in this case satisfy (1.22). This shows that arbitrarily small perturbations in K can cause drastically different behaviours in positive solutions of (1.5). (The "radial" version of this result is even more striking as we have seen in the Introduction.)

The proofs of both Propositions 1.20 and 1.21 are interesting. The first makes use of a global bifurcation result of P. Rabinowitz[**R**] and the second one includes an application of the Rellich-Pohozaev identity in estimating those quantities in (1.22). We refer the reader to [**DN2**] for further details.

For the rest of this section, we shall restrict ourselves to radial K's and radial solutions.

First, we recall the result of [**DN2**] stated in the Introduction (pp. 2-3). The proof is a bit tricky and we shall not go into that here. We only remark that the Rellich-Pohozaev identity again plays an important role in the proof. This result (and its proof) was recently extended to equation (1.5′) with $1 < \tau < n^*$ under appropriate hypothesis on K by Kusano and Naito [**KNa**].

There have been lots of works on radial K's and it is simply not possible to review all of them here. We shall only mention a few of them which we are familiar with. (Many of the other papers, such as [**LL**], are not yet published.)

Our basic assumptions on $K = K(r)$ for the rest of this section are that $K \geq 0$ in $[0, \infty)$ is Hölder continuous and $K \sim Cr^\ell$ at ∞ for some $\ell \in [-2, 0)$, and, we shall be dealing with the equation (1.5′) with $\tau > 1$. The following simple lemma is a special case of a result obtained in [**N1**].

LEMMA 1.23. *A positive radial solution u of (1.5′) must satisfy the following estimates at $r = \infty$:*

$$(1.24)\qquad Cr^{-\frac{\ell+2}{\tau-1}} \geq u(r) \geq Cr^{2-n} \text{ if } \ell > -2,$$

$$(1.25)\qquad C(\log r)^{-\frac{1}{\tau-1}} \geq u(r) \geq Cr^{2-n} \text{ if } \ell = -2.$$

It is not hard to see that both (1.24) and (1.25) are sharp (see [**NY2**]). Our next result (obtained by Li and Ni [**LN**]) says that only two extremes of (1.24) and (1.25) are possible under additional reasonable hypothesis on K.

THEOREM 1.26.

(i) *Suppose that $\ell > -2$ and $\tau > \frac{n+2+2\ell}{n-2}$. If $(r^{-\ell}K)_r \leq 0$ at ∞, then any positive radial solution u of (1.5′) in $\mathbb{R}^n$ must behave like either $\sim Cr^{2-n}$ or $\sim Cr^{-\frac{\ell+2}{\tau-1}}$ at ∞.*

(ii) *Suppose that $\ell = -2$. If*

$$(1.27)\qquad \limsup_{r\to\infty} r(\log r)^2(r^2K)_r \leq 0,$$

then any positive radial solution u of (1.5′) in $\mathbb{R}^n$ must behave like either $\sim Cr^{2-n}$ or $\sim C(logr)^{-\frac{1}{\tau-1}}$ at ∞.

The condition on K above (e.g. (1.27)) seems to be necessary.

THEOREM 1.28. *Suppose that $K(r) = O(r^\sigma)$ at $r = 0$ for some $\sigma \geq 0$, and that $\frac{n+2+2\sigma}{n-2} > \tau > \frac{n+2+2\ell}{n-2}$.*

(i) *If $0 > \ell > -2$ and $(r^{-\ell}K)_r \leq 0$ at ∞, then equation (1.5′) possesses a positive entire radial solution in $\mathbb{R}^n$ of which the asymptotic behaviour is $\sim Cr^{2-n}$ at ∞, and for every $\alpha > 0$ sufficiently small, the initial value problem*

$$\begin{cases} u'' + \dfrac{n-1}{r}u' + Ku^\tau = 0 \\ u'(0) = 0, \quad u(0) = \alpha > 0 \end{cases}$$

possesses a solution $u(r;\alpha) > 0$ for all $r > 0$ and $u(r,\alpha) \sim Cr^{-\frac{\ell+2}{\tau-1}}$ at ∞.

(ii) *If $\ell = -2$ and (1.27) holds with $r^2K \to c > 0$ at ∞, then the same conclusions as in (i) hold with $Cr^{-\frac{\ell+2}{\tau-1}}$ replaced by $C(\log r)^{-\frac{1}{\tau-1}}$.*

This theorem is proved in [**LN**] by using an existence result of Ni and Yotsutani in [**NY2**].

REMARKS:

(i) Note that, for instance, $K = r(r+1)/(r^3+1)$ satisfies all the hypotheses on K in Theorem 1.28. Thus, for this K, we have two and only two kinds of solutions, one behaves like Cr^{2-n} at $r = \infty$ which gives rise to incomplete conformal metrics in $\mathbb{R}^n$, the other behaves like $Cr^{-\frac{1}{\tau-1}}$ ($\ell = -1$, $\tau > \frac{n}{n-2}$) which gives rise to complete noncompact conformal metrics in $\mathbb{R}^n$ (when $\tau = n^*$, $\frac{1}{\tau-1} = \frac{n-2}{4}$). That is, for this particular K, we have two kinds of metrics both are conformal to the usual one in $\mathbb{R}^n$, having K as their scalar curvature, and yet one is complete, the other is incomplete.

(ii) Various other results and examples concerning existence, nonexistence and asymptotic behaviours were obtained in [**N1,DN2**].

(iii) Other interesting works (in the case K is radial) include some very recent joint work of C.-S. Lin and S.-S. Lin [**LL**]. (The solutions they obtained have finite "energy" and can be pulled back to S^n).

(iv) The non-radial versions (or analogues) of the above theorems remain open. These seem to be hard problems and worth pursuing.

§2. Conformal Gaussian curvature equations.

Let (M, g) be a Riemannian manifold of dimension 2 and K be a given smooth function on M. Similarly as before, we may consider the following question: can we find a new metric g_1 on M such that K is the Gaussian curvature of (M, g_1) and g_1 is conformal to g (i.e. $g_1 = \varphi g$ for some function $\varphi > 0$)? Again, we compute the Gaussian curvature of φg and we obtain a nonlinear equation in φ. To simplify the equation (i.e. to eliminate the first order term of φ in the equation), this time we set $\varphi = e^{2u}$ (compare to the higher dimensional case) and the equation becomes

$$\Delta_g u - k + K e^{2u} = 0 \tag{2.1}$$

on M, where Δ_g is the Laplace-Beltrami operator on M in the g metric and k is the Gaussian curvature of M in the original metric g. Note that since $e^{2u} > 0$ for any $u \in \mathbb{R}$, any solution (not necessarily positive) of (2.1) would give rise to such a new metric g_1. In case M is compact, much has been done (see the recent monograph by J. Kazdan [**K**]). In case M is complete, non-compact, little was known until recently due to similar difficulties we encountered in higher dimensional case (see the previous section). The first natural case is of course $M = \mathbb{R}^2$ and g is the standard metric. In this case, the equation (2.1) reduces to

$$\Delta u + Ke^{2u} = 0 \tag{2.2}$$

in $\mathbb{R}^2$, where Δ is the usual Laplacian. In this section, we shall mostly concentrate on discussing the existence (and nonexistence) of solutions (not necessarily positive) of (2.2) and of (2.1) in case g is uniformly elliptic. We should remark that other cases of (2.1) have also been considered; for instance, the case where (M, g) is the Poincaré disc has been considered by [**AM**] and [**CTY**]. However, we shall not discuss that here.

Returning to equation (2.2), we are interested in studying its entire solutions in $\mathbb{R}^2$ with no positivity requirement imposed on the solution u.

The first result on (2.2) seems due to L. V. Ahlfors [**A**] in 1938 - it was a nonexistence theorem. Ahlfors's result was reproved and improved by several authors including Wittich [**W**] in 1944, Osserman [**O**] in 1956, Sattinger [**Sa**] in 1972 and Oleinik [**Ol**] in 1978. Sattinger's theorem may be stated as follows.

THEOREM 2.3. *If $K \le 0$ and $K(x) \le -C|x|^{-2}$ at ∞, then equation (2.2) possesses no entire solutions in $\mathbb{R}^2$.*

This nonexistence result was slightly improved by Ni [**N2**] in 1982 to include "oscillatory" K's. For instance, Ni's theorem implies that for

$$\tilde{K}(R) = -\frac{1+\sin r}{r^2}, \text{ or } \tilde{K}(R) = -(1+\sin r), \text{ at } r = \infty.$$

where $\tilde{K}$ is some kind of average of K on the sphere $|x| = r$, equation (2.2) does not have any (entire) solutions in $\mathbb{R}^2$. Other extensions were also obtained recently in [**CL**].

The proof of Theorem 2.3 is somewhat similar to that of the nonexistence results in §1 above, is hence omitted here.

It seems that the first existence theorem for (2.2) was not known until the publication of Ni's paper [**N2**] in 1982. Ni proved the following theorem which is complementary to the non-existence result above.

THEOREM 2.4. *If $K \leq 0$ and $\not\equiv 0$ in $\mathbb{R}^2$, $K(x) \geq -C|x|^{-\ell}$ at ∞ for some $\ell > 2$, then equation (2.2) possesses infinitely many solutions. More precisely, for every sufficiently small $\alpha > 0$, there exists an entire solution u of (2.2) such that*

$$u(x) - \alpha \log |x| = O(1) \text{ at } |x| = \infty.$$

It is easy to see that the above solutions u (with logarithmic asymptotics) give rise to new conformal metrics which are complete. Theorem 2.4 was subsequently improved by McOwen [**M2**] in 1984 as follows.

THEOREM 2.5. *Under the same hypotheses of K in the above theorem (assuming that $K(0) < 0$), then for every $\alpha \in (0, \frac{\ell-2}{2})$, there exist a solution u of (2.2) in $\mathbb{R}^2$ and a constant u_∞ such that*

$$u(x) - \alpha \log |x| - u_\infty = O(|x|^\gamma) \text{ at } |x| = \infty \tag{2.6}$$

for every $\gamma > \max\{-1, 2 - \ell + 2\alpha\}$.

In proving Theorem 2.4 Ni used the barrier method which we described in the previous section. McOwen's method is, however, totally different. He used the "weighted Sobolev space" technique which also enables him to obtain various results when K is positive somewhere or K changes sign in $\mathbb{R}^2$ (we shall briefly describe those later). At this point, we would like to point out that the barrier method, although gives less precise information about the solutions obtained, is more elementary and easier to adapt in different situations. For instance, if we know more about the asymptotic behaviour of K, then it is possible to obtain solutions of (2.2) with different asymptotic behaviours at ∞. To illustrate this point, we include the following result of Ni [**N2**]. Other similar results may be obtained by exactly the same arguments except that different barriers should be used.

THEOREM 2.7. *Suppose that $K < 0$ in $\mathbb{R}^2$ and that there exist two positive constants A_1, A_2 such that*

$$-A_1 e^{-2|x|^2} \geq K(x) \geq -A_2 e^{-2|x|^2} \text{ at } |x| = \infty,$$

then besides the solutions with logarithmic asymptotics given by (2.6), equation (2.2) also possesses a solution u in $\mathbb{R}^2$ with

$$u(x) - |x|^2 = O(1) \text{ at } |x| = \infty.$$

The second advantage of the barrier method is that it applies to the more general equation (2.1) (see Kenig and Ni [**KN2**]) while the weighted normed space method seems hard to generalize due to the restrictions on the linear theory (see, e.g. McOwen [**M1**]).

On the other hand, J. Moser studied equation (2.1) on S^2 in [**Mo**] by variational arguments. Following Moser's approach and using weighted Sobolev spaces instead of the classical ones, McOwen [**M3**] established the following theorems.

THEOREM 2.8. *If $K(x_0) > 0$ for some $x_0 \in \mathbb{R}^2$ and $K(x) = O(|x|^{-\ell})$ at $x = \infty$ for some $\ell > 0$, then for every $\alpha \in (-2, \min\{0, \ell - 2\})$, there exists a solution u of (2.2) with a constant u_∞ such that*

$$(2.9) \qquad u(x) - \alpha \log |x - x_0| - u_\infty = O(|x|^\gamma) \text{ at } \infty.$$

for every $\gamma > \max\{-1, 2 - \ell + 2\alpha\}$.

THEOREM 2.10. *If $K(x) = O(|x|^{-\ell})$ at ∞ for some $\ell > 2$, then (2.2) has a solution satisfying (2.9) with $\alpha = 0$ if and only if K changes sign and $\int_{\mathbb{R}^2} K\, dx < 0$.*

The proofs of these two theorems are similar to that of Moser's results which are rather well-known and are thus omitted here.

We now come to equation (2.1) on $(\mathbb{R}^2, g)$ where g is a uniformly elliptic Riemannian metric. To study (2.1), it seems best to free ourselves from geometry and just consider

$$(2.11) \qquad Lu - k + Ke^{2u} = 0$$

in $\mathbb{R}^2$, where $L = \partial_i(a_{ij}(x)\partial_j)$ with $a_{ij}(x)\xi_i\xi_j \geq \lambda|\xi|^2$ for all x, $\xi \in \mathbb{R}^2$, for some constant $\lambda > 0$ and that $\Lambda \geq \|A(x)\|_{L^\infty}$ holds for all $x \in \mathbb{R}^2$ where $A(x)$ is the $n \times n$ matrix $(a_{ij}(x))$.

REMARK: Observe that equation (2.1) may be reduced to equation (2.2) via the Uniformization Theorem, under appropriate conditions on the metric g. Since the reduced equation takes the form

$$(2.12) \qquad \Delta v + (K \circ \varphi^{-1}) \cdot e^{2v} = 0$$

in $\mathbb{R}^2$, where φ is the conformal diffeomorphism guaranteed by the Uniformization Theorem. In order to apply our results for (2.2) to the equation (2.12), it is essential to know the asymptotic behaviour of φ near ∞. For instance, the estimates

$$(2.13) \qquad C_1|x| \leq |\varphi(x)| \leq C_2|x|$$

for $|x|$ large, would suffice. However, this kind of estimates only hold for metrics of the form $g_{ij}(x) = \lambda(x)\delta_{ij}$ outside a compact set in $\mathbf{R}^2$ for some positive function $\lambda(x)$. (This is due to Ahlfors and Bers [**AB**] in 1960.) In general, even for uniformly elliptic metrics, (2.13) does not hold. Instead, one has the following sharp behaviour

$$C_1|x|^{1/\sigma} \leq |\varphi(x)| \leq C|x|^{\sigma} \tag{2.14}$$

for $|x|$ large, where σ is the ellipticity constant of g at ∞. ((2.14) clearly extends (2.13) and was obtained in [**LnN**].)

With this behaviour, one could still obtain theorems concerning (2.1) by reducing it to (2.12). However, this would not give the optimal results. Furthermore, if we replace Δ_g by L in (2.1), this approach does not seem to give anything. Therefore it seems worthwhile to study (2.11) directly. We shall follow the paper by Kenig and Ni [**KN2**].

The method used in [**KN2**] is again the barrier method and our approach is very close to the higher dimensional case for the equation (1.11) except that in the present case it is more complicated. To state our results, we let $g(x)$ be the Green's function for L outside the unit ball with pole at ∞, i.e.

$$\begin{cases} Lg = 0 \text{ in } |x| > 1, \\ \ \ g = 0 \text{ on } |x| = 1, \end{cases}$$

and $g(x) \sim \log|x|$ at ∞. (This may be done by an inversion, i.e. the Kelvin transform.) Our main result is

THEOREM 2.15. *If $|k(x)| \leq C|x|^{-2-\varepsilon}$ at ∞ for some $\varepsilon > 0$, $C > 0$, and $K \leq 0$ in $\mathbf{R}^2$ with*

$$|K(x)| \leq A\frac{\exp[\alpha g(x)]}{|x|^4} \text{ at } x = \infty,$$

for some constant $A > 0$ and $\alpha < 4\pi\lambda - 2\int_{\mathbf{R}^2} k\,dx$. Then for every a with $0 < a < 4\pi\lambda$ and $a < \left[4\pi\lambda - 2\int_{\mathbf{R}^2} k - \alpha\right]/2$, equation (2.11) has an entire solution u with

$$u(x) = \left(\int_{\mathbf{R}^2} k\,dx + \frac{a}{2}\right) g(x) + O(1) \text{ at } \infty.$$

REMARKS:

(i) When $L = \Delta$, we have $\lambda = 1$, $g(x) = \frac{1}{2\pi}\log|x|$, $k \equiv 0$ and

$$\frac{\exp[\alpha g(x)]}{|x|^4} = |x|^{\frac{\alpha}{2\pi}-4}.$$

Also observe that $\alpha < 4\pi$ if and only if $\frac{\alpha}{2\pi} - 4 < -2$, and this reduces to Theorem 2.4. This shows that Theorem 2.15 is in general the best possible. This can also be seen from the following nonexistence result with suitable examples (see [**KN2**]).

PROPOSITION 2.16. *If $k \geq 0$ in $\mathbb{R}^2$ with*

$$k(x) \geq C_0 |x|^{-2-\varepsilon} \text{ for } |x| \geq R_0,$$

where C_0, $\varepsilon > 0$, then equation (2.11) does not have any entire solution in $\mathbb{R}^2$ for $K \leq 0$ in $\mathbb{R}^2$ and $|K(x)| \geq C|x|^{-2-\alpha}$ at ∞, for some $\alpha < \frac{2C_0}{\varepsilon R_0^\varepsilon}$.

(ii) Note that the total curvature $\int_{\mathbb{R}^2} k$ plays an important role in the results above.

As in the higher dimensional case, the crucial step in proving Theorem 2.15 is to solve a linear equation with appropriate estimates.

THEOREM 2.17. *Suppose that $f \in L^1(\mathbb{R}^2)$ and is locally bounded in $\mathbb{R}^2$ with*

$$\int_{|x|>1} |f(x)|^p |x|^{4(p-1)} dx < \infty$$

for some $p > 1$. Then there exists a solution u of $Lu = f$ in $\mathbb{R}^2$ with the asymptotic behavior

$$u(x) = \left(\int_{\mathbb{R}^2} f\, dx \right) \cdot g(x) + O(1) \text{ at } \infty.$$

REMARK: The proof of this theorem may be used to establish the existence of a (global) fundamental solution of L in $\mathbb{R}^2$ with logarithmic behavior both at the pole and at ∞ (as we have pointed out in the Introduction). For the details of these proofs, we refer the reader to [**KN2**].

Open problems. It is not hard to see that if $K = K(r)$ is radial and $K \geq 0$ in $\mathbb{R}^2$, then there is essentially no obstruction for existence of a solution of (2.2). However, if we drop the "radial" assumption on K, little is known. Thus we pose the following

PROBLEM. *Does (2.2) possess a solution in $\mathbb{R}^2$ for every $K \geq 0$ in $\mathbb{R}^2$?*

Other problems have to do with the case K changes sign. In this direction we do have Theorems 2.8 and 2.10. However, the situation is not at all clear if K changes sign and does not decay to 0 at ∞. I believe this is also worth pursuing.

§3. Matukuma equation, Eddington equation and related topics.

In 1930, Matukuma, an astrophysicist, proposed the following mathematical model to describe the dynamics of a globular cluster of stars,

$$\Delta u + \frac{u^p}{1+|x|^2} = 0, \quad x \in \mathbb{R}^3, \tag{3.1}$$

where $p > 1$, u represents the gravitational potential (therefore $u > 0$), $\rho = -(4\pi)^{-1}\Delta u = \{4\pi(1+|x|^2)\}^{-1}u^p$ represents the density and $\int \rho\, dx$ represents the total mass (for details, see [**NY1**,**NY2**]). His aim was to improve the model proposed earlier (in 1915) by Eddington [**E**]

$$\Delta u + \frac{e^{2u}}{1+|x|^2} = 0, \quad x \in \mathbb{R}^3. \tag{3.2}$$

Since the globular cluster has radial symmetry, positive radial entire solutions of (3.1) or (3.2) (i.e. solutions of (3.1) or (3.2) with $u(x) = u(|x|) > 0$ for all $x \in \mathbb{R}^3$) are of particular interest, and equations (3.1) and (3.2) reduce to the following ordinary differential equations respectively

$$\begin{cases} u_{rr} + \dfrac{2}{r}u_r + \dfrac{u^p}{1+r^2} = 0, \\ u(0) = \alpha, \quad u_r(0) = 0; \end{cases} \tag{3.1$_\alpha$}$$

$$\begin{cases} u_{rr} + \dfrac{2}{r}u_r + \dfrac{e^{2u}}{1+r^2} = 0, \\ u(0) = \alpha, \quad u_r(0) = 0; \end{cases} \tag{3.2$_\alpha$}$$

where $\alpha > 0$. For each $\alpha > 0$, we denote the (global) unique solution of (3.1$_\alpha$) (or (3.2$_\alpha$)) by $u = u(r;\alpha)$. Studying the structure of solutions of (3.1$_\alpha$), Matukuma conjectured:

(i) *if $p < 3$, then $u(r;\alpha)$ has a finite zero for every $\alpha > 0$,*

(ii) *if $p = 3$, then $u(r;\alpha)$ is a positive entire solution with finite total mass for every $\alpha > 0$,*

(iii) *if $p > 3$, then $u(r;\alpha)$ is a positive entire solution with infinite total mass for every $\alpha > 0$.*

In 1938, Matukuma found an interesting exact solution

$$u(r;\sqrt{3}) = \left(\frac{3}{1+r^2}\right)^{1/2} \quad (p=3), \tag{3.3}$$

which confirms part of his conjecture.

Equations (3.1) and (3.2) have been around for a long time, at least in Japan. However, there seems to be very little mathematical contribution in the literature on these two equations. Very recently, equations (3.1) and (3.2) have been studied by Ni and Yotsutani [**NY1,NY2**]. It turns out that the equation (3.1) is more delicate than Matukuma had expected. In answer to his conjecture, Ni and Yotsutani proved that for a solution $u(r;\alpha)$ of the equation (3.1) the following theorems hold:

(iv) *if* $1 < p < 5$, *then* $u(r;\alpha)$ *has a finite zero for every sufficiently large* $\alpha > 0$,

(v) *if* $1 < p < 5$, *then* $u(r;\alpha)$ *is a positive entire solution with infinite total mass for every sufficiently small* $\alpha > 0$,

(vi) *if* $p \geq 5$, *then* $u(r;\alpha)$ *is a positive entire solution with infinite total mass for every* $\alpha > 0$.

Theorem (iv) follows from an easy scaling argument while the existence part in (v) is more delicate. The crucial idea involved here is to combine an extended version of Rellich-Pohozaev identity and some a priori estimates for solutions of (3.1_α) *with small initial value* $\alpha > 0$. The proof of the fact that those solutions obtained in (v) and (vi) have infinite total mass consists of two steps: first, reduce the problem to the limiting behaviour of $r^{n-2}u(r)$ at $r=\infty$; then, show that $r^{2-n}u(r) \to \infty$ as $r \to \infty$. It is rather interesting to note that not only the exponent p but also the initial value α has vital influence on the behaviour of the solution $u(r;\alpha)$. The methods and conclusions described above admit generalizations (see [**NY2**]) which in particular include the following equation

$$\Delta u + K(r)u^p = 0, \tag{3.4}$$

where Δ is the Laplace operator in $\mathbb{R}^n$, $n \geq 3$, $K \geq 0$ on $[0,\infty)$ and $p > 1$. When K decays faster than Cr^{-2} or grows faster than Cr^2 at ∞, equation (3.4) was studied by Ni [**N1**] in 1982 in connection to a problem in Riemannian geometry (where $p = (n+2)/(n-2)$). When K decays slower than or equal to Cr^{-2} at ∞, various existence and nonexistence results, including Theorems (iv)–(vi) above, for positive entire solutions of (3.4) were established in [**NY2**]. (We refer the readers to §1 for other works on this interesting equation.)

The existence of positive entire solutions of (3.1) with finite total mass was, however, not treated in [**NY2**], but was obtained in a later paper of Li and Ni [**LN**]. It was first proved in [**N1**] that a positive entire solution u of the equation (3.1) in $\mathbf{R}^n$ must satisfy the following estimates at ∞ (as we have mentioned in §1 above)

$$\frac{C_1}{(\log r)^{p-1}} \geq u(r) \geq \frac{C_2}{r} \tag{3.5}$$

and this is sharp (see [**NY2**,**LN**]). Furthermore, recall that it was established in [**LN**] that the asymptotics of a positive entire solution u of (3.1) must be either

$$\frac{C_1}{r} \geq u(r) \geq \frac{C_2}{r} \tag{3.6}$$

or

$$\frac{C_1}{(\log r)^{p-1}} \geq u(r) \geq \frac{C_2}{(\log r)^{p-1}} \tag{3.7}$$

at $r = \infty$. (See §1 also.) Therefore, all the solutions in (v) and (vi) above (found in [**NY2**]) must behave like $C(\log r)^{1-p}$ at ∞ since they have infinite total mass. Now, a finite total mass solution u^* of (3.1) may be found in case $1 < p < 5$ by minimizing the Dirichlet integral

$$\int_{\mathbf{R}^3} |\nabla u|^2 \, dx$$

among all the radial functions on $\mathbf{R}^3$ subject to the constraint

$$S(u) \equiv \int_{\mathbf{R}^3} \frac{|u|^{p+1}}{1 + |x|^2} \, dx = 1.$$

Noting that a positive entire solution with asymptotic behaviour (3.7) does not have finite value $S(u)$, we conclude that u^* must satisfy (3.6) and thus the corresponding total mass must be finite. Denote the initial value of u^* by $u^*(0) \equiv \alpha^*$. We would then like to know the relationship between α^*

and those α's in (iv) and (v) above. In [**NY2**] it was conjectured that if $1 < p < 5$ then

$$\begin{aligned}\alpha^* &= \inf\{\alpha > 0 \mid u(r;\alpha) \text{ is given by (iv)}\}\\ &= \sup\{\alpha > 0 \mid u(r;\alpha) \text{ is given by (v)}\}.\end{aligned}$$

Note that in particular this would imply that the equation (3.1) has a unique positive radial entire solution with finite total mass. This conjecture still seems to be open. As before, all the theorems and estimates above can be extended to include the equation (3.4) in $\mathbf{R}^n$, $n \geq 3$, with appropriate decaying conditions on K (see [**N1**] and [**LN**] for details), and we would like to just mention that if we compare the results above to our previous analysis on the equation of prescribed conformal scalar curvatures, we see immediately that the quadratic decay of K in (3.1) seems to be critical.

We now come to the Eddington equation (3.2) and (3.2_α); the story here is quite different. It was observed in [**NY2**] that (3.2) does not have any positive radial entire solutions. That is, for every $\alpha > 0$, the solution $u(r;\alpha)$ of (3.2_α) must have a finite zero. (This perhaps indicates that Matukuma equation is a better physical model than Eddington equation.) To understand the Eddington equation a little more, we shall discuss a generalized version of it; namely,

$$\Delta u + K(x)e^{2u} = 0 \tag{3.8}$$

in $\mathbf{R}^n$, $n \geq 3$, where $K \geq 0$. Although our analysis does cover the non-radial case, we shall limit our discussions, for simplicity, to the radial case only; that is, to the following initial value problem

$$\begin{cases} u_{rr} + \dfrac{n-1}{r}u_r + K(r)e^{2u} = 0, \\ u(0) = \alpha > 0, \quad u_r(0) = 0. \end{cases} \tag{3.8_α}$$

Again we denote the solution of (3.8_α) by $u(r;\alpha)$. It seems that the following quantity k plays an important role in our analysis

$$k \equiv \frac{1}{n-2}\int_{\mathbf{R}_+} rK(r)\,dr$$

where $\mathbf{R}_+ = (0,\infty)$. (In the non-radial case the corresponding quantity is

$$k \equiv \sup_{x\in\mathbf{R}^n}\int_{\mathbf{R}^n} G(x,y)K(y)\,dy$$

where $G(x,y)$ is the fundamental solution of Δ on $\mathbb{R}^n$.) It was proved in [**NY2**] that *if* $k = \infty$ *then* $u(r;\alpha)$ *must have a finite zero for every* $\alpha > 0$ *(and thus* (3.8_α) *does not possess any positive entire solutions); and if* $k \leq \frac{1}{2e}$, *then there exist* $0 < \alpha_1 < \alpha_2 \leq \infty$ *such that* $u(r;\alpha)$ *is a positive entire solution for every* $\alpha \in (\alpha_1, \alpha_2)$. The sufficient condition $k = \infty$ for nonexistence was later improved to $k \geq \frac{1}{2}$ by Li and Ni in [**LN**]. The situation when k falls into the range $(\frac{1}{2e}, \frac{1}{2})$ is, however, not yet entirely understood. (In $\mathbb{R}^2$, $u(r,\alpha)$ must have a finite zero for every $\alpha > 0$ as long as $K \geq 0$.) We should remark that the crucial point which causes those differences between equations (3.1) and (3.2) seems to be the fact that at $u = 0$, $e^u = 1 \neq 0 = u^p$, while the rapid growth of the nonlinearity e^u in (3.8) does not seem to be important. For instance, the above results for (3.8) can be easily generalized to include the more general equation

$$\Delta u + K(x) f(u) = 0,$$

where $f(0) > 0$, f is non-decreasing in $\mathbb{R}_+$ and $f^q(u)$ is convex for some $0 < q < 1$ (see [**NY2,LN**] for details).

§4. Scalar field equation and its generalizations.

In early 1950's several physicists considered the following equation in $\mathbb{R}^3$

$$\Delta u - u + u^3 = 0 \tag{4.1}$$

in their studies in nuclear physics. In 1961, a similar equation with a different exponent arose in Takahashi's work concerning the structure of nucleon core; namely,

$$\Delta u - u + u^2 = 0. \tag{4.2}$$

Physical considerations suggest that equation (4.1) may be expected to possess a positive radial solution in $\mathbb{R}^3$ which tends to 0 at infinity. Since then, equations of the form

$$\Delta u - u + u^p = 0 \tag{4.3}$$

in $\mathbb{R}^n$, where $p > 1$, or more generally,

$$\Delta u - u + f(u) = 0 \tag{4.4}$$

in $\mathbb{R}^n$, with appropriate hypotheses on f, have been studied extensively. Equations (4.3) and (4.4) also arise in other branches of applied mathematics, e.g. in phase transitions in the van der Waal fluid (see e.g. [**MS**]).

The first mathematical proof of existence concerning (4.3) is due to Z. Nehari [**Ne**] in 1963. He showed that in $\mathbb{R}^3$ equation (4.3) has a positive radial solution which tends to 0 at infinity provided that $1 < p \leq 4$. He also proved that in case $p = 5$, such a solution does not exist. Nehari's results may be extended to general p and n with $p < n^* \equiv (n+2)/(n-2)$ for existence and $p \geq n^*$ for nonexistence (see e.g. Berger [**B**]). In 1977, W. Strauss [**S**] studied (4.4) in connection to nonlinear Klein-Gordon equations. He considered $f(u) = \alpha u^s + \beta u^q$ with $s > 1$, $q > 1$ and $\alpha, \beta \in \mathbb{R}$, and he obtained various existence and nonexistence theorems for the problem

$$(4.5) \qquad \begin{cases} \Delta u - u + f(u) = 0 \\ u > 0 \text{ in } \mathbb{R}^n, \text{ and } u \to 0 \text{ at } \infty, \end{cases}$$

with constants α, β, s and q belonging to various intervals. However, the case $\alpha < 0 < \beta$ and $q > s > n^*$ was left open, and even though Strauss's result were later generalized by Berestycki and Lions [**BL**], this case was still left untouched. The fact that in this case (4.5) does not possess any solution was only recently established by Ni and Serrin [**NS1**] (see [**NS3**] also).

In a different direction, one could consider the following "non-autonomous" generalization of (4.3)

$$(4.6) \qquad \begin{cases} \Delta u - u + Q(x)u^p = 0 \\ u > 0 \text{ in } \mathbb{R}^n, \\ u \to 0 \text{ at } \infty, \end{cases}$$

where $Q \geq 0$ is a given "potential". Problem (4.6) itself does not seem to have any physical meaning. Nonetheless, (4.6) seems to be an easier model for other more significant and more difficult equations in entire space, such as the equation of conformally prescribed scalar curvatures. Indeed, recent works of Ding and Ni [**DN1,DN2**] showed some similarities in the methods and techniques involved. Furthermore, the solvability of (4.6) itself seems to be a mathematically interesting question. Existence results under various assumptions on Q have been obtained by several authors including C. A. Stuart [**St**], Ding and Ni [**DN1**], Y. Li [**L**], and P.-L. Lions [**L**] (who used the so-called "concentration-compactness" method). However, very simple examples show that in general one cannot hope to solve (4.6) even for bounded Q's. For instance, when $Q(x) = Q(x_1)$ where $x = (x_1, ..., x_1)$ and Q is strictly increasing in x_1, an easy argument involving integration by parts shows that (4.6) does not have any solution. (Note that any solution of (4.6) must decay exponentially at infinity.) Therefore it seems a little surprising to know that in case Q is radial and bounded, there is no obstruction for the existence of (4.6). In fact the following result was recently established by Ding and Ni [**DN1**, p. 302, Corollary 4.8].

THEOREM 4.7. *If* $1 < p < n^*$ *and*

$$Q(x) = Q(|x|) \leq C|x|^\sigma \text{ at } \infty,$$

for some $\sigma < (p-1)(n-1)/2$, *then (4.6) possesses a radial solution.*

This result seems to contain all the previous ones (e.g. [**Ne**,**B**,**BLP**]) in the radial case as far as the existence is concerned. Furthermore, it is also proved by Y. Li [**Li**] that *this result is optimal.*

As we have just mentioned that in extending this very clean result for radial potentials to non-radial ones the difficulties we encounter are genuine - similar statements for non-radial Q's are simply not true. In this connection, we should mention that the example above disproves a "theorem" announced in [**NoS**] which claims that, in particular, (4.6) is solvable for all bounded $Q \geq 0$.

In direct variational approaches (e.g. the "concentration-compactness" technique), roughly speaking, one tries to find the maximum of the functional

$$\int_{\mathbf{R}^n} Q|u|^{p+1}\,dx$$

subject to the constraint

$$\int_{\mathbf{R}^n} (|\nabla u|^2 + |u|^2)\,dx = 1.$$

We shall construct a simple non-radial potential Q for which the above constrained maximum is not assumed while the corresponding problem (4.6) still has a solution. We start with a radial potential $Q^*(r)$ which is bounded and strictly increasing in $(0, \infty)$ (e.g. take $Q^*(r) = (1+r^2)/(2+r^2)$). Then Ding and Ni's theorem above implies that there exists a radial solution v of the problem

$$\begin{cases} \Delta v - v + Q^* v^p = 0, \\ v > 0 \text{ in } \mathbf{R}^n \text{ and } v \to 0 \text{ at } \infty. \end{cases}$$

In other words,

$$Q^*(r) = \big[v(r) - \Delta v(r)\big]v^{-p}(r).$$

Now set

$$u(x) = \begin{cases} v(|x|) \text{ in } |x| \geq 1 \\ \text{a small non-radial } C^2\text{-perturbation of } v \text{ in } |x| < 1. \end{cases}$$

Since $Q^* \leq \frac{2}{3}$ in $|x| \leq 1$, we can choose the perturbation in the definition of u so small that the new potential $Q(x)$ defined by

$$Q(x) \equiv \big[u(x) - \Delta u(x)\big]u^{-p}(x)$$

satisfies that $Q(x) \leq \frac{3}{4}$ in $|x| \leq 1$. Since $Q \equiv Q^*$ outside $|x| \leq 1$, we see that $Q(x) \to 1$ as $|x| \to \infty$. Thus it is easy to see that there exists an $R_0 > 0$ such that

$$\operatorname*{Min}_{2R \leq |x| \leq 4R} Q(x) \geq \operatorname*{Max}_{|x| \leq R} Q(x)$$

for all $R \geq R_0$. (In fact, we may choose $R_0 = 1$.) Now Proposition 5.12 in [**DN1**, p. 306] implies that the constrained supremum

$$\operatorname*{Sup}_{\|u\|_{H^1(\mathbf{R}^n)}=1} \int_{\mathbf{R}^n} Q(x)|u|^{p+1}\,dx$$

is *not* attained. This example shows that the possible achievement of a direct variational approach would perhaps be rather limited.

The following alternative approach seems to be natural: we first solve the corresponding Dirichlet problem on the ball B_R with radius R centered at the origin

$$(4.6_R) \qquad \begin{cases} \Delta u - u + Q(x)u^p = 0 & \text{in } B_R, \\ u > 0 & \text{in } B_R, \\ u = 0 & \text{on } \partial B_R, \end{cases}$$

where $1 < p < n^*$. Denoting the solution we obtained (by e.g. the Mountain-Pass Lemma) by u_R and letting R tend to ∞, we then show that the sequence of solutions u_R converges to a nontrivial solution of (4.6). The crucial step in this approach is that in proving the convergence of the sequence u_R, we need to get a positive a priori lower estimate of the L^∞-norm of u_R in a fixed compact subset of $\mathbf{R}^n$, uniformly in R. Using this approach,

Ding and Ni [**DN1**] succeeded in solving (4.6) under various circumstances; for instance Theorem 4.7 stated above is obtained this way.

The uniqueness question (up to translations) concerning (4.5), even in the special case $f(u) = u^p$, $p > 1$, seems to be very hard and has received much attention since early 1960's. It was proved by [**GNN2**] that any solution of (4.5) must be radial (provided that $f(u) = O(u^p)$ near $u = 0$ for some $p > 1$). Therefore we only have to deal with the corresponding ordinary differential equation of (4.5). In 1972, C. V. Coffman [**Co**], using the method of comparison identities, proved that (4.5) has a unique radial solution if $f(u) = u^3$ and $n = 3$. The following extension was recently obtained in [**MS**]: *for each* $n < 8$, *there exists* $1 < p_0(n) < n^*$, *with* $p_0(n) = \frac{n}{n-2}$ *if* $n \leq 4$ *and* $p_0(n) = \frac{8}{n}$ *if* $4 < n < 8$, *such that if* $f(u) = u^p$ *and* $p < p_0(n)$ *then (4.5) has a unique radial solution.* Note that there is a gap between the ranges of existence and uniqueness even in the simplest case $f(u) = u^p$.

§5. Remarks.

A number of open problems concerning equations (1)–(5) have been mentioned in §1–§4 already. In this section, I shall compile a short list of comments and open problems concerning equations related to those we have discussed in previous sections.

(I) Concerning radial symmetry of positive solutions of elliptic equations on $\mathbf{R}^n$, one may ask the following question:

(i) Is it true that *all positive* solutions of the problem

$$(5.1) \qquad \begin{cases} \Delta u + f(u) = 0 \text{ in } \mathrm{R}^n, \\ u \to 0 \text{ at } \infty, \end{cases}$$

where f is C^1, are necessarily radial?

If we drop the positivity requirement on the solutions of (5.1), then, it was proved by W.-Y. Ding [**D**] that the problem

$$(5.2) \qquad \begin{cases} \Delta u + |u|^{n^*-1} u = 0 \text{ in } , \\ u \to 0 \text{ at } \infty \end{cases}$$

has infinitely many distinct solutions which change sign on $\mathbb{R}^n$. Moreover, none of those solutions can be radial. Nonetheless, the following question remains open:

(ii) Does there exist a solution (*not* positive) of the problem

$$(5.3) \qquad \begin{cases} \Delta u - u + |u|^{p-1}u = 0 \text{ in } \mathbb{R}^n, & p > 1, \\ u \to 0 \text{ at } \infty, \end{cases}$$

which is not radially symmetric?

(II) Due to the homogeneity of the nonlinearity, at least the radial solutions of the classical Lane-Emden equations are well-understood. The following equation is an immediate generalization:

$$(5.4) \qquad \Delta u + |u|^{p-1}u + |u|^{q-1}u = 0 \text{ in } \mathbb{R}^n,$$

where $1 < p < q$. However, the structure of the set of all radial solutions of (5.4) is still not clear. For instance, does (5.4) have a *positive* entire solution if $\frac{n}{n-2} < p < n^* < q$? Partial answers have been obtained in [**LiN**]. In an "easier" range $\frac{n}{n-2} < p < q < n^*$, it is known [**NS2**] that (5.4) does not have any positive radial solutions on $\mathbb{R}^n$; but, is it true that (5.4) possesses no positive entire solutions (not just radial ones) on $\mathbb{R}^n$ at all?

(III) In [**PS**], Peletier and Serrin established the uniqueness of the problem

$$(5.5) \qquad \begin{cases} \Delta u + f(u) = 0 & \text{in } \mathbb{R}^n, \\ u > 0 & \text{in } \mathbb{R}^n, \\ u \to 0 & \text{at } \infty. \end{cases}$$

where $f(u)/(u - \beta)$ is non-increasing in (β, ∞) with $\beta \equiv \inf D$ and $D = \{u > 0 \mid \int_0^u f(t)dt > 0\}$. Recently, this work was generalized by Kaper and Kwong [**KK**]. (See the last paragraph of §4 for other works and open problems concerning uniqueness of positive entire solutions.)

(IV) *Singular* solutions of the problem

$$\begin{cases} \Delta u + f(u) = 0 & \text{in } \mathbb{R}^n \backslash \{0\}, \\ u > 0 & \text{in } \mathbb{R}^n \backslash \{0\}, \\ u \to 0 & \text{at } \infty, \end{cases}$$

have been discussed in [**NS3**] for nonlinear terms $-u+u^p$ and others. In a recent paper [**S2**], Schoen discussed the existence of solutions of $\Delta u+u^{n^*}=0$ with singularities prescribed at given points and sets.

(**V**) The prescribed mean curvature equation

$$\begin{cases} \operatorname{div}\left(\dfrac{Du}{\sqrt{1+|Du|^2}}\right)+f(u)=0 \text{ in } \mathbb{R}^n, \\ u>0 \text{ in } \mathbb{R}^n, \\ u\to 0 \text{ at } \infty, \end{cases}$$

and the generalized Laplace equation

$$\begin{cases} \operatorname{div}\left(|Du|^{m-2}Du\right)+f(u)=0 \text{ in } \mathbb{R}^n, \\ u>0 \text{ in } \mathbb{R}^n, \\ u\to 0 \text{ at } \infty, \end{cases}$$

where $1<m<n$ and $f(u)\sim u^p$, or $-u+u^p$, $p>1$, have also been considered in a series of papers of Atkinson, Franchi, Lanconnelli, Ni, Peletier, Serrin, and others.

This research supported in party by NSF Grant DMS-8601246

School of Mathematics, University of Minnesota, Minneapolis MN 55455

REFERENCES

[A] L. V. Ahlfors, *An extension of Schwarz's lemma*, Trans. Amer. Math. Soc. **43** (1938), 359–364.

[AB] L. V. Ahlfors and L. Bers, *Riemann's mapping theorem for variable metrics*, Ann. Math. **72** (1960), 385–404.

[AK] K. Ako and T. Kusano, *On bounded solutions of second order elliptic differential equations*, J. Fac. Sci. Univ. Tokyo (Sect. I) **11** (1964), 29–37.

[AM] P. Aviles and R. McOwen, *Conformal deformations of complete manifolds with negative curvature*, J. Diff. Geom. **21** (1985), 269–281.

[BFH] J. Batt, W. Faltenbacher, and E. Horst, *Stationary spherically symmetric models in stellar dynamics*, Arch. Rat. Mech. Anal. **93** (1986), 159–183.

[BL] H. Berestycki and P. L. Lions, *Nonlinear scalar field equations I*, Arch. Rat. Mech. Anal. **82** (1983), 313–345.

[BLP] H. Berestycki, P. L. Lions and L. A. Peletier, *An ODE approach to the existence of positive solutions for semilinear problems in* $\mathbb{R}^n$, Indiana Univ. Math. J. **30** (1981), 141–157.

[B] M. S. Berger, *On the existence and structure of stationary states for a nonlinear Klein-Gordon equation*, J. Funct. Anal. **9** (1972), 249–261.

[C] E. Calabi, *An extension of E. Hopf maximum principle with an application to Riemannian geometry*, Duke Math. J. **25** (1958), 45–56.

[CCY] C.-Y. Chen, K.-S. Cheng, and W.-N. Yu, *Conformal deformations of metrics on* $H^n(-1)$ *with prescribed scalar curvatures*, preprint.

[CL] K.-S. Cheng and J. T. Lin, *On the elliptic equations* $\Delta u = K(x)u^\sigma$ *and* $\Delta u = K(x)e^{2u}$, Trans. Amer. Math. Soc., to appear.

[CTY] K.-S. Cheng, F.-S. P. Tsen, and W.-N. Yu, *Conformal deformations of metrics on Poincare disk*, preprint.

[Co] C. V. Coffman, *Uniqueness of the ground state solution for* $\Delta u - u + u^3 = 0$ *and a variational characterization of other solutions*, Arch. Rat. Mech. Anal. **46** (1972), 81–95.

[D] W.-Y. Ding, *On a conformally invariant elliptic equation on* $\mathbb{R}^n$, Comm. Math. Phys. **107** (1986), 331–335.

[DN1] W.-Y. Ding and W.-M. Ni, *On the existence of positive entire solutions of a semilinear elliptic equation*, Arch. Rat. Mech. Anal. **91** (1986), 283–308.

[DN2] W.-Y. Ding and W.-M. Ni, *On the elliptic equation* $\Delta u + Ku^{(n+2)/(n-2)} = 0$ *and related topics*, Duke Math. J. **52** (1985), 485–506.

[E] A. S. Eddington, *The dynamics of a globular stellar system*, Monthly Notices of the Royal Astronomical Society **75** (1915), 366–376.

[F] A. Friedman, *Bounded entire solutions of elliptic equations*, Pacific J. Math. **44** (1973), 497–507.

[GNN1] B. Gidas, W.-M. Ni and L. Nirenberg, *Symmetry and related properties via the maximum principle*, Comm. Math. Phys. **68** (1979), 209–243.

[GNN2] B. Gidas, W.-M. Ni and L. Nirenberg, *Symmetry of positive solutions of nonlinear elliptic equations in* $\mathbb{R}^n$, Advances in Math., Supplementary Studies **7A** (1981), 369–402.

[KK] H. G. Kaper and M.-K. Kwong, *Uniqueness of nonnegative solutions of a class of semilinear elliptic equations*, preprint.

[Ka] N. Kawano, *On bounded entire solutions of semilinear elliptic equations*, Hiroshima Math. J. **14** (1984), 125–158.

[K] J. Kazdan, *Prescribing the Curvature of a Riemannian Manifold*, NSF-CBMS Regional Conference Lecture Notes **57** (1985).

[KN1] C. E. Kenig and W.-M. Ni, *An exterior Dirichlet problem with applications to some nonlinear equations arising in geometry*, Amer. J. Math. **106** (1984), 689–702.

[**KN2**] C. E. Kenig and W.-M. Ni, *On the elliptic equation* $Lu - k + Ke^{2u} = 0$, Ann. Scuo. Norm. Sup. Pisa (Series IV) **12** (1985), 191–224.

[**KNa**] T. Kusano and M. Naito, *Oscillation theory of entire solutions of second order superlinear elliptic equations*, Funkcial Ekvac., to appear.

[**Li**] Y. Li, *Remarks on a semilinear elliptic equation in* $\mathbf{R}^n$, to appear in J. Diff. Eqns.

[**LN**] Y. Li and W.-M. Ni, *in preparation.*

[**L**] P. L. Lions, *The concentration-compactness principle in the calculus of variations, the limit case*, Revista Mat. Iberoamericana **1** (1985), 145-201; *II*, 45–121.

[**LL**] C.-S. Lin and S.-S. Lin, in preparation.

[**LiN**] C.-S. Lin and W.-M. Ni, *A counterexample to the Nodal Domain Conjecture and a related semilinear equation*, Proc. Amer. Math. Soc., to appear..

[**Ln**] F.-H. Lin, *On the elliptic equation* $D_i[A_{ij}(x)D_jU] - k(x)U + K(x)U^p = 0$, Proc. Amer. Math. Soc. **95** (1985), 219–226.

[**LnN**] F.-H. Lin and W.-M. Ni, *On the least growth of harmonic functions and the boundary behaviour of Riemann mappings*, Comm. PDE **10** (1985), 767–786.

[**LSW**] W. Littman, G. Stampacchia, and H. Weinberger, *Regular points for elliptic equations with discontinuous coefficients*, Ann. Scuo. Norm. Sup. Pisa (Series III) **17** (1963), 45–79.

[**LoN**] C. Loewner and L. Nirenberg, *Partial differential equations invariant under conformal and projective transformations*, in "Contributions to analysis," Academic Press, 1974, pp. 245–272.

[**MS**] K. McLeod and J. Serrin, *Uniqueness of positive radial solutions of* $\Delta u + f(u) = 0$ *in* $\mathbf{R}^n$, Arch. Rat. Mech. Anal., to appear.

[**MNS**] McLeod, J.B., W.-M. Ni, and J. Serrin, *in preparation.*

[**M1**] R. McOwen, *The behavior of the Laplacian on weighted Sobolev spaces*, Comm. Pure Appl. Math. **32** (1979), 783–795.

[**M2**] R. McOwen, *On the equation* $\Delta u + Ke^{2u} = f$ *and prescribed negative curvature in* $\mathbf{R}^2$, J. Math. Anal. Appl. **103** (1984), 365–370.

[**M3**] R. McOwen, *Conformal metric in* $\mathbf{R}^2$ *with prescribed Gaussian curvature and positive total curvature*, Indiana Univ. Math. J. **34** (1985), 97–104.

[**Mo**] J. Moser, *On a nonlinear problem in differential geometry*, in "Dynamical systems," M. Peixoto, ed., Academic Press, New York, 1973.

[**Na**] M. Naito, *A note on bounded positive entire solutions of semilinear elliptic equations*, Hiroshima Math. J. **14** (1984), 211–214.

[**Ne**] Z. Nehari, *On a nonlinear differential equation arising in nuclear physics*, Proc. Royal Irish Acad. **62** (1963), 117–135.

[**N1**] W.-M. Ni, *On the elliptic equation* $\Delta u + K(x)u^{(n+2)/(n-2)} = 0$, *its generalization and applications in geometry*, Indiana Univ. Math. J. **31** (1982), 493–529.

[**N2**] W.-M. Ni, *On the elliptic equation* $\Delta u + Ke^{2u} = 0$ *and conformal metrics with prescribed Gaussian curvatures*, Invent. Math. **66** (1982), 343–352.

[**NS1**] W.-M. Ni and J. Serrin, *Nonexistence theorems for quasilinear partial differential equations*, Rend. Circolo Mat. Palermo. (Centenary Supplement), Series II **8** (1985), 171–185.

[**NS2**] W.-M. Ni and J. Serrin, *Existence and nonexistence theorems for ground states for quasilinear partial differential equations*, Accad. Naz. dei Lincei **77** (1986), 231–257.

[**NS3**] W.-M. Ni and J. Serrin, *Nonexistence theorems for singular solutions of quasilinear partial differential equations*, Comm. Pure Appl. Math. **39** (1986), 379–399.

[**NY1**] W.-M. Ni and S. Yotsutani, *On Matukuma's equation and related topics*, Proc. Japan Acad. (Series A) **62** (1986), 260–263.

[**NY2**] W.-M. Ni and S. Yotsutani, *Semilinear elliptic equations of Matukuma-type and related topics*, Japan J. Appl. Math., to appear.

[NW] L. Nirenberg and H. F. Walker, *The null spaces of elliptic partial differential operators in* $\mathbb{R}^n$, J. Math. Anal. Appl. **42** (1973), 271–301.

[NoS] E. S. Noussair and C. A. Swanson, *Existence theorems for generalized Klein-Gordon equations*, Bull. Amer. Math. Soc. (New Series) **8** (1983), 333–336.

[Ol] O. A. Oleinik, *On the equation* $\Delta u + k(x)e^u = 0$, Russian Math. Surveys **33** (1978), 243–244.

[O] R. Osserman, *On the inequality* $\Delta u \geq f(u)$, Pacific J. Math. **7** (1957), 1641–1647.

[PS] L. A. Peletier and J. Serrin, *Uniqueness of positive solutions of semilinear equations in* $\mathbb{R}^n$, Arch. Rat. Mech. Anal. **81** (1983), 181–197.

[R] P. Rabinowitz, *Some aspects of nonlinear eigenvalue problems*, Rocky Mountain J. Math. **3** (1973), 161–202.

[Sa] D. H. Sattinger, *Conformal metrics in* $\mathbb{R}^2$ *with prescribed curvature*, Indiana Univ. Math. J. **22** (1972), 1–4.

[S1] R. Schoen, *Conformal deformation of a Riemannian metric to constant scalar curvature*, J. Diff. Geom. **20** (1984), 479–495.

[S2] R. Schoen, *The existence of weak solutions with prescribed singular behavior for a conformally invariant scalar equation*, preprint.

[S] W. A. Strauss, *Existence of solitary waves in higher dimensions*, Comm. Math. Phys. **55** (1977), 149–162.

[St] C. A. Stuart, *Bifurcation for Dirichlet problems without eigenvalues*, Proc. London Math. Soc. (3) **45** (1982), 169–192.

[W] H. Wittich, *Ganze Losungen der Differentialgleichung* $\Delta u = e^u$, Math. Z. **49** (1944), 579–582.

Global Existence Results for a Strongly Coupled Quasilinear Parabolic System

M.A. POZIO AND A. TESEI

1. The Problem.

We are interested in existence, uniqueness and *a priori* estimates of nonnegative solutions to the following problem:

$$(1.0)\qquad \begin{aligned} &u_t = \Delta[(1+\alpha v)u] + uf(u,v), \\ &v_t = \Delta v + vg(u,v) && \text{in } (0,\infty)\times\Omega, \\ &u = v = 0 && \text{in } (0,\infty)\times\partial\Omega, \\ &u(0) = u_0, v(0) = v_0 && \text{in } \Omega. \end{aligned}$$

Here $\Omega \subset \mathbb{R}^n$ is an open bounded set with smooth boundary $\partial\Omega$; $\alpha > 0$ is a parameter, u_0 and v_0 are given nonnegative functions and f, g satisfy the following assumptions:

$$(H)\qquad \begin{aligned} &\text{(i)} \quad f,g \in C^1; \quad f_u, f_v, g_u, g_v \text{ are locally Lipschitz continuous and negative;} \\ &\text{(ii)} \quad f(u,0) \to -\infty \text{ for } u \to +\infty, \quad g(0,v) \to -\infty \text{ for } v \to +\infty; \\ &\text{(iii)} \quad f(0,0) > 0, \quad g(0,0) > 0. \end{aligned}$$

Using the unknown $w := (1+\alpha v)u$, the problem can be rewritten as follows:

$$(1.1)\qquad \begin{aligned} &v_t = \Delta v + vG(v,w), \\ &w_t = (1+\alpha v)\Delta w + \frac{\alpha w}{1+\alpha v}\Delta v + wH(v,w), && \text{in } (0,\infty)\times\Omega, \\ &v = w = 0 && \text{in } (0,\infty)\times\partial\Omega, \\ &v(0) = v_0, w(0) = w_0 := (1+\alpha v_0)u_0 && \text{in } \Omega, \end{aligned}$$

where $G(v,w) := g\left(\frac{w}{1+\alpha v}, v\right)$ and

$$H(v,w) := f\left(\frac{w}{1+\alpha v}, v\right) + \frac{\alpha v}{1+\alpha v} g\left(\frac{w}{1+\alpha v}, v\right).$$

The above problem is suggested by a class of competition models in population dynamics (which also motivates assumption (H)) [**SKT**]. Existence results for the case $n = 1$ and Neumann homogeneous boundary conditions are quoted in [**MaMi**]. Related results, concerning existence and attractivity of stationary solutions, can be found in [**MaMi, Mi, ATW**].

Our interest in (1.1) is motivated here by its strongly coupling character. The existence results we shall prove hold in fact for a class of strongly coupled parabolic quasilinear systems. Local existence results for systems of this kind were proved in [**Am**].

2. Statement of the results.

Let us consider the Hölder space $C^{k+\sigma}(\bar{\Omega})$ (k integer, $\sigma \in (0,1)$) endowed with the usual norm, namely

$$u \to |u|_{k+\sigma} := |u|_k + \sum_{|\alpha|=k} \sup_{x,y\in\bar{\Omega},x\neq y} \frac{|D^\alpha u(x) - D^\alpha u(y)|}{|x-y|^\sigma}$$

(where $\alpha = (\alpha_1, \ldots, \alpha_n)$ is any multi-index). We shall denote by $h^{k+\sigma}(\bar{\Omega})$ the subspace of $C^{k+\sigma}(\bar{\Omega})$ consisting of all functions whose derivatives of k-th order satisfy the following condition:

$$\lim_{\tau\to 0} \sup_{x,y\in\bar{\Omega},|x-y|\leq\tau} \frac{|D^\alpha u(x) - D^\alpha u(y)|}{\tau^\sigma} = 0 \qquad (|\alpha| = k).$$

The so-called little-Hölder space $h^{k+\sigma}(\bar{\Omega})$ is clearly a Banach space under the $|.|_{k+\sigma}$-norm. We also denote by $h_0^\sigma(\bar{\Omega})$ the Banach space of $u \in h^\sigma(\bar{\Omega})$ such that $u = 0$ on $\partial\Omega$ and by $h_0^{2+\sigma}(\bar{\Omega})$ the Banach space of $u \in h^{2+\sigma}(\bar{\Omega})$ such that $u = \Delta u = 0$ on $\partial\Omega$.

Concerning local existence and uniqueness of solutions to (1.1), the following result can be proved.

THEOREM 2.1. *Let (H) be satisfied. For any nonnegative* $v_0 \in h_0^{2+\sigma}(\bar{\Omega})$ *and* $w_0 \in h_0^\sigma(\bar{\Omega})$ *there exists a unique nonnegative local solution* (v, w) *of (1.1); here* $v \in C(0,T; h_0^{2+\sigma}(\bar{\Omega})) \cap C^1(0,T; h_0^\sigma(\bar{\Omega}))$ *and* $w \in C(0,T; h_0^\sigma(\bar{\Omega}))$ $(T = T(v_0, w_0) > 0; \sigma \in (0,2), \sigma \neq 1)$.

The proof makes use of fixed point techniques in the following way. For any $\bar{w}$ in some closed set S_T consider the problem

$$
(2.1)\qquad \begin{aligned}
&v_t = \Delta v + vG(v,\bar{w}) && \text{in } (0,T)\times\Omega,\\
&v = 0 && \text{in } (0,T)\times\partial\Omega,\\
&v(0) = v_0 && \text{in } \Omega.
\end{aligned}
$$

Let there exist a unique solution $\bar{v}$ of (2.1). Consider again the problem

$$
(2.2)\qquad \begin{aligned}
&w_t = (1+\alpha\bar{v})\Delta w + \bar{w}\left(\frac{\alpha\Delta\bar{v}}{1+\alpha\bar{v}} + H(\bar{v},\bar{w})\right) && \text{in } (0,T)\times\Omega,\\
&w = 0 && \text{in } (0,T)\times\partial\Omega,\\
&w(0) = w_0 && \text{in } \Omega.
\end{aligned}
$$

Suppose there exists a unique solution $\hat{w}$ of (2.2); define the map B : $\bar{w} \to \hat{w}$. Under a suitable choice of the function spaces which v_0, w_0 belong to, we can prove that the map B

(i) is well-defined and maps S_T onto itself,

(ii) is a contraction in S_T,

provided that T is suitably small.

Denote by w the unique fixed point of B in S_T, call v the unique solution of (2.1) with $\bar{w} = w$. Clearly, (v, w) is the unique solution of (1.1). It follows easily from (1.1) and the maximum principle that $v_0 \geq 0$, $w_0 \geq 0$ entail $v(t) \geq 0$, $w(t) \geq 0$ for any $t \in [0,T]$.

Concerning the global existence of solutions to problem (1.1), the following can be proved.

THEOREM 2.2. *Let (H) be satisfied. For any nonnegative* $v_0 \in h_0^{2+\sigma}(\bar{\Omega})$ *and* $w_0 \in h_0^{\sigma}(\bar{\Omega})$ *there exists* $\bar{\alpha} = \bar{\alpha}(v_0, w_0) > 0$ *with the following property: if* $\alpha \in (0, \bar{\alpha})$, *then the (unique, nonnegative) solution of (1.1) is global (namely,* $T = +\infty$*).*

Let us mention that the *a priori* estimates underlying Theorem 2.2 are uniform in $(0, +\infty)$.

3. Mathematical Framework.

Our approach to problem (1.1) makes use of methods for abstract parabolic problems — namely, for problems like

$$v'(t) = Av(t) + f(t) \quad (t > 0), \qquad (3.1)$$
$$v(0) = 0.$$

Here $A : D_A \subset E \to E$ is the infinitesimal generator of a strongly continuous analytic semigroup e^{At} in some Banach space E and f is continuous with values in E.

The solution of (3.1) is given by

$$v(t) = (e^A * f)(t) = \int_0^t ds\; e^{A(t-s)} f(s) \quad (t \geq 0). \qquad (3.2)$$

However, it is well-known [**DPG1, Lu**] that the convolution with the semigroup e^{At} does not take continuous functions with values in E into continuous functions with values in D_A. In other words, no "maximal regularity" result holds for problem (3.1).

Since maximal regularity is needed in several steps of our procedure, we find it convenient to work in continuous interpolation spaces intermediate between D_A and E, where such a maximal regularity result is known to hold [**DPG1**]. Let us therefore introduce such spaces.

Let E be a Banach space with norm $|.|_E$. For any $T > 0$ we denote by $C^k(0,T;E)$ the space of k times continuously differentiable functions from $[0,T]$ to E $(k \geq 0)$, endowed with the usual norm; we set $C(0,T;E) := C^0(0,T;E)$.

Let $A : D_A \subset E \to E$ be the infinitesimal generator of a strongly continuous analytic semigroup e^{At} in E. Observe that thus, in particular, D_A is dense in E; however, this is not necessary for several further developments (see [**Si**] for details).

We can now define a family of intermediate spaces between D_A (thought of as endowed with the graph norm) and E. Such spaces proved to be useful in the theory of abstract parabolic equations (see e.g. [**DPG1, DPSi, Lu**]). They can also be regarded as real interpolation spaces [**DPG1**], but such characterization is not needed for the present purposes.

DEFINITION 3.1. *For each $\theta \in (0,1)$ we define the Banach space*

$$D_A(\theta) := \left\{ u \in E \text{ s.t. } \lim_{t \to 0} \left| t^{1-\theta} A e^{At} u \right|_E = 0 \right\}$$

endowed with the norm

$$|u|_{D_A(\theta)} := |u|_E + \sup_{t>0} \left|t^{1-\theta} A e^{At} u\right|_E .$$

Since A generates an analytic semigroup, for any $k = 0, 1, 2 \ldots$ there is $M_k > 0$ such that (see e.g. [**Yo**])

$$\left\|t^k A^k e^{At}\right\|_{\ell(E)} \leq M_k \quad (t > 0);$$

here $\ell(E)$ denotes the Banach algebra of continuous linear operators in E, with norm $\|.\|_{\ell(E)}$. Hence we have the estimate

$$\left|t^{1-\theta} A e^{At} u\right|_E \leq \frac{M_1 \, |u|_E}{t^\theta},$$

which shows that the map $t \to t^{1-\theta} A e^{At} u$ is bounded in $[a, +\infty)$ for any $a > 0$. It follows that

$$\sup_{t>0} \left|t^{1-\theta} A e^{At} u\right|_E < +\infty$$

for any $u \in D_A(\theta)$, so that the above definition of $|.|_{D_A(\theta)}$ makes sense.

Let us also consider the following family of intermediate spaces between D_{A^2} and D_A.

DEFINITION 3.2. *For any $\theta \in (0,1)$ we define the Banach space*

$$D_A(\theta + 1) := \{u \in D_A \text{ s.t. } Au \in D_A(\theta)\}$$

endowed with the norm

$$|u|_{D_A(\theta+1)} := |u|_E + |Au|_{D_A(\theta)} .$$

We shall also consider the following spaces:

$$C^k(0, T; D_A(\theta)), \quad C^k(0, T; D_A(\theta + 1)) \quad (k \geq 0; \theta \in (0,1)),$$

endowed with the natural norms.

Many properties of the spaces $D_A(\theta)$, $D_A(\theta + 1)$ are discussed at length, and proved, in [**Si**]. In connection with what was said before, let us only mention the following maximal regularity result.

THEOREM 3.1. *Let $\theta \in (0,1)$. If $f \in C(0,T;D_A(\theta))$, then $(e^A * f)'$, $A(e^A * f) \in C(0,T;D_A(\theta))$. Moreover, there is $c = c(\theta,T) > 0$ such that the following inequalities hold:*

$$|(e^A * f)'|_{C(0,T;D_A(\theta))} \leq (c+1)\,|f|_{C(0,T;D_A(\theta))}; \tag{3.3}$$

$$|(e^A * f)|_{C(0,T;D_A(\theta))} \leq c\,|f|_{C(0,T;D_A(\theta))}. \tag{3.4}$$

Throughout the following we shall make the choice $E = C_0(\bar{\Omega})$, where $C_0(\bar{\Omega})$ denotes the Banach space of continuous functions on $\bar{\Omega}$ which vanish on $\partial\Omega$, endowed with the maximum norm.

Define the operator A in E as follows:

$$\begin{aligned} D_A &:= \{u \in C_0(\bar{\Omega}) \text{ s.t. } u \in W^{2,q}(\Omega) \\ &\qquad\qquad \text{for some } q > n \text{ and } \Delta u \in C_0(\bar{\Omega})\}, \\ Au &:= \Delta u \quad (u \in D_A). \end{aligned} \tag{3.5}$$

It was proved in [**St**] that A generates a strongly continuous analytic semigroup in E. The intermediate spaces $D_A(\theta)$, $D_A(\theta+1)$ were given in [**Lu**] the following concrete characterization, of interest in the present case.

THEOREM 3.2. *Let $\Omega \subset \mathbf{R}^n$ be an open bounded set with boundary $\partial\Omega$ of class C^∞. Let A be a uniformly elliptic operator in $C_0(\bar{\Omega})$ defined as follows:*

$$D_A := \{u \in C_0(\bar{\Omega}) \text{ s.t. } Au \in C_0(\bar{\Omega})\},$$

$$(Au)(x) := \sum_{i=1}^{n} a_{ij}(x)(\partial^2_{ij}u)(x) + \sum_{i=1}^{n} b_i(x)(\partial_i u)(x) + c(x)u(x) \quad (x \in \Omega),$$

where $a_{ij}, b_i, c \in C(\bar{\Omega})$ $(i,j = 1,\dots,n)$. Then $D_A(\theta)$ is isomorphic to the Banach space $h^{2\theta}_0(\bar{\Omega})$ of $f \in h^{2\theta}(\bar{\Omega})$ which vanish on $\partial\Omega$ $(\theta \in (0,1), \theta \neq 1/2)$. If moreover $a_{ij}, b_i, c \in h^{2\theta}(\bar{\Omega})$ and $\partial\Omega$ is of class $h^{2\theta+2}$, $D_A(\theta+1)$ is isomorphic to the Banach space $h^{2\theta+2}_0(\bar{\Omega})$ of $f \in h^{2\theta+2}(\bar{\Omega})$ such that both f and Af vanish on $\partial\Omega$ $(\theta \in (0,1), \theta \neq 1/2)$.

It should now be clear which rôle the little-Hölder spaces play in our approach. Let us just mention that the statements of Theorems 2.1 and 2.2 remain true in v_0, w_0 are taken in little-Hölder spaces of higher index; this is clear enough from Theorem 3.2 and the proofs sketched in the following section.

4. Proofs.

Let us sketch the proof of Theorem 2.1; the proof of Theorem 2.2 makes use of similar techniques (see [**PT**]).

Consider the closed set

$$S_T = S_T(\theta) := \Big\{ w \in C(0,T; D_A(\theta)) \text{ s.t. } w(0) = w_0, w(t) \geq 0 \quad \forall t \in [0,T], \quad |w|_{C(0,T;D_A(\theta))} \leq W_\theta \Big\};$$

here $T, W_\theta > 0$ will be fixed in the following.

According to the ideas outlined in Section 3, let us rewrite (2.1) as an abstract problem in E:

$$(4.1) \qquad \begin{aligned} &v_t = Av + vG(v, \bar{w}) \text{ in } (0,T), \\ &v(0) = v_0. \end{aligned}$$

The operator A was defined in (3.5) and

$$\big[v(t)G(v,\bar{w})(t)\big](x) := v(t,x)G(v(t,x), \bar{w}(t,x)) \quad (t \in (0,T); x \in \Omega),$$

where $\bar{w} \in S_T$. The initial datum v_0 is taken in the intermediate space $D_A(\theta + 1)(\theta \in (0,1))$, which was characterized in Theorem 3.2 for the present situation. Then the existence of a unique local solution

$$\bar{v} \in C(0,T; D_A(\theta+1)) \cap C^1(0,T; D_A(\theta))$$

of (4.1) follows by [**DPG1**].

Similarly, let us think of (2.2) as an abstract problem in E of the following form:

$$(4.2) \qquad \begin{aligned} &w_t = A_0 w + M(t)w + \bar{w}H_1(\bar{v}, \bar{w}) \text{ in } (0,T), \\ &w(0) = w_0. \end{aligned}$$

Here the operator A_0 and the family $\{M(t)\}$ of operators in E are defined as follows:

$$\begin{aligned} &D_{A_0} := D_A, \\ &A_0 u := (1 + \alpha v_0)\Delta u \quad (u \in D_A); \\ &D(M(t)) := D_A, \\ &M(t)w := \alpha(\bar{v}(t) - v_0)\Delta w \quad (w \in D_A; t \in (0,T)). \end{aligned}$$

Moreover,

$$[\bar{w}(t)H_1(\bar{v},\bar{w})(t)](x) := \bar{w}(t,x)\left(\frac{\alpha\Delta\bar{v}}{1+\alpha\bar{v}} + H(\bar{v},\bar{w})\right)(t,x)$$
$$(t \in (0,T); x \in \Omega).$$

As pointed out in the definition of S_T, the initial datum w_0 is taken in the intermediate space $D_A(\theta)$ $(\theta \in (0,1))$. Then the existence of a unique local solution $\hat{w} \in C(0,T; D_A(\theta))$ of (4.2) follows by [**DPG2**].

Thus the map $\mathsf{B} : \bar{w} \to \hat{w}$ is well-defined. In order to prove that B is a contraction of S_T onto itself (modulo a proper choice of T and W_θ) we need *a priori* estimates on $\bar{v}$ and $\hat{w}$. This in turn requires representation formulas for such quantities.

Concerning $\bar{v}$ we can write

$$\bar{v}(t) = e^{At}v_0 + \int_0^t ds\, e^{A(t-s)}\bar{v}(s)G(\bar{v},\bar{w})(s) \quad (t \in [0,T]). \tag{4.3}$$

From (4.3) we get by a bootstrap argument *a priori* estimates for $\bar{v}$ in more and more regular spaces. We end up with the estimate

$$|\bar{v}|_{C(0,T;D_A(\theta+1))} \leq V_{\theta+1} \quad (\theta \in (0,1)), \tag{4.4}$$

with a suitable constant $V_{\theta+1} > 0$.

We can now prove the following continuity properties (relevant to deal with the family $\{M(t)\}$):

$$(4.5_1)\quad |\bar{v}(t) - v_0|_{D_A(\theta)} \to 0 \quad \text{for } t \to 0 \quad (\theta \in (0,1));$$
$$(4.5_2)\quad |\bar{v}(t) - v_0|_{D_A(\theta'+1)} \to 0 \quad \text{for } t \to 0 \quad (\theta' \in (0,\theta); \theta \in (0,1)).$$

In order to get a representation for $\hat{w}$, consider the operator Q defined as follows:

$$h \to Qh, (Qh)(t) := -M(t)(e^A * h)(t) \quad (t \in [0,T]).$$

Due to maximal regularity results (see Theorem 3.1), Q maps $C(0,T; D_A(\theta))$ into itself. As is easily checked, were $(I+Q)$ invertible, we could write $\hat{w}$ as follows:

$$\hat{w}(t) = e^{A_0 t}w_0 + e^{A_0} * \left\{(I+Q)^{-1}\left[\bar{w}H_1(\bar{v},\bar{w}) + M(\cdot)e^{A_0\cdot}w_0\right]\right\}(t) \tag{4.6}$$

for $t \in [0,T]$ (here I denotes the identity operator).

Due to (4.5_1), $(I+Q)$ is shown to be invertible in $C(0,T;D_A(\theta))$ for any T small enough. Then (4.6) is given a sound meaning (possibly taking T even smaller) using (4.5_2) and the extrapolation methods in [**DPG2**]. From (4.6), (4.4) and a proper choice of W_θ it follows that $\mathbb{B}$ maps S_T into itself.

The contractivity of $\mathbb{B}$ is proved along the same path. In particular, the proof makes use of

a) *a priori* estimate of $\bar{v}$ in the supremum norm, uniformly in $(0,+\infty)$ (this follows easily from assumption (H));

b) regularizing properties of the semigroup e^{At}. Then the conclusion follows as in [**ScT**, Theorem 1].

Dipartimento di Matematica, II Universitá di Roma
Via O. Raimondo, Roma 00173, ITALY

Research at MSRI supported in part by NSF Grant DMS 812079-05.

REFERENCES

[**Am**] H. Amann, *Quasilinear evolution equations and parabolic systems*, Trans. Amer. Math. Soc. **293** (1986), 191–227.

[**ATW**] D.G. Aronson, A. Tesei and H. Weinberger, *A density dependent diffusion system with stable discontinuous stationary solutions*, Quaderni dell'Istituto per le Applicazioni del Calcolo "M. Picone", CNR, Roma (1987).

[**DPG1**] G. Da Prato and P. Grisvard, *Equations d'évolution abstraites nonlinéaires de type parabolique*, Ann. Mat. Pura Appl. **120** (1979), 329–396.

[**DPG2**] G. Da Prato and P. Grisvard, *Maximal regularity for evolution equations by interpolation and extrapolation*, J. Funct. Anal. **58** (1984), 107–124.

[**DPSi**] G. Da Prato and E. Sinestrari, *Hölder regularity for non-autonomous abstract parabolic equations*, Israel J. Math. **42** (1982), 1–19.

[**Lu**] A. Lunardi, *Interpolation spaces between domains of elliptic operators and spaces of continuous functions with applications to nonlinear parabolic equations*, Math. Nachr. **121** (1985), 295–318.

[**MaMi**] H. Matano and M. Mimura, *Pattern formation in competition-diffusion systems in nonconvex domains*, Publ. RIMS, Kyoto Univ. **19** (1983), 1049–1079.

[**Mi**] M. Mimura, *Stationary problems of some density-dependent diffusion systems with competitive dynamics*, Hiroshima Math. J. **11** (1981), 621–635.

[**PT**] M.A. Pozio and A. Tesei, forthcoming.

[**ScT**] A. Schiaffino and A. Tesei, *Linearized stability results in continuous interpolation spaces*, Ann. Scuola Normale Superiore-Pisa, Serie IV **11** (1984), 281–301.

[**Si**] E. Sinestrari, *On the abstract Cauchy problem of parabolic type in spaces of continuous functions*, J. Math. Anal. Appl. **107** (1985), 16–66.

[**SKT**] N. Shigesada, K. Kawasaki and E. Teramoto, *Spatial segregation of interacting species*, J. Theor. Biol. **79** (1979), 83–99.

[**St**] B. Stewart, *Generation of analytic semigroups by strongly elliptic operators*, Trans. Amer. Math. Soc. **199** (1974), 141–162.

[**Yo**] K. Yosida, "Functional Analysis," Springer, Berlin, 1968.

A Survey of Some Superlinear Problems

PAUL H. RABINOWITZ

Introduction.

Consider the semilinear elliptic partial differential equation:

$$(0.1) \qquad \begin{cases} -\Delta u = p(x,u) & x \in \Omega \\ u = 0 & x \in \partial\Omega \end{cases}$$

Here and in what follows, Ω always denotes a bounded domain in $\mathbb{R}^n$ with a smooth boundary. Suppose that $p(x,\xi)$ is *superlinear* as $|\xi| \to \infty$, i.e.

$$(0.2) \qquad p(x,\xi)\xi^{-1} \to \infty \text{ as } |\xi| \to \infty.$$

The goal of this lecture is to survey some of the more recent literature concerning the existence and multiplicity of solutions of (0.1) using variational methods. Such studies begin with the ingenious work of Nehari [**1**] for $n = 1$ where he exploited nodal properties is an essential way.

We will also survey certain evolution counterparts of (0.1) where one has a finite or infinite dimensional Hamiltonian system and time periodic solutions are sought. As examples we have:

(a) a finite dimensional Hamiltonian system:

$$(0.3) \qquad \dot{z} = JH_z(t,z), \quad J = \begin{pmatrix} 0 & -\mathrm{id} \\ \mathrm{id} & 0 \end{pmatrix}$$

where $H : \mathbb{R} \times \mathbb{R}^{2n} \to \mathbb{R}$ and $H(t,z)$ is T-periodic in t, id denotes the $n \times n$ identity matrix, and we seek T-periodic in t solutions;

(b) semilinear wave equations:

$$(0.4) \qquad \begin{aligned} &u_{tt} - \Delta u + g(x,t,u) = 0, && x \in \Omega, t \in \mathbb{R} \\ &u(x,t) = 0, && x \in \partial\Omega \\ &u(x,t+T) = u(x,t), && x \in \Omega \end{aligned}$$

where g is T-periodic in t;

(c) semilinear Schrödinger equations:

$$
(0.5) \qquad \begin{aligned} & iu_t - \Delta u + h(x,t,|u|)u = 0, && x \in \Omega, t \in \mathbb{R} \\ & u(x,t) = 0, && x \in \partial\Omega \\ & u(x,t+T) = u(x,t) && \end{aligned}
$$

where h is T-periodic in t and u is complex valued.

In §1, the elliptic case (0.1) will be discussed and the evolution cases will be treated in §2.

1. The Elliptic Case.

Attention here will be restricted to

$$
(1.1) \qquad \begin{aligned} -\Delta u &= p(x,u) && x \in \Omega \\ u &= 0 && x \in \partial\Omega. \end{aligned}
$$

There are analogues of most of the results that will be stated for (1.1) with $-\Delta$ replaced by a more general divergence structure elliptic operator. Also, for simplicity, we will always assume that

$$
(p_1) \qquad p \in C^1(\bar{\Omega} \times \mathbb{R}, \mathbb{R}).
$$

It has been conjectured that if p is superlinear, (1.1) possesses multiple solutions. However, this is not known, even for $n = 1$, without further hypotheses on p. To apply variational methods to (1.1), a stronger version of (0.2) is required. Thus we assume

$$
(p_2) \qquad \begin{cases} \text{there are constants } \mu > 2 \text{ and } R > 0 \text{ such that} \\ 0 < \mu P(x,\xi) \le \xi p(x,\xi) \text{ for } |\xi| \ge R \text{ where} \\ P(x,\xi) = \int_0^{\xi} p(x,\xi)\,dt. \end{cases}
$$

This hypothesis implies the existence of constants $\alpha_1, \alpha_2 > 0$ such that

$$
(1.2) \qquad P(x,\xi) \ge \alpha_1 |\xi|^{\mu} - \alpha_2 \text{ for all } \xi \in \mathbb{R} \text{ and } x \in \bar{\Omega}
$$

Hence by (p_2) and (1.2), p satisfies (0.2).

To treat (1.1) via the calculus of variations requires studying the corresponding functional

$$I(u) \equiv \int_\Omega \left[\tfrac{1}{2}|\nabla u|^2 - P(x,u)\right]\,dx. \tag{1.3}$$

A natural space in which to work with (1.2) is $E = W_0^{1,2}(\Omega)$ for then the first term in $I(u)$ is smooth. However, in order for the second term to be well-defined on E, further growth conditions are required for p. To motivate them, consider the special case of

$$P(\xi) = \frac{1}{s+1}|\xi|^{s+1}.$$

If $1 < s < (n+2)(n-2)^{-1} \equiv n^*$ and $n \geq 3$ (with milder restrictions if $n = 1$ or 2), the P term in (1.3) belongs to $C^1(E, \mathbb{R})$. Moreover, I satisfies the Palais-Smale condition (PS), an important technical condition which when satisfied permits the use of certain standard variational tools. If $s = n^*$, the so-called *critical exponent*, the P term is defined for all $u \in E$ but the (PS) condition does not hold. If $s > n^*$, the P term is not even defined for all $u \in E$. Accordingly we divide our discussion of what is known about (1.1) into three cases corresponding to the above three possibilities for s.

We begin with the "subcritical case" where p satisfies

$$|p(x,\xi)| \leq a_1 + a_2|\xi|^s \text{ where } 1 < s < n^* \text{ and } n \geq 3. \tag{p_3}$$

(A milder restriction holds if $n = 1$ or 2.) As was mentioned above, (p_1)–(p_3) imply that (PS) holds for (1.3). This permits the use of minimax methods from the calculus of variations to treat (1.3). In particular as an easy application of the Mountain Pass Theorem one gets:

THEOREM 1.4 [**2**]. *Suppose p satisfies (p_1)–(p_3) and*

$$p(x,\xi) = o(|\xi|) \text{ as } \xi \to 0. \tag{p_4}$$

Then (1.1) possesses a positive and a negative solution in Ω.

REMARK 1.5:

(i) The Mountain Pass Theorem actually gives a critical point of (1.3) in $W_0^{1,2}(\Omega)$. Then, (p_1), (p_3) and results from elliptic regularity theory show that it is a classical solution of (1.1). Further tricks involving the maximum principle yield a positive and negative solution of (1.1).

(ii) The Mountain Pass Theorem provides a minimax characterization of a positive critical value of I. Such minimax characterizations can be useful since they lead to a priori bounds for the corresponding critical points.

(iii) Earlier versions of Theorem 1.4 under more restrictive hypotheses are due to Coffman [**3**] and Hempel [**4**].

By (p_4), the linearization of the right-hand side of (1.1) vanishes when $u = 0$. What happens if $p(x,\xi) = \lambda a(x)\xi + \hat{p}(x,\xi)$ where $\hat{p}$ satisfies (p_1)–(p_4), $\lambda \in \mathbb{R}$, and $a(x) > 0$ in $\bar{\Omega}$? If $\lambda < \lambda_1$, the smallest eigenvalue of

$$(1.6)\qquad \begin{aligned} -\Delta v &= \lambda a(x) v, & x \in \Omega \\ v &= 0, & x \in \partial\Omega, \end{aligned}$$

the proof of Theorem 1.4 works equally well to give a positive and negative solution of (1.1) for this case. Suppose however that $\lambda \geq \lambda_1$. Multiplying (1.1) by v_1, a positive eigenfunction of (1.6) corresponding to λ_1, and integrating over Ω readily yields:

$$(1.7)\qquad (\lambda - \lambda_1)\int_\Omega a u v_1\, dx = -\int_\Omega \hat{p}(x,u) v_1\, dx.$$

If e.g. $\hat{p}$ satisfies

$$(p_5)\qquad \begin{aligned} &\xi\hat{p}(x,\xi) \geq 0 \text{ for all } \xi \in \mathbb{R} \text{ with equality} \\ &\text{if and only if } \xi = 0, \end{aligned}$$

we see from (1.7) that (1.1) does not possess a positive or negative solution. Indeed if e.g. u were a positive solution of (1.1), the left-hand side of (1.7) would be a nonnegative while the right-hand side would be negative. Hence in general when $\lambda \geq \lambda_1$, (1.1) does not possess a positive or a negative solution.

On the other hand a generalized version of the Mountain Pass Theorem yields:

THEOREM 1.8 [**5**]. *If $p(x,\xi) = \lambda a(x)\xi + \hat{p}(x,\xi)$ where $\hat{p}$ satisfies (p_1)–(p_5), then for all $\lambda \in \mathbb{R}$, (1.1) possesses at least one nontrivial solution.*

REMARK 1.9:

(i) By above remarks, if $\lambda \geq \lambda_1$ this solution cannot be of one sign in Ω.

(ii) The generalized Mountain Pass Theorem gets the solution as a critical point of (1.3) having a positive critical value.

A natural question to ask next is what happens in the setting of Theorem 1.8 if (p_5) is dropped? Recently S. Li [**6**] and V. Benci [**7**] independently have shown:

THEOREM 1.10. *Suppose* $p(x,\xi) = \lambda a(x)\xi + \hat{p}(x,\xi)$ *where* $\hat{p}$ *satisfies* (p_1)–(p_4). *If* λ *is not an eigenvalue of (1.6), then (1.1) possess a nontrivial solution.*

Both Li and Benci use variational arguments which are not of a minimax nature and give no information about whether T has a positive critical value. With such additional information, it may be possible to eliminate the restriction on λ.

If p satisfies

$$(p_6) \qquad p(x,\xi) \text{ is odd in } \xi,$$

then the behavior of p near 0 is not important and one can make much stronger assertions about the number of solutions of (1.1). Indeed, if (p_6) holds, I is even in u and a Z_2 symmetric version of the Mountain Pass Theorem then yields:

THEOREM 1.11 [**11**]. *If p satisfies (p_1)–(p_6), (1.1) possesses an unbounded (in $L^\infty(\Omega)$) sequence of solutions.*

REMARK 1.12: As in Theorems 1.4 and 1.8, one has a minimax characterization of the corresponding critical values of I.

A natural question that arises in the case just treated is what happens if it is subjected to a perturbation that destroys the Z_2 symmetry? The simplest such perturbation would be to add an inhomogeneous term $f(x)$ to $p(x,\xi)$:

$$(1.13) \qquad \begin{aligned} -\Delta u &= p(x,u) + f(x), & x &\in \Omega \\ u &= 0, & x &\in \partial\Omega. \end{aligned}$$

The next result provides a partial answer:

THEOREM 1.14 [**5**]. *If p satisfies (p_1)–(p_3), (p_6), and $f \in C^1(\bar{\Omega})$, then (1.13) possesses an unbounded (in $L^\infty(\Omega)$) sequence of solutions provided that s in (p_3) satisfies*

$$(1.15) \qquad s < s^* = \frac{\mu n + (\mu - 1)(n+2)}{\mu n + (\mu - 1)(n-2)}.$$

REMARK 1.16:

(i) The proof of Theorem 1.14 employs some of the machinery of Theorem 1.11 together with additional arguments. The proof does not provide minimax characterizations of the corresponding critical values, a defect which may well be related to the inability to capture the full range of subcritical values of s in Theorem 1.14.

(ii) Slightly less general versions of Theorem 1.14 can be found in Bahri-Berestycki [**8**], Struwe [**9**], and Dong-Li [**10**] while a more general result with s^* replaced by $n/n-2$ has been announced by Bahri and P. L. Lions [**11**].

(iii) Bahri [**12**] has also obtained an interesting version of Theorem 1.14 for $p(x,\xi) = |\xi|^{s-1}\xi$ and any $s < n^*$ in which he shows (1.13) has infinitely many solutions for an open dense set of f in $W^{-1,2}(\Omega)$.

Next we turn to the case of the critical exponent $s \equiv n^*$ in (p_2). Much current effort is being expended on questions involving the critical exponent. Since Haim Brezis will give a detailed discussion of this case in his talk, we will only make a few remarks here.

If p depends only on u, solutions of (1.1) satisfy an identity due to Pohozaev [**13**] which states that if $\nu = \nu(x)$ is the outward pointing normal to $\partial\Omega$,

$$\int_\Omega [2nP(u) - (n-2)up(u)]\,dx = \int_{\partial\Omega} x\cdot\nu(x)\left|\frac{\partial u}{\partial\nu}\right|^2 dS. \tag{1.17}$$

(Some interesting extensions of this identity have been obtained recently by Pucci and Serrin [**14**].) In particular if $x\cdot\nu(x) > 0$, i.e. Ω is star-shaped with respect to the origin, and $p(\xi) = |\xi|^{n^*-1}\xi$, (1.17) shows $\frac{\partial u}{\partial\nu} = 0$ on $\partial\Omega$. But if u is of one sign in Ω, e.g. positive, the maximum principle implies $\frac{\partial u}{\partial\nu} < 0$ on $\partial\Omega$. Hence u cannot be of one sign in Ω. A more elaborate argument (for which we only know of an unpublished proof of Brezis) implies $u \equiv 0$ in Ω. Hence (1.1) possesses no nontrivial solutions for this particularly simple case. Thus one cannot hope to obtain interesting existence results for the critical exponent unless the nonlinearities p or domain Ω are more complicated. Some progress has been made in both of these directions. Brezis and Nirenberg studied problems of the form $p(x,\xi) = |\xi|^{n^*-1}\xi +$ lower order terms [**15**]. In particular for

$$\begin{aligned} -\Delta u &= \lambda u + |u|^{n^*-1}u, && x \in \Omega \\ u &= 0, && x \in \partial\Omega, \end{aligned} \tag{1.18}$$

they proved:

THEOREM 1.19. *If* $\lambda \in (0, \lambda_1)$ *and* $n \geq 4$, *(1.18) possesses a positive solution.*

If $n = 3$, the situation is more complicated [**15**]. On the basis of this success and comparing this result with Theorem 1.11, one might conjecture that (1.18) possesses multiple solutions, even infinitely many. Some partial answers have recently been given. Cerami, Solimini, and Struwe [**16**] and independently Z. Dong [**17**] have showed that if $n \geq 6$ and $\lambda \in (0, \lambda_1)$, there exists at least two pairs of nontrivial solutions to (1.18). Capozzi, Fortunato, and Palmieri [**18**] and independently Z. Dong [**17**] have proved that for any $\lambda > 0$, (1.18) possesses at least one pair of nontrivial solutions. Finally, in the direction of more complicated domains, a deep result of Bahri and Coron [**19**] shows that if Ω has a "hole" in it, the equation

$$(1.20) \qquad \begin{aligned} -\Delta u &= |u|^{n^*-1}u, \quad && x \in \Omega \\ u &= 0, && x \in \partial\Omega \end{aligned}$$

has a positive solution.

It is clear from this brief description that many open questions remain for $s = n^*$. Even less is known about (1.1) for $s > n^*$. As was mentioned earlier, if $s > n^*$, $I(u)$ in (1.3) is no longer defined for all $u \in W_0^{1,2}(\Omega)$ and (PS) is not satisfied. Thus one loses the compactness structure which is generally required for a variational treatment of (1.1). Nevertheless, the Schauder or L^p theories of linear elliptic boundary value problems allow (1.1) to be converted to an operator equation of the form

$$(1.21) \qquad u + K(u) = 0$$

where K is the compact operator on an appropriate Hölder or L^p space. Thus despite the loss of the usual variational compactness machinery, there is still some compactness inherent in the problem.

There are at least three types of results for (1.1) that apply to the case $s > n^*$. They are not however restricted to this case. The first relies on the compactness structure of (1.21). Suppose as in Theorem 1.10, $p(x, \xi) = \lambda a(x)\xi + \hat{p}(x, \xi)$ where $\hat{p}$ satisfies (p_4). We will now intrepret a solution of (1.1) as a pair (λ, u) satisfying this equation. Let λ_j be an eigenvalue of (1.16). Then by a global bifurcation theorem we get:

THEOREM 1.22 [20]. *If $p(x,\xi) = \lambda a(x)\xi + \hat{p}(x,\xi)$ where $\hat{p}$ satisfies (p_1) and (p_4) and λ_j is an eigenvalue of odd multiplicity of (1.6), then there exists a connected set C_j of nontrivial solutions of (1.1) in $\mathbb{R} \times C(\bar{\Omega})$ which meets $(\lambda_j, 0)$ and is either unbounded or meets $(\lambda_k, 0)$ where $k \neq j$.*

The odd multiplicity condition of Theorem 1.22 is satisfied in particular if $j = 1$ since λ_1 is a simple eigenvalue of (1.6). Moreover elementary arguments based on the maximum principle further imply if $(\lambda, u) \in C_1 \setminus \{(\lambda_1, 0)\}$, then u is of one sign in Ω and C_1 is unbounded in $\mathbb{R} \times C(\bar{\Omega})$ [20]. However even though C_1 is unbounded in $\mathbb{R} \times C(\bar{\Omega})$, it may be bounded in $\mathbb{R} \times W_0^{1,2}(\Omega)$. Indeed consider the special case of

$$(1.23)\qquad \begin{aligned} -\Delta u &= (\lambda + |u|^{s-1})u, & x &\in \Omega \\ u &= 0, & x &\in \partial\Omega \end{aligned}$$

where $s > n^*$.

By the Pohozaev identity (1.17), any solution of (1.23) for star-shaped Ω satisfies

$$2n \int_\Omega \left(\frac{\lambda u^2}{2} + \frac{|u|^{s+1}}{s+1} \right) dx \geq (n-2) \int_\Omega (\lambda u^2 + |u|^{s+1})\, dx$$

or

$$(1.24)\qquad \left(1 - \frac{2n}{(n-2)(s+1)}\right) \int_\Omega |u|^{s+1} \leq \frac{n}{n-2} \int_\Omega u^2\, dx.$$

Thus, using the Hölder inequality, (1.24) yields an a priori bound for u in $L^{s+1}(\Omega)$ and then from (1.23) for u in $W_0^{1,2}(\Omega)$. In particular if $(\lambda, u) \in C_1$ for (1.23) and therefore u is positive or negative, a simple comparison argument implies $\lambda \in (0, \lambda_1)$. Hence C_1 is bounded in $\mathbb{R} \times W_0^{1,2}(\Omega)$.

The second type of result we know of for (1.1) where s does not play a role involves using the variational structure of (1.1) more strongly in the setting of Theorem 1.22. In so doing, we can weaken the requirement that λ_j be of odd multiplicity, but at the expense of less information about the structure of the solution set near $(\lambda_j, 0)$. More precisely we have:

THEOREM 1.25. *If $p(x,\xi) = \lambda a(x)\xi + \hat{p}(x,\xi)$ where $\hat{p}$ satisfies (p_1) and (p_4), then for each eigenvalue λ_j of (1.6), $(\lambda_j, 0)$ is a bifurcation point for (1.1), i.e. there exists a sequence of (nontrivial) solutions (μ_k, u_k) of (1.1) such that $\mu_k \to \lambda_j$ and $u_k \to 0$ (e.g. in $C^2(\Omega)$).*

The proof of Theorem 1.25 just involves redefining $\hat{p}(x,\xi)$ outside of a neighborhood of $\xi = 0$, invoking a variational bifurcation theorem of

Krasnoselski [**21**] or related results, and making some simple estimates. See e.g. [**5**] for details. It is not known whether this family of solutions one has near $(\lambda_j, 0)$ continues into the large.

The final type of result we want to mention for (1.1) occurs when Ω is a ball and $p(x,\xi) = p(|x|,\xi)$. Then one can seek solutions which depend solely on $|x| = r$ and (1.1) reduces to a nonlinear ordinary differential equation. For example, for an annular type domain, $a < r < b$, it is not difficult to show that (1.20) with n^* replaced by s has a positive solution for any choice of s. C. Budd, jointly with J. Norbury, has made an interesting study of the effect of nonradial perturbations of p on the radial case and we refer to the written version of Budd's talk in this Proceedings for more details.

2. Superlinear evolution problems.

This section concerns superlinear evolution problems that have been treated by variational methods. In particular we will discuss (finite dimensional) Hamiltonian systems, semilinear wave equations, and semilinear Schrödinger equations. The Hamiltonian system case is the best understood and we will begin with it. Thus consider

$$\dot{z} = JH_z(z), \quad J = \begin{pmatrix} 0 & -\mathrm{id} \\ \mathrm{id} & 0 \end{pmatrix} \tag{2.1}$$

where $z = (p, q)$ with $p, q \in \mathbb{R}^n$, id the $n \times n$ identity matrix, and $H : \mathbb{R}^{2n} \to \mathbb{R}$.

Assume:

$$(H_1) \qquad H \in C^1(\mathbb{R}^{2n}, \mathbb{R})$$

and by analogy with (p_2),

(H_2) there are constants $\mu > 2$ and $R > 0$ such that $0 < \mu H(z) = z \cdot H_z(z)$ for $|z| \geq R$.

As earlier (H_2) implies there are constants $a_1, a_2 > 0$ such that

$$H(z) \geq a_1 |z|^\mu - a_2$$

so H is a "superquadratic" function.

The problem we pose for (2.1) is: given $T > 0$, does (2.1) possess a T periodic solution. The variational formulation of (2.1) is to find critical points of

$$I(z) = \int_0^T (p \cdot \dot{q} - H(z))\, dt \tag{2.2}$$

is an appropriate class of T-periodic functions. By making a change of variables in (2.1) which replaces H by $\frac{2\pi}{T}H$, it can be assumed that $T = 2\pi$ in (2.2). The term

$$A(z) \equiv \int_0^{2\pi} p \cdot \dot{q}\, dt \tag{2.3}$$

in (2.2) is called the action integral. An natural space to work in so that (2.3) is defined and smooth is $X = W^{1/2,2}(S^1, \mathbb{R}^{2n})$, i.e. the space of $2n$-tuples of 2π-periodic functions under the "$\frac{1}{2}$ derivative" norm. See e.g. [**5**] for details. Unfortunately the H term in (2.2) is not defined on all of X unless H satisfies hypotheses such as a power growth condition. However, one can get around this complication by certain truncation arguments that we will not describe here but can be found, e.g., in [**5**]. Thus for simplicity suppose $I \in C^1(X, \mathbb{R})$. The functional I possesses a natural S^1 symmetry: namely if $z(t) \in X$, $I(z(t+\theta)) = I(z(t))$ for all $\theta \in [0, 2\pi]$. Hence one can use an S^1 symmetry version of the Mountain Pass Theorem to prove an analogue of Theorem 1.11:

THEOREM 2.4 [**22**]. *If H satisfies (H_1)–(H_2), then for any $T > 0$, (2.1) possesses an unbounded sequence of T-periodic solutions.*

One might further ask whether for each $T > 0$, there is a solution having *minimal* period T. This cannot be expected in the generality of Theorem 2.4 since there are simple examples of Hamiltonians satisfying (H_1)–(H_2) and for which there is an upper bound on the (minimal) period. However, using more refined arguments, Ekeland and Hofer [**23**] have proved:

THEOREM 2.5. *Suppose $H \in C^2(\mathbb{R}^{2n}, \mathbb{R})$, $H_{zz}(z)$ is positive definite for all $z \neq 0$, $H(z)|z|^{-2} \to 0$ as $|z| \to 0$, and (H_2) holds for all $z \in \mathbb{R}^{2n} \setminus \{0\}$. Then for any $T > 0$, (2.1) possesses a solution having minimal period T.*

The same sort of ideas that go into the proof of Theorem 2.4 can be used to prove a similar theorem for second order Hamiltonian systems of the form

$$\ddot{q} + V'(q) = 0. \tag{2.6}$$

Here $q \in \mathbb{R}^n$, $V \in C^1(\mathbb{R}^n, \mathbb{R})$, and V satisfies (H_2). Indeed (2.6) is technically much simpler to work with than (2.1). No truncations are necessary in treating the corresponding functional [**24**].

If H in (2.1) also depends on T in a time periodic fashion, one can also get an existence result, but at the expense of additional assumptions.

THEOREM 2.7 [**25**]. *Suppose H satisfies (H_2),*

(H_1') $\quad H \in C^1(\mathbb{R} \times \mathbb{R}^{2n}, \mathbb{R})$ *and is T periodic in t,*

(H_3) $\quad H(t,z) = o(|z|^2)$ as $|x| \to 0$,

(H_4) $\quad H \geq 0,$

and

(H_5) $\quad |H_z(t,z)| \leq \alpha z \cdot H_z(t,z) + \beta$ *for some constants* $\alpha, \beta > 0$.

Then

$$\dot{z} = JH_z(t,z) \tag{2.8}$$

possesses at least one nonconstant T-periodic solution.

Theorem 2.7 follows most simply from a generalized version of the Mountain Pass Theorem [**5**].

If $H(z,t)$ is T-periodic in t, it is also kT-periodic for any $k \in \mathbb{N}$. One can therefore ask for kT-periodic solutions of (2.8) different from the T-periodic solution determined already. Such solutions are called *subharmonics* for (2.8). Using the ideas in the proof of Theorem 2.7 together with some additional comparison arguments, one gets:

THEOREM 2.9 [**26**]. *Under the hypotheses of Theorem 2.7, there exists a sequence $(k_i) \subset \mathbb{N}$, $k_i \to \infty$ as $i \to \infty$, and corresponding district k_iT-periodic solutions of (2.8).*

Ekeland and Hofer [**27**] also have a result in this spirit where one gets subharmonics having minimal period kT.

Returning to (2.1), there are analogues for this equation of Theorem 1.14. Consider

$$\dot{z} = JH_z(z) + f(t). \tag{2.10}$$

THEOREM 2.11. *Suppose $H \in C^2(\mathbb{R}^{2n}, \mathbb{R})$, $f \in C([0,T], \mathbb{R})$ and is T-periodic, (H_2) holds, and*

$$H(z) \leq a_3|z|^\sigma + a_4 \text{ where } \sigma < 2\mu. \tag{H_6}$$

Then (2.10) possesses an unbounded sequence of T-periodic solutions.

This result is due to Bahri and Berestycki [**28**] via a complicated variational argument. It is not known whether (H_6) is essential. Indeed for the analogous second order Hamiltonian system

$$\ddot{q} + V'(q) = \phi(t) \tag{2.12}$$

one has a much better result without any upper growth restrictions on V:

THEOREM 2.13. *Suppose $V \in C^1(\mathbb{R}^n, \mathbb{R})$, $\phi \in C([0,T], \mathbb{R})$ and is T-periodic, and V satisfies (H_2). Then (2.12) possesses an unbounded sequence of T-periodic solutions.*

Theorem 2.13 was proved for $V \in C^2(\mathbb{R}^n, \mathbb{R})$ by Bahri and Berestycki [**29**] using an argument related to that of Theorem 2.11. The C^1 case is due recently to Y. Long [**30**] who used an argument more in spirit with that of Theorem 1.14.

Another interesting result for (2.10) was proved recently by Y. Long. Note that we can also write this equation as

$$J\dot{z} + H_z(z) = g(t). \tag{2.14}$$

Let $L^2([0,T], \mathbb{R}^{2n})$ denote the space of square integrable T-periodic functions with values in $\mathbb{R}^{2n}$ and $W^{1,2}([0,T], \mathbb{R}^{2n})$ the subspace of L^2 of function having square integrable first derivatives.

THEOREM 2.15 [**30**]. *Suppose H satisfies (H_1),*

$$\frac{z \cdot H_z(z)}{|z|^2} \to \infty \text{ as } |z| \to \infty \tag{H_7}$$

and

$$H(z) \leq a_1 z \cdot H_z + a_2 \text{ for all } z \in \mathbb{R}^{2n}. \tag{H_8}$$

Then for any $T > 0$, the range of $J\frac{d}{dt} + H_z(\cdot)$ acting on $W^{1,2}([0,T], \mathbb{R}^{2n})$ is dense in $L^2([0,T], \mathbb{R}^{2n})$.

Long also has some results in this spirit when H depends on t, e.g.

THEOREM 2.16. *Suppose H satisfies (H_1),*

$$(H_7') \qquad \text{there is a } \mu > 2 \text{ such that } \frac{z \cdot H_z}{|z|^\mu} \to \infty \text{ as } |z| \to \infty,$$

and

$$(H_9) \qquad H_t(t,z) \geq -a|z|^{\mu/2} - b \text{ for some } a, b > 0.$$

Then the range of $J\frac{d}{dt} - H_z(t,\cdot)$ acting on $W^{1,2}([0,T],\mathbb{R}^{2n})$ is dense in $L^2([0,T],\mathbb{R}^{2n})$.

Theorems 2.15 and 2.16 were motivated by work of Tanaka [**31**] on (0.4) that we will describe shortly. The proofs of these results are actually fairly elementary and do not involve variational methods. It is natural to conjecture that there is a version of Theorem 2.15 for (1.1).

Now we turn to semilinear wave equations. Consider

$$(2.17) \qquad \begin{aligned} &u_{tt} - \Delta u + g(x,u) = 0, && x \in \Omega, t \in \mathbb{R}, \\ &u(x,t) = 0, && x \in \partial\Omega \\ &u(x,t+T) = u(x,t). \end{aligned}$$

The corresponding functional is

$$(2.18) \qquad I(u) = \int_0^T \int_\Omega \left[\tfrac{1}{2}(u_t^2 - u_x^2) - G(x,u)\right] dx\, dt$$

where G is the primitive of g. Note that the same S^1 symmetry that one finds in (2.2) is present here. However (2.18) is much more difficult to treat than our earlier cases due to a lack of compactness or roughly equivalently of the (PS) condition. Thus far the only real success for this variational approach has been when $n = 1$, $\Omega = (0,\ell)$, and $\ell T^{-1} \in \mathbb{Q}$. This last condition is also related to compactness via the spectrum of the wave operator in the class of functions satisfying the periodicity and boundary conditions. As a sample existence result here we have:

THEOREM 2.19 [**32**]. *Suppose that $g \in C^2([0,\ell] \times \mathbb{R},\mathbb{R})$, satisfies (p_2), and is strictly monotone increasing in u. If further $T\ell^{-1} \in \mathbb{Q}$, then (2.17) possesses an unbounded (e.g. in L^∞) sequence of (classical) solutions.*

REMARKS 2.20:

(i) It is not known whether the requirement that $T\ell^{-1} \in \mathbb{Q}$ is essential but see Theorem 2.22 below.

(ii) If g is independent of x, the strict monotonicity assumption can be dropped. See Coron [**33**] or [**32**].

(iii) When g depends on t in a T-periodic fashion, there are versions of Theorem 2.7 for this setting. See e.g. [**34,35**].

(iv) There are also some results on perturbation from symmetry here in the spirit of Theorems 1.14, 2.11 and 2.13. See e.g. [**36,37**]. They involve further growth conditions on g which we suspect will prove to be unnecessary.

(v) Some results concerning subharmonics such as Theorem 2.9 have been found for (2.17) — see e.g. [**38**] — but nothing is known about solutions having a minimal period.

(vi) See also the survey paper of Brezis [**39**] for more results concerning (2.17).

As was mentioned earlier, Tanaka [**31**] has obtained density results like Theorem 2.15 for (2.17). To be more precise, he treated

$$\begin{aligned} & u_{tt} - u_{xx} + g(u) = \phi(x,t), \quad x \in (0,\ell), t \in \mathbb{R}, \\ (2.21) \qquad & u(0,t) = 0 = u(\ell,t) \\ & u(x,t+T) = u(x,t). \end{aligned}$$

working in the class X of functions satisfying the periodicity and boundary conditions under the norm

$$\|u\| = \sup_t \left(\int_0^\ell (u_t^2 - u_x^2)\,dx \right)^{\frac{1}{2}}.$$

Tanaka proved:

THEOREM 2.22. *Suppose g is continuous,*

$$g(\xi)\xi^{-1} \to \infty \text{ as } |\xi| \to \infty, \text{ and } G(\xi) \le a(1 + \xi f(\xi))$$

for all ξ. Then for a dense set of $\phi \in L^2([0,\ell] \times [0,T])$, there is a weak solution of (2.21), i.e. the range of $\frac{\partial^2}{\partial t^2} - \frac{\partial^2}{\partial x^2} + g(\cdot)$ acting (weakly) on X is dense in $L^2([0,\ell] \times [0,T])$.

REMARKS:

(i) Replacing $(0,\ell)$ by Ω and u_{xx} by Δu, Tanaka [**31**] proved a similar result for the higher dimensional case under the additional growth restriction $|g(\xi)| \le a_1 + a_2|\xi|^s$, $s < n(n-2)^{-1}$.

(ii) Y. Long [**30**] has generalized Tanaka's results somewhat, permitting g to depend on x and t.

Lastly we turn very briefly to the case of semilinear Schrödinger equations. Consider

$$
(2.24)\qquad \begin{aligned} & iu_t - \Delta u + h(x,t,|u|^2)u = 0, && x \in \Omega, t \in \mathbb{R}, \\ & u(x,t) = 0, && x \in \partial\Omega \\ & u(x,t+T) = u(x,t). && \end{aligned}
$$

Here u is complex valued. Almost nothing is known for (2.24) in the spirit of the results presented above except for some recent work on density, such as Theorems 2.15, 2.16 and 2.22. In fact, Y. Long [**30**] has obtained such results for special cases of (2.24) by formulating them as infinite dimensional Hamiltonian systems and using ideas such as those that go into the proofs of Theorems 2.15, 2.16 and 2.22.

Mathematics Department and Mathematics Research Center, University of Wisconsin, Madison WI 53706

This research was sponsored in part by NSF Grant MCS-8110556 and by the United States Army under Contract No. DAAG29-80-C-0041.

Research at MSRI supported in part by NSF Grant DMS 812079-05.

REFERENCES

1. Z. Nehari, *Characteristic values associated with a class of nonlinear second order differential equations*, Acta Math. **105** (1961), 141–175.
2. A. Ambrosetti and P. H. Rabinowitz, *Dual variational methods in critical point theory and applications*, J. Funct. Anal. **14** (1973), 349–381.
3. C. V. Coffman, *On a class of nonlinear elliptic boundary value problems*, J. Math. Mech. **19** (1970), 351–356.
4. J. A. Hempel, *Superlinear variational boundary value problems and nonuniqueness*, Thesis, University of New England, Australia (1970).
5. P. H. Rabinowitz, *Minimax Methods in Critical Point Theory with Applications to Differential Equations*, AMS Regional Conference Series in Math., vol. 65 (1986).
6. S. Li, *Some aspects of critical point theory*, preprint, Institute of Math. Academia Sinica.
7. V. Benci, private communication.
8. A. Bahri and H. Berestycki, *A perturbation method in critical point theory and applications*, Trans. Amer. Math. Soc. vol 267 (1981), 1–32.
9. M. Struwe, *Infinitely many critical points for functionals which are not even and applications to superlinear boundary value problems*, Manus. Math. **32** (1980), 335-364.
10. G. C. Dong and S. Li, *On the existence of infinitely many solutions of the Dirichlet problem for some nonlinear elliptic equations*, Sci. Sinica, Ser. A. **25** (1982), 468–475.
11. A. Bahri and P. L. Lions, *Remarks on the variational theory of critical points and applications*, C. R. Acad. Sci. Paris, Ser. I Math. **301** (1985), 145–148.
12. A. Bahri, *Topological results on a certain class of functionals and applications*, J. Funct. Anal. **41** (1981), 397–427.
13. S. I. Pohozaev, *Eigenfunctions of the equation* $\Delta u + \lambda f(u) = 0$, Soviet Math. **5** (1965), 1408–1411.
14. P. Pucci and J. Serrin, *A general variational identity*, preprint.
15. H. Brezis and L. Nirenberg, *Positive solutions of nonlinear elliptic equations involving critical Sobolev exponents*, Comm. Pure Appl. Math. **36** (1983), 437–477.
16. G. Cerami, S. Solimini and M. Struwe, *Some existence results for superlinear elliptic boundary value problems involving critical exponents*, preprint.
17. Z. Dong, to appear in Nonlinear Analysis TMA.
18. A. Capozzi, D. Fortunato and G. Palmieri, *An existence result for nonlinear elliptic problems involving the critical Sobolev exponent*, to appear in Ann. Inst. H. Poincaré: Analyse Nonlinéaire.
19. A. Bahri and J.-M. Coron, *Sur un équation elliptique nonlinéaire avec l'exposant critique de Sobolev*, C. R. Acad. Sci. Paris **301, Ser. 1** (1985), 345–348.
20. P. H. Rabinowitz, *Some global results for nonlinear eigenvalue problems*, J. Funct. Anal. **7** (1971), 487–513.
21. M. A. Krasnosel'skii, "Topological Methods in the Theory of Nonlinear Integral Equations," Macmillan, New York, 1964.
22. P. H. Rabinowitz, *Periodic solutions of large norm of Hamiltonian systems*, J. Diff. Eq. **50** (1983), 33–48.
23. I. Ekeland and H. Hofer, *Periodic solutions with prescribed minimal period for convex autonomous Hamiltonian systems*, Invent. Math. **81** (1985), 155–188.
24. P. H. Rabinowitz, *On large norm periodic solutions of some differential equations*, in "Ergodic Theory and Dynamical Systems," Proc. Sp. Yr. Maryland 1979-1980, A. Katok, ed., Birkhäuser, 1982, pp. 193–210.
25. P. H. Rabinowitz, *Periodic solutions of Hamiltonian systems*, Comm. Pure Appl. Math. **31** (1978), 157–184.
26. P. H. Rabinowitz, *Subharmonic solutions of Hamiltonian systems*, Comm. Pure Appl. Math. **33** (1980), 609–633.

27. I. Ekeland and H. Hofer, *Subharmonics for convex nonautonomous Hamiltonian systems*, preprint.
28. A. Bahri and H. Berestycki, *Forced vibrations of superquadratic Hamiltonian systems*, Acta Math. **151** (1984), 143–197.
29. A. Bahri and H. Berestycki, *Existence of forced oscillations for some nonlinear differential equations*, Comm. Pure Appl. Math. **37** (1984), 403–442.
30. Y. Long, Thesis, University of Wisconsin, Madison, in progress.
31. K. Tanaka, *Density of the range of a wave operator with nonmonotone superlinear nonlinearity*, preprint.
32. P. H. Rabinowitz, *Large amplitude time periodic solutions of a semilinear wave equation*, Comm. Pure Appl. Math. **37** (1984), 189–206.
33. J.-M. Coron, *Periodic solutions of a nonlinear wave equation without assumption of monotonicity*, preprint.
34. P. H. Rabinowitz, *Free vibrations of a semilinear wave equation*, Comm. Pure Appl. Math. **31** (1978), 31–68.
35. H. Brezis, J.-M. Coron and L. Nirenberg, *Free vibrations for a nonlinear wave equation and a theorem of P. Rabinowitz*, Comm. Pure Appl. Math. **33** (1980), 667–689.
36. J. P. Ollivry, *Vibrations forcées pour une équation d'onde nonlineaire*, C. R. Acad. Sci. Paris, Ser. I **297** (1983), 29–32.
37. K. Tanaka, *Infinitely many periodic solutions for a superlinear forced wave equation*, to appear in Nonlinear Analysis TMA.
38. P. H. Rabinowitz, *Subharmonic solutions of a forced wave equation*, in "Contributions to Analysis and Geometry," D. N. Clark, G. Pecelli and R. Sacksteder, eds.,, Johns Hopkins University Press, 1982, pp. 285-292.
39. H. Brezis, *Periodic solutions of nonlinear vibrating strings and duality principle*, Bull Amer. Math. Soc. **8** (1983), 410–426.

A Priori Estimates for Reaction-Diffusion Systems

FRANZ ROTHE

Abstract. An a priori estimate of some L_p-norm is easy to obtain for many reaction-diffusion systems from applications. We state conditions for which an L_p-estimate implies an L_∞-estimate. As an application we prove that the solutions of a predator-prey system and a nuclear reactor model are bounded uniformly in space and time.

This note surveys the results of [**10,11**], where complete proofs are given. Related results were obtained by Alikakos, Ball, Henry, Kielhöfer and Massat [**1,2,5,6,8**].

Consider reaction-diffusion systems of the general type

$$u_t - D\Delta u = F(x,t,u) \quad \text{for } x \in \Omega, t > 0. \tag{1}$$

Here $x \in \Omega \subset \mathbb{R}^n$ where Ω is a bounded region, $t \in [0,\infty)$ denotes the time and $u = (u_1,\ldots,u_m) \in \mathbb{R}^m$. $D = \operatorname{diag}(d_1,\ldots,d_m)$ is the nonnegative diagonal matrix of diffusion coefficients and the function $F = (F_1,\ldots,F_m)$ models the reaction terms. For diffusing components we adopt boundary conditions

$$d_i u_i = 0 \text{ or } d_i(b_i u_i + \partial u_i/\partial n) = 0 \text{ for } x \in \partial\Omega, t > 0, i = 1\ldots m. \tag{2}$$

Note that components without diffusion ($d_i = 0$) and different types of boundary conditions may occur in one system. Finally take initial data

$$u(x,0) = u_0(x) \quad \text{for } x \in \Omega. \tag{3}$$

Assume sufficient smoothness, e.g. for some $a \in (0,1)$ let $\partial\Omega \in C^{2+\alpha}$, $b_i \in C^{1+\alpha}(\partial\Omega)$, $u_0 \in C^{2+\alpha}(\bar{\Omega}, \mathbb{R}^m)$, $F \in C^\alpha(\bar{\Omega} \times [0,\infty) \times \mathbb{R}^m, \mathbb{R}^m)$ and locally Lipschitz in u. For compatibility of (2) and (3) assume that for $x \in \partial\Omega$, u_0 satisfies (2) and that for $x \in \partial\Omega$, $t = 0$, (1) holds in case of Dirichlet boundary conditions.

Construction of a solution of (1)–(2)–(3) uses an integrated form as (10), i.e. a mild solution. It exists on a maximum time interval $(0, T_{\max})$ and

$$\lim_{t \to T_{\max}} \|u(\cdot,t)\|_\infty = \infty \text{ if } T_{\max} < \infty. \tag{4}$$

By semigroup theory (see [**2**,**5**]) one gets $u \in C((0,T_{\max}), C^\alpha(\bar{\Omega}, \mathbb{R}^m))$ and the Schauder estimates for linear equations [**7**] imply $u \in C([0,T_{\max}), C^{2+\alpha}(\bar{\Omega}, \mathbb{R}^m))$. Here compatibility conditions are needed.

For special systems occuring in applications elementary calculations involving partial integrations or comparison arguments often yield estimates of some functionals. We want to give sufficient (and also essentially necessary) conditions under which these primary weak a priori estiamtes imply uniform estimates.

To this end we focus attention on a single component $v = u_i$ with diffusion coefficient $d_i \neq 0$ and apply a "bootstrap" argument. Application to all components in some convenient order yields a uniform estimate for the vector u. For ecological systems satisfying a food pyramid condition (see [**14**]) there exists a natural order convenient for this process. But the Propositions below apply to more general situations. They remain valid for vector-valued v and f. Secondly, systems involving nondiffusing components as the examples below and in [**10**] can also be handled. Finally as shown in [**11**] one gets better results by adapting the "feedback" argument used in Proposition 1 to the structure of the special system considered.

Let $v = u_1$ be, for example, the first component of the vector u and let $b = b_1$, $v_0 = u_{01}$ and $d_1 = 1$. $L \in [0,\infty)$ will be specified below. Considering the other components of u as known functions of (x,t) let

$$f(x,t,v) = Lv + F_1(x,t,(v,u_2(x,t),\dots u_m(x,t))). \tag{5}$$

The main object is the following initial-boundary value problem for v:

$$v_t - \Delta v + Lv = f(x,t,v) \quad \text{for } x \in \Omega, t > 0; \tag{6}$$

$$v = 0 \text{ or } bv + \partial v/\partial n = 0 \quad \text{for } x \in \partial\Omega, t > 0; \tag{7}$$

$$v(x,0) = v_0(x) \quad \text{for } x \in \Omega. \tag{8}$$

The solution of the corresponding linear system with $f \equiv 0$ is formally given by $v(x,t) = (S(t)v_0)(x)$. The formal semigroup $S(t)$ can be extended to an analytic semigroup $S_p(t)$ in $L_p(\Omega)$ for $p \in (1,\infty)$ as shown in [**3**]. As proved in [**11**], S_p can be extended to $L_1(\Omega)$. We denote the extension again by $S(t)$. For $t > 0$, $v_0 \in L_1(\Omega)$ clearly $(S(t)v_0)(x)$ satisfies (6)–(7). Choose $L \geq 0$ large enough such that for some $\lambda > 0$ independent of p

$$\|S(t)v_0\|_p \leq Ke^{-\lambda t} \|v_0\|_p \text{ for all } v_0 \in L_p.$$

In case of Dirichlet boundary condition or if $b \geq 0$, $b \not\equiv 0$, $L = 0$ is sufficient.

DEFINITION. *Let $T \in (0,\infty) \cup \{\infty\}$. A mild solution of the initial-boundary value problem (6)–(7)–(8) on $(0,T)$ is a measurable function $v : (x,t) \in \Omega \times (0,T) \to \mathbb{R}$ such that*

$$\sup\{\|v(\cdot,\tau)\|_\infty \mid \tau \in (s,t)\} < \infty \text{ and} \tag{9}$$
$$\int_0^t \|f(\cdot,\tau,v(\cdot,\tau))\|_1 \, d\tau < \infty \text{ for all } s,t \in (0,T).$$

$$v(\cdot,t) = v_0 + \int_0^t S(t-\tau) f(\cdot,\tau,v(\cdot,\tau))\, d\tau. \tag{10}$$

Clearly classical solutions are also mild solutions.

For $r_1 \in (0,\infty)$, $r_2 \in (0,\infty]$ let $L_{r_1,r_2,T}$ be the space of all measurable functions $v : (x,t) \in \Omega \times (0,T) \to \mathbb{R}$ such that for $r_2 < \infty$, $r_2 = \infty$ the norm

$$\|v\|_{r_1,r_2,T} = \sup\left\{\left(\int_{t_1}^{t_2} \left(\frac{1}{|\Omega|}\int_\Omega |v(x,t)|^{r_1}\, dx\right)^{r_2/r_1} dt\right)^{1/r_2} \;\middle|\; |t_2 - t_1| < 1, \quad t_i < T\right\}$$

$$\|v\|_{r_1,\infty,T} = \operatorname*{sup\ ess}_{t\in(0,T)} \left(\frac{1}{|\Omega|}\int_\Omega |v(x,t)|^{r_1}\, dx\right)^{1/r_1} \quad \text{respectively, is finite.}$$

In the assumptions for f we introduce a function c motivated by (5). We assume that for some $q_1, q_2 \in [1,\infty]$ and $\gamma \in [1,\infty)$ there exists a function $c \in L_{q_1,q_2,T}$ such that

(f0) $f(\cdot,\cdot,v) \in L_{q_1,q_2,T}$ for all $v \in \mathbb{R}$,

(f1) $|f(x,t,v)| \le c(x,t)(1+|v|)^\gamma$ for $v \in \mathbb{R}$ and almost all $(x,t) \in \Omega \times (0,T)$.

If additionally we have $v \ge 0$ the two-sided estimate (f1) can be replaced by the one-sided estimate

(f2) $f(x,t,v) \le c(x,t)(1+v)^\gamma$ for $v \ge 0$ and almost all $(x,t) \in \Omega \times (0,T)$.

Sometimes the local Lipschitz condition is needed: there exists a continuous function $\ell : [0,\infty) \to [0,\infty)$ such that

(f3) $|f(x,t,v) - f(x,t,w)| \le c(x,t)\ell(|v|+|w|)|v-w|$ for $v,w \in \mathbb{R}$ and almost all $(x,t) \in \Omega \times (0,T)$.

In Propositions 1 through 3 let v be a classical or mild solution of the initial-boundary value problem (6)–(7)–(8) in the time interval $(0,T)$ with initial data $v_0 \in L_{p_0}(\Omega)$ for some $p_0 \in [1,\infty]$ and let the function f satisfy (f0) and the two-sided bound (f1) or the one-sided bound (f2) if it is additionally known that $v \geq 0$.

PROPOSITION 1. *Let the exponents $r_1 \in (0,\infty)$, $r_2 \in (0,\infty]$, $q_1, q_2 \in [1,\infty]$ and $\gamma \in [1,\infty)$ satisfy*

$$1/q_1 + (\gamma-1)/r_1 < \min(1, 2/n), \tag{11}$$

$$n/2q_1 + 1/q_2 + (\gamma-1)/(n/2r_1 + 1/r_2) < 1. \tag{12}$$

Then there exists an exponent α depending only on the exponents $r_1, r_2, p_0, q_1, q_2, \gamma$ and a constant K depending on these exponents and on the domain Ω and the boundary conditions (not on T, v_0 or v) such that for all $t \in (0,T)$

$$\|v(\cdot,t)\|_\infty \leq K \min(1,t)^{-n/2p_0}(1 + \|c\|_{q_1,q_2,t} + \|v\|_{r_1,r_2,t} + \|v_0\|_{p_0})^\alpha. \tag{13}$$

PROPOSITION 2. *Let the exponents $p_0, q_1, q_2 \in [1,\infty]$ and $\gamma \in [1,\infty)$ satisfy*

$$n/2q_1 + 1/q_2 + (\gamma-1)n/2p_0 < 1. \tag{14}$$

Then for some exponent α and constant K as in Proposition 1

$$\|v(\cdot,t)\|_\infty \leq K \min(1,t)^{-\alpha}(1 + \|c\|_{q_1,q_2,t} + \|v\|_{r_1,r_2,t} + \|v_0\|_{p_0})^\alpha. \tag{15}$$

for all $t \in (0,T)$.

If the local Lipschitz condition (f3) holds, v can be extended to a maximal time interval $(0,T_{\max})$ and

$$\lim_{t\to T_{\max}} \|v(\cdot,t)\|_{p_0} = \infty \text{ if } T_{\max} < \infty.$$

Complete proofs are given in [**11**]. We take an example and sketch the proof of Proposition 1 in a diagram. Let $u = u(x,t)$ satisfy

$$u_t - \Delta u \leq u^\gamma \text{ and } 0 \leq u \text{ for } x \in \Omega, t > 0;$$

$$u = 0 \text{ for } x \in \partial\Omega, t > 0 \text{ and for } x \in \Omega, t = 0;$$

$$\|u\|_r = \sup_{t>0} \|u(\cdot,t)\|_r < \infty \text{ where } (\gamma-1)n/2r < 1.$$

We get the diagram:

$$\|u\|_r < \infty$$

$$\downarrow$$

$$\text{(I)} \quad \left\{ \begin{array}{c} \|u^\gamma\|_p \leq \\ \|u\|_r^{r/p} \, \|u\|_\infty^{\gamma - r/p} \end{array} \right\}$$

$$\downarrow$$

feedback with $\gamma - r/p < 1$

$$\|u^\gamma\|_p < \infty$$

$$\downarrow$$

$$\text{(II)} \quad \left\{ \begin{array}{c} u_t - \Delta u \leq u^\gamma, 0 \leq u, \\ n/2p < 1 \text{ imply} \\ \|u\|_\infty < K(\Omega, p) \, \|u^\gamma\|_p \end{array} \right\}$$

$$\downarrow$$

$$\|u\|_\infty < \infty$$

(I) is an elementary inequality. To show (II) denote by $S(t)$ the analytic semigroup in $L_p(\Omega)$ generated by $-\Delta$ with Dirichlet boundary conditions. Then

$$0 \leq u(\cdot, t) \leq \int_0^t S(t-s) u(\cdot, s)^\gamma \, ds. \tag{16}$$

Using the explicit kernel for the heat equation one shows

$$\|S(t)u\|_\infty \leq K(\Omega, p) \exp(-\lambda t) t^{-n/2p} \, \|u\|_p \, , \tag{17}$$

where $\lambda > 0$ is the principal eigenvalue of $-\Delta$ in Ω. Hence by (16)–(17)

$$\|u(\cdot, t)\|_\infty \leq K \left(\int_0^\infty e^{-\lambda s} s^{-n/2p} \, ds \right) \sup_{0 < s < t} \|u(\cdot, s)\|_p$$

which implies (II). (I) and (II) together yield

$$\|u\|_\infty \leq K \, \|u^\gamma\|_p \leq K \, \|u\|_r^{r/p} \, \|u\|_\infty^{\gamma - r/p} .$$

Since $\gamma - r/p < 1$ one solves for $\|u\|_\infty$ and gets the desired estimate

$$\|u\|_\infty \leq \left(K \, \|u\|_r^{r/p} \right)^{1/(1 - (\gamma - r/p))} .$$

For this "feedback" argument to work we have to choose $p \in (1,\infty)$ such that

$$(\gamma - 1)/r < 1/p \text{ and } 1/p < \min(1, 2/n).$$

This is possible if and only if $(\gamma - 1)n/2r < 1$. □

The Lipschitz condition usually assumed in existence proofs is not needed in Proposition 1. For equality in (11) or (12) the feedback argument breaks down. In the case of sublinear equations with $\gamma = 1$ conditions (11)–(12) do not restrict r_1 or r_2. As shown in [**11**], in this case an estimate of $\|v(t)\|_\infty$ follows already from the weaker a priori estimate

$$\sup\left\{\int_{t_1}^{t_2}\int_\Omega \Phi(|v(x,s)|)\,dx\,ds \mid t_2 - t_1 < 1, t_i \in (0,T)\right\} < \infty,$$

where $\Phi = \Phi(v)$ is a continuous function $[0,\infty) \to [0,\infty)$ such that $\lim_{v\to\infty}\Phi(v) = \infty$.

The proof of Proposition 2 uses the variation of constants formula

$$v(\cdot, t + \Delta t) = v(\cdot, t) + \int_0^{\Delta t} S(\Delta t - s) f(\cdot, t + s, v(\cdot, t + s))\,ds$$

for arbitrary $t \in (0,T)$ and Δt small enough. The advantage of this method is that assumption (11) can be dropped, which improves the result for dimension $n = 1$. More importantly, it is possible to get a weakened result for the limit case of equality in (12).

PROPOSITION 3. *Let f satisfy the Lipschitz condition (f3) and extend the solution v to the maximal time interval $(0, T_{\max})$. Let the exponents $q_1 \in [1,\infty]$, $q_2 \in (1,\infty], p_0, \gamma \in (1,\infty)$ occuring in (f1) satisfy*

$$n/2q_1 + 1/q_2 + (\gamma - 1)n/2p_0 = 1. \tag{18}$$

If $T_{\max} < \infty$ and for some (hence all) $s \in (0, T_{\max})$ the set

$$k(\Pi) = \left\{v(\cdot,\tau)/\|v(\cdot,\tau)\|_{p_0} \mid \tau \in (s, T_{\max})\right\}$$

is precompact in the space $L_{p_0}(\Omega)$, then

$$\lim_{t\to T_{\max}} \|v(\cdot,t)\|_{p_0} = \infty.$$

In the limit case no functional relation like (13) or (15) holds. For the case that (14) holds with inequality sign reversed Weissler [**13**] gives an example of nonexistence of solutions and Haraux [**4**] gives for $\Omega = \mathbb{R}^n$ an example of multiple solutions. The nontrivial solution is bounded in L_{p_0} but unbounded in L_∞ and hence violates (13). Propositions 1 and 2 can not only be exploited to characterize the behavior at $T_{\max} < \infty$ but also to prove global existence of solutions if some a priori estimate is available.

PROPOSITION 4. *Let f satisfy the two-sided bound (f1) and the Lipschitz condition (f3). Let the exponents $r_1 \in (0,\infty)$, $r_2 \in (0,\infty], p_0, q_1, q_2 \in [1,\infty]$, $\gamma \in [1,\infty)$ satisfy (11). Assume that there exists a function φ such that for all $v_0 \in C_0^\infty(\bar{\Omega})$ the (unique classical) solution of the initial-boundary value problem (6)–(7)–(8) on any time interval $(0,T)$ satisfies an a priori estimate*

$$\|v_1\|_{r_1,r_2,T} \leq \varphi(\|v_0\|_{p_0}).$$

Then for initial data $v_0 \in L_{p_0}(\Omega)$ the initial-boundary value problem (6)–(7)–(8) has a global solution v on the interval $[0,\infty)$ satisfying

$$\lim_{t\to 0} \|v(\cdot,t) - S(t)v_0\|_{p_0} = 0,$$

$$\sup_{t>0} \min(1,t)^{n/2p_0} \|v(\cdot,t)\|_\infty < \infty.$$

This solution is not necessary unique. If $f = f(x,t,v)$ is α-Hölder continuous with respect to (x,t), for $t > 0$ this is a classical solution satisfying (6)–(7).

Application to Special Examples.

1. The predator-prey system with one diffusing and one sedentary species:

$$\begin{aligned}
u_t &= \pm u(1-v) && \text{for } x \in \Omega, t > 0;\\
v_t &= \Delta v \mp v(1-u) && \text{for } x \in \Omega, t > 0;\\
\partial v/\partial n &= 0 && \text{for } x \in \partial\Omega, t > 0;\\
u(x,0) &= u_0(x) && \text{for } x \in \Omega;\\
v(x,0) &= v_0(x) && \text{for } x \in \Omega.
\end{aligned}$$

According to the interpretation of u and v as densities of species (see [**9**]) we take smooth initial data $u_0 > 0$, $v_0 > 0$. By the maximum principle the solution satisfies $u(x,t) > 0$, $v(x,t) > 0$ for $x \in \Omega$, $t > 0$. Define the function $s = s(u) = u - \log u - 1$. Clearly $s \geq 0$, $s'' \geq 0$ for $u > 0$. An elementary calculation shows that the functional

$$\Phi(t) = \frac{1}{|\Omega|}\int_\Omega [s(u(x,t)) + s(v(x,t))]\,dx \geq 0 \text{ satisfies}$$

$$d\Phi/dt = \frac{1}{|\Omega|}\int_\Omega (\nabla v/v)^2\,dx \geq 0.$$

Hence Φ is bounded on the maximal existence interval $[0, T_{\max})$. By

$$s\left(\frac{1}{|\Omega|}\int u(x,t)\,dx\right) \le \frac{1}{|\Omega|}\int s(u(x,t))\,dx \text{ (Jensen's inequality)}$$

we get the global L_1-bound

$$\sup\{\|u(\cdot,t)\|_1 + \|v(\cdot,t)\|_1 \mid t \in [0, T_{\max})\} < \infty. \tag{19}$$

We have to distinguish the cases (a) and (b) (lower and upper signs).

(a) Consider first the simpler case of diffusing prey (lower signs). The equation for v implies

$$v_t - \Delta v \le v. \tag{20}$$

Using (19)–(20), Proposition 1 can be applied with $\gamma = 1$, $q_1 = q_2 = \infty$ (weight function $c \equiv 1$) and $r_1 = 1$, $r_2 = \infty$. Hence we get

$$\sup\{\|v(\cdot,t)\|_\infty \mid t \in [0, T_{\max})\} < \infty. \tag{21}$$

Elementary considerations about the ordinary differential equation for u show $T_{\max} = \infty$. Note that we cannot get a global L_∞-bound for u.

(b) Take the case of diffusing predator (upper signs). To the equation for v:

$$v_t - \Delta v = v(-1 + u),$$

we apply Proposition 2 with $\gamma = 1$, $c = (-1 + u)$ as weight function and $q_1 = 1$, $q_2 = \infty$, $r_1 = 1$, $r_2 = \infty$. Condition (14) turns out to be satisfied only in the one-dimensional case $n = 1$. Then we conclude as in case (a) that (21) holds and $T_{\max} = \infty$. Note that Proposition 1 cannot be used.

2. Consider the following system from nuclear engineering (see [**12**]):

$$\begin{aligned}
u_t &= \Delta u + u(\alpha - v) && \text{for } x \in \Omega, t > 0;\\
v_t &= -v + u && \text{for } x \in \Omega, t > 0;\\
u &= 0 && \text{for } x \in \partial\Omega, t > 0;\\
u(x,0) &= u_0(x) && \text{for } x \in \Omega;\\
v(x,0) &= v_0(x) && \text{for } x \in \Omega.
\end{aligned}$$

Take smooth initial data $u_0, v_0 > 0$. Then the maximum principle implies $u(x,t) > 0$ for all $x \in \Omega, t > 0$. For $\alpha > \alpha_0$ there exists a unique positive equilibrium $U = V = U(x)$. Here α_0 is the principal eigenvalue of $-\Delta$ in Ω. One shows that the functional Φ is nonincreasing:

$$\Phi(t) = \int_\Omega U^2(x)s(u(x,t)/U(x)) + U(x)(v(x,t) - U(x))^2/2\, dx.$$

Hence Φ is bounded and Jensen's inequality implies

$$\sup\left\{\int_\Omega U(x)u(x,t) + U(x)v^2(x,t)\, dx \mid t \in [0, T_{\max})\right\} < \infty.$$

Let $s \in (0,1)$, $r \in (0,1/2)$ and $1 + 1/s = 1/r$. Then $\int_\Omega U^{-s}\, dx < \infty$. Hölder's inequality implies

$$\left(\int_\Omega u^r\, dx\right)^{1/r} \leq \int_\Omega Uu\, dx \left(\int_\Omega U^{-s}\, dx\right)^{1/s} \quad \text{and hence}$$
$$\sup\{\|u(\cdot,t)\|_r + \|v(\cdot,t)\|_{2r} \mid t \in [0, T_{\max})\} < \infty.$$

Proposition 1 can be applied to the equation for $u > 0$ taking $c = \alpha$ as weight function and $\gamma = 1$, $q_1 = \infty$, $r_1 = r$, $q_2 = r_2 = \infty$. Hence (11)–(12) can be fulfilled for any space dimension n. We conclude

$$\sup\{\|u(\cdot,t)\|_\infty \mid t \in [0, T_{\max})\} < \infty$$

and by considering the differential equation for v that $T_{\max} = \infty$. Using compactness arguments and the functional Φ, it is shown in [**11**] that the solution $u(x,t)$ converges to the equilibrium $U(x)$ uniformly in x for $t \to \infty$.

Universität der Bundeswehr München, Institut für Mathematik,
Werner-Heisenberg-Weg 39, D8014 Neubiberg WEST GERMANY

References

1. N. D. Alikakos, *L_p-bounds of solutions of reaction-diffusion equations*, Comm. Part. Diff. Equ. **4** (1979), 827-868.
2. J. M. Ball, *Remarks on blow-up and nonexistence theorems for nonlinear evolution equations*, Quart. J. Math. Oxford **28** (1977), 473–486.
3. A. Friedman, "Partial Differential Equations," Holt, Rinehart and Winston, 1969.
4. A. Haraux and F. B. Weissler, *Nonuniqueness for a semilinear initial value problem*, preprint (1981).
5. D. Henry, *Geometric Theorem of Semilinear Parabolic Equations*, Springer-Verlag, New York, Springer Lecture Notes in Math. **840** (1981).
6. H. Kielhöfer, *Global solutions of semilinear evolution equations satisfying an energy inequality*, J. Diff. Equ. **36** (1980), 188–222.
7. O. A. Ladyzenskaja, V. A. Solonnikov and N. N. Ural'cera, "Linear and Quasilinear Equations of Parabolic Type," Transl. of Math. Monographs, vol. 23, Am. Math. Soc., Providence, R. I., 1968.
8. P. Massatt, *Obtaining L_∞-bounds from L_p-bounds for solutions of various parabolic partial differential equations*, Conference Proceedings, Intern. Symposium on Dynamical Systems, University of Florida (1981).
9. R. May, "Stability and Complexity in Model Ecosystems," Princeton University Press, Princeton, N. J., 1974.
10. F. Rothe, *Uniform bound from bounded L_p-functions in reaction-diffusion equations*, J. Diff. Equ. **45** (1982), 207–233.
11. F. Rothe, *Global Solutions of Reaction-Diffusion Systems*, Springer-Verlag, New York, Springer Lecture Notes in Math. **1072** (1984).
12. E. T. Rumble and W. E. Kastenberg, *On the application of eigenfunction expansion to problems in nonlinear space-time reactor dynamics*, Nucl. Sci. Eng. **49** (1972), 172–187.
13. F. B. Weissler, *Local existence and nonexistence for semilinear parabolic equations in L_p*, Indiana Univ. J. **29** (1980), 79–102.
14. S. Williams and P. Chow, *Nonlinear reaction-diffusion models for interacting populations*, J. Math. Anal. Appl. **62** (1978), 157–169.

Qualitative Behavior for a Class of Reaction-Diffusion-Convection Equations

PAUL SACKS

Consider the initial and boundary value problem

$$v_t = v_{xx} + (g(v))_x + f(v) \quad 0 < x < L \quad t > 0 \tag{1}$$

$$v(0,t) = v(L,t) = 0 \qquad t > 0 \tag{2}$$

$$v(x,0) = v_0(x) \geq 0 \qquad 0 < x < L \tag{3}$$

where f and g are in general nonlinear functions on R^+, satisfying certain structure conditions. We are interested in describing the structure of the (semi)-dynamical system governed by (1)–(2), namely, what are the equilibrium states, what is the behavior of solutions close to a given equilibrium state, and what else can be said concerning the global description of the set of all possible solutions of (1)–(2).

If $g \equiv 0$ the equation is of reaction-diffusion type, and much is known under various conditions on f; see e.g. Henry [**3**], Matano [**7**], or Smoller [**10**]. On the other hand if $f \equiv 0$ then a well-known special case is Burger's equation ($g(v) = v^2$) which is easy to analyze because of the Hopf-Cole transformation (Hopf [**5**]). In any case if $f = 0$ (or more generally if $f(v) \leq 0$ for $v \geq 0$) and g is any reasonable function, the dynamics are trivial, namely all solutions tend to zero exponentially as $t \to \infty$. In order to obtain some interesting structure one must consider source terms f which are somewhere positive.

A notable difference between the cases $g \equiv 0$ and $g \not\equiv 0$ is that in the former case the dynamical system has gradient structure, while this is in general false for the latter case. As is true for systems of O.D.E.'s the gradient-like property can be exploited in many ways. For example, when $g \equiv 0$ the steady states of (1)–(2) may be obtained as solutions of variational problems. Another interesting consequence is that the ω-limit set of a precompact orbit must be contained in the set of equilibrium states; the fact that in certain cases the ω-limit set must reduce to a single point is also tied up with this structure (Matano [**8**], Hale and Massat [**2**]). Analogous results about the stabilization of solutions of (1)–(2)–(3) have been proved by Zelenyak [**11**] who constructs a Lyapunov function for a fairly general quasilinear parabolic equation in one space variable. However such results are definitely false in more than one space dimension. For example, in two or more dimensions one can write down examples of semilinear autonomous problems that have time-periodic solutions which are not steady states. Thus no general result about the stabilization of bounded orbits is possible. See also Matano [**9**] and Hirsch [**4**] for analyses of a class of dynamical systems which includes those generated by equations like (1).

In this article we will discuss some special cases of (1)–(2)–(3), namely

$$f(v) = v^p \quad p > 1 \tag{4}$$

$$g(v) = \varepsilon v^m \quad m \geq 1 \tag{5}$$

where $\varepsilon \in \mathsf{R}$ is a parameter. Our point of view is that the situation is well understood for $\varepsilon = 0$, and we want to understand how the structure of the dynamical system changes as ε varies. We need only consider $\varepsilon \geq 0$ since the change of variables $x \to L - x$ takes ε to $-\varepsilon$.

Details of the proofs and discussion of more general nonlinearities can be found in Chen, Levine and Sacks [**1**] and Levine, Payne, Sacks and Straughan [**6**]. The organization is as follows. We first discuss existence and multiplicity results for the steady state problem. Second, we discuss their local stability properties. Finally, we make further remarks about the global structure of the solutions of (1)–(2).

Steady state solutions.

We consider the boundary value problem

$$u'' + \varepsilon(u^m)' + u^p = 0 \quad 0 < x < L \tag{6}$$

$$u(0) = u(L) = 0 \qquad u > 0 \qquad 0 < x < L. \tag{7}$$

For $\varepsilon = 0$ it is well-known that for any $p > 1$ there is a unique solution of (6)–(7). When $\varepsilon \neq 0$ the answer will depend in some way on m, p and ε. For purposes of discussion we will consider m and p to be fixed; we may then speak of the solution diagram for (6)–(7), namely a representation of $\|u\|_{L^\infty}$ versus ε whenever u is a solution of (6)–(7) for a given ε.

To begin with there are certain things which are true for any $p > 1$, $m \geq 1$. First, there is a curve of solutions $u(\cdot, \varepsilon)$ of (6)–(7) emanating from the known $\varepsilon = 0$ solution. Also $\|u(\cdot,\varepsilon)\|_{L^\infty}$ is a function of ε near $(0, \|u(\cdot,0)\|_{L^\infty})$, hence this curve cannot return to the $\varepsilon = 0$ axis for any positive value of $\|u\|_{L^\infty}$, by uniqueness for the $\varepsilon = 0$ problem. Next, it can be shown that there exists $\delta > 0$ independent of ε such that $\|u\|_{L^\infty} \geq \delta$ for any solution of (6)–(7). Now an argument using topological degree theory shows that this curve of solutions cannot be bounded in the $(\varepsilon, \|u\|_{L^\infty})$ plane, hence either there exists a solution (6)–(7) for every ε or else there is a branch of solutions along which $\|u\|_{L^\infty}$ tends to ∞.

Further analysis involves mainly the derivation of a priori estimates, i.e. identifying portions of the $(\varepsilon, \|u\|_{L^\infty})$ plane in which solutions cannot be found. The results now depend on p and m; we find that there are at least four distinct possibilities.

Case 1. $1 < p \leq m$. In this case there exists $\varepsilon_0 > 0$ such that there are no solutions for $\varepsilon > \varepsilon_0$, at least one solution for $\varepsilon = \varepsilon_0$ and at least two solutions for $0 < \varepsilon < \varepsilon_0$. As $\varepsilon \to 0^+$ there is a maximal solution of (6)–(7) tending to ∞ in L^∞. Based on numerical evidence we conjecture that there is always exactly one solution for $\varepsilon = \varepsilon_0$ and exactly two for $0 < \varepsilon < \varepsilon_0$.

Case 2. $m < p < 2m - 1$. There still exists $\varepsilon_0 > 0$ such that (6)–(7) has no solution for $\varepsilon > \varepsilon_0$, but now there also exists $\varepsilon_1 \in (0, \varepsilon_0]$ such that for $0 < \varepsilon < \varepsilon_1$ there is exactly one solution. It is not true in general that $\varepsilon_0 = \varepsilon_1$; for example if p is sufficiently close to m there exists an interval of ε for which there are at least two solutions. We conjecture that $\varepsilon_1 < \varepsilon_0$ for all $p \in (m, 2m-1)$ and that (6)–(7) has two solutions for $\varepsilon \in (\varepsilon_1, \varepsilon_0)$. In any case there must be a branch of solutions along which ε remains bounded but $\|u\|_{L^\infty}$ tends to ∞. However it is also true that u remains bounded in $L^\infty_{\text{loc}}(0, L)$ along this branch. It can be shown that there exists a function $U^* > 0$ such that U^* satisfies (6) on $(0, L)$, $U^*(L) = 0$, and U^* is the limit (at least along a subsequence) of this unbounded branch of solutions.

Case 3. $p = 2m - 1$. In this case there exists $\varepsilon_0 > 0$ such that (6)–(7) has no solution for $\varepsilon \geq \varepsilon_0$, and exactly one solution for $0 \leq \varepsilon < \varepsilon_0$. As $\varepsilon \to \varepsilon_0^-$ there is a singular limit as in Case 2.

Case 4. $p > 2m - 1$. There is exactly one solution for every $\varepsilon \geq 0$.

In Figure 1 we illustrate these four situations with solution diagrams taken from [**1**] computed by T. F. Chen. In each case $m = 2$ and the representative values of p are 1.5 (Fig. 1a), $p = 2.6$ (Fig. 1b), $p = 3$

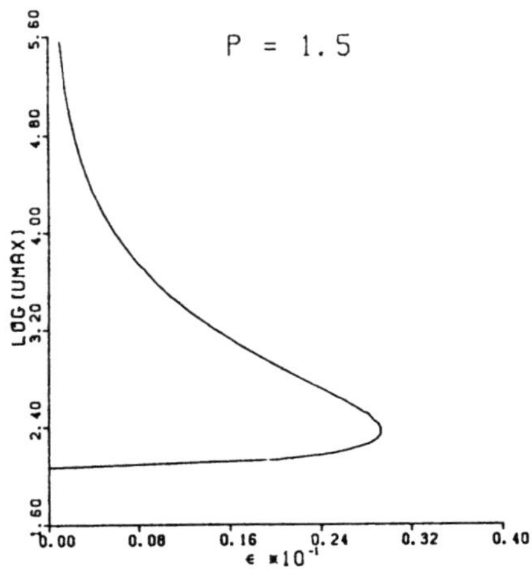

Figure 1a

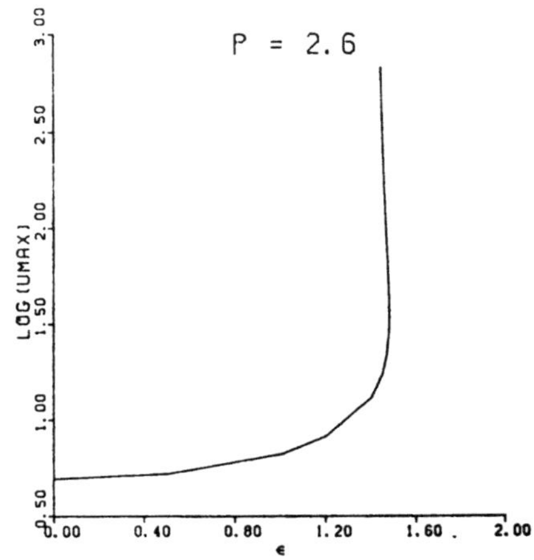

Figure 1b

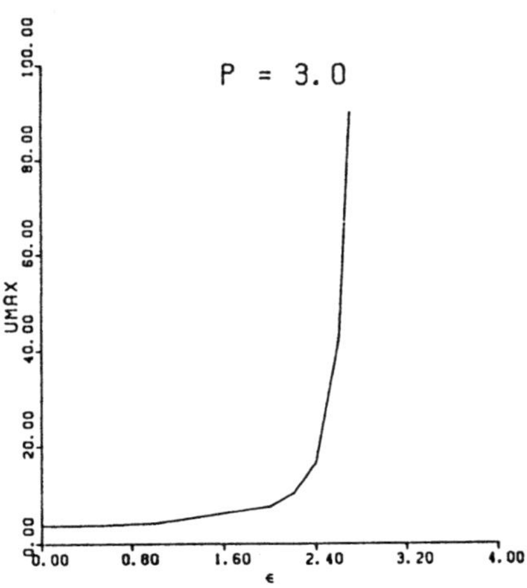

Figure 1c

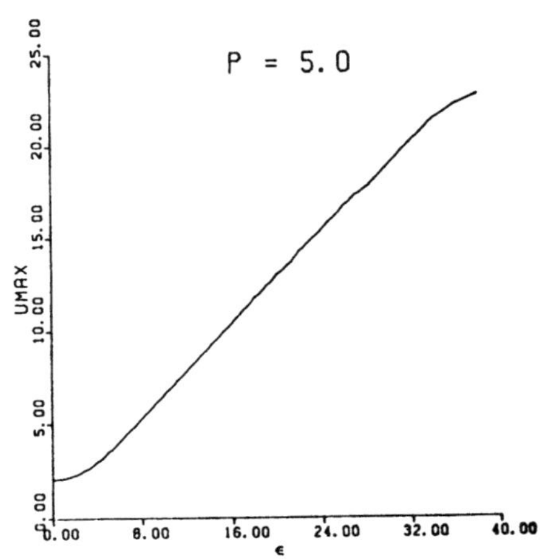

Figure 1d

(Fig. 1c), and $p = 5$ (Fig. 1d).

Now we will briefly indicate some of the argument used in the proofs of the above facts.

For the case $1 < p \leq m$ we have an a priori estimate

$$\|u\|_{L^\infty} \leq C(\varepsilon) < \infty \quad \text{for } \varepsilon > 0 \tag{8}$$

where u is any solution of (6)–(7). Here is a proof for $1 < p < m$. We integrate the equation from 0 to L to get

$$u'(L) - u'(0) = \int_0^L u^p(x)\, dx. \tag{9}$$

Next, if we multiply by u', integrate from 0 to L, integrate by parts and use the equation in a straightforward way, there follows

$$u'(L)^2 - u'(0)^2 = -2\varepsilon \int_0^L u^{m+p}(x)\, dx. \tag{10}$$

Eliminating $u'(0)$ between (8) and (9) we obtain

$$u'(L) = \frac{\varepsilon \int_0^L u^{m+p}(x)\, dx}{\int_0^L u^p(x)\, dx} - \frac{1}{2} \int_0^L u^p(x)\, dx. \tag{11}$$

If u is a solution of (6)–(7) then $u'(L) < 0$, hence

$$\int_0^L u^{m+p}(x)\, dx \leq \frac{1}{2\varepsilon} \left(\int_0^L u^p(x)\, dx \right)^2 \leq \frac{C}{\varepsilon} \int_0^L u^{2p}(x)\, dx \tag{12}$$

from which it follows easily that

$$\|u\|_{L^{m+p}} \leq \frac{C}{\varepsilon^{\frac{1}{m-p}}}. \tag{13}$$

Since we also have (multiplying the equation by u and integrating)

$$\int_0^L (u'(x))^2\, dx = \int_0^L u^{p+1}(x)\, dx \tag{14}$$

the estimate (8) follows from Sobolev's inequality.

Another useful tool in the analysis of (6)–(7) is the following change of variables. Let

$$y = \int_0^x u^{m-1}(s)\, ds \tag{15}$$

and

$$h(y) = u^m(x(y)). \tag{16}$$

If $u(x)$ solves (6)–(7) then $h(y)$ satisfies

$$h'' + m\varepsilon h + mh^{\frac{p+1-m}{m}} = 0 \qquad 0 < x < L \tag{17}$$

$$h(0) = h(R) = 0 \qquad h > 0 \quad 0 < y < R. \tag{18}$$

where $R = \int_0^L u^{m-1}(x)\,dx$.

From the equation (17) we see that it is plausible that some change of behavior might occur at $p = 2m - 1$. In the case that $p = 2m - 1$ the equation is linear and hence easily analyzable. For example a solution h of (16)–(17) is impossible for $\varepsilon \geq \sqrt{\frac{4}{m}}$ no matter what R is, while if $\varepsilon < \sqrt{\frac{4}{m}}$, R is uniquely determined by ε. One can also show that solutions of (6)–(7) are necessarily ordered, hence no more than one solution can correspond to a given R.

Behavior near an equilibrium solution.

To analyze the behavior of solutions of (1)–(2)–(3) near an equilibrium solution one can proceed, at least formally, by linearization. Rigorous theorems may be obtained by application of results in Henry [**3**].

Aside from the solutions of (6)–(7) we have also to consider the behavior near $u \equiv 0$; we will also discuss the "stability of ∞". This is really the blow-up question. If $v_0(x)$ is sufficiently large, does $v(x,t)$ tend to ∞ in finite or infinite time?

First of all, $u \equiv 0$ is always a strongly stable equilibrium solution of (1)–(2)–(3) for any choice of $m \geq 1$, $p > 1$ and $\varepsilon \in \mathbb{R}$. Indeed if $m > 1$ then the linearization about $u \equiv 0$ is just the heat equation, while if $m = 1$ it is a simple calculation to show that the corresponding linear problem still has zero asymptotically stable.

Now aside from the steady state $u \equiv 0$ there may or may not exist other steady states. We will denote by $u_\varepsilon > 0$ the minimal solution (6)–(7) whenever it exists. Recalling the earlier remarks (i) that solutions of (6)–(7) are necessarily ordered and (ii) there exists $\delta > 0$ such that $\|u\|_{L^\infty} \geq \delta$

for any solution of (6)–(7), we see that if there exists any solution of (6)–(7) then a minimal solution u_ε exists.

Now employing some ideas from Matano [9] we can show that u_ε is always unstable from below, whenever it exists. Indeed the map $v_0 \to v(\cdot, t)$ can be shown to be a strongly order-preserving local semidynamical system; see [9] for definitions. Hence the arguments of [9] (e.g. Lemma 3.1) show that if u_ε were stable from below then since 0 is stable from above there would have to exist another equilibrium solution between 0 and u_ε, contradicting the minimality of u_ε. Now in general we cannot assert that u_ε is unstable from above, since for example at the turning point in Figure 1a the solution should be expected to be stable from above and unstable from below. In any case it is clear that 0 is an eigenvalue of the linearized operator in this situation. We can, however, show that the set of ε for which u_ε is strongly unstable is open and its complement is nowhere dense.

Turning now to the behavior near ∞ we have the following: if $p \le m$ then any solution of (1)–(2)–(3) with $v_0 \in L^\infty$ is bounded uniformly for all t. If $p > m$ then for v_0 sufficiently large, $v(x,t) \to \infty$ in finite time.

The proof of uniform boundedness for $p \le m$ is achieved by exhibiting arbitrarily large time-independent supersolutions. There may be found in the form Ae^{-Bx} for constants A and B properly chosen. The proof of finite time blow-up for $p > m$ is a variation of the eigenfunction method. One examines the behavior of

$$h(t) = \int_0^L v(x,t) \left(\sin \tfrac{\pi}{L} x\right)^n \, dx \qquad n = \frac{p}{p-m}$$

to see that $h(t) \to \infty$ in finite time if $h(0)$ is large enough.

Finally, if $p \le m$ then because of the estimate (8) there must exist a maximal solution u^ε of (6)–(7) for $0 < \varepsilon < \varepsilon_0$, $u^\varepsilon \neq u_\varepsilon$. Using the fact that all solutions of (1)–(2)–(3) must be uniformly bounded in this case, we can show that u^ε is stable from above wherever it exists. Again, we cannot expect that u^ε is always strongly stable, but the set of ε for which u^ε fails to be strongly stable is nowhere dense.

Other remarks.

(i) If $p \le m$ and $\varepsilon > \varepsilon_0$ then $u \equiv 0$ is the only nonnegative steady state and all solutions of (1)–(2)–(3) with $v_0 \in L^\infty(0, L)$ are uniformly bounded.

By standard regularity theory $\|v(\cdot,t)\|_{C^{2+\alpha}(0,L)}$ is then uniformly bounded as $t \to \infty$ and so by Theorem 1 of [**11**] and the maximum principle we must have $v(\cdot,t)$ converging to a nonnegative steady state as $t \to \infty$. Hence $u \equiv 0$ is a global attractor for positive solutions in this case.

(ii) In the case that $m < p \leq 2m-1$ and $\varepsilon > \varepsilon_0$ we have the situation that $u \equiv 0$ is a strongly stable equilibrium, solutions of (1)–(2)–(3) with v_0 sufficiently large tend to infinity in finite time, and there are no positive equilibrium states. In the case that $g \equiv 0$ such a state of affairs is not possible, as is pointed out by Matano [**7**].

(iii) Results like those described here may be obtained if u^p and u^m are replaced by more general nonlinearities f and g respectively; see [**1**] and [**6**] for details. Roughly speaking, the condition that $p \leq m$ can be replaced either by

$$\varlimsup_{s\to\infty} \frac{f(s)}{g(s)} = 0 \text{ and } \int_0^\infty \sqrt{g'(s)}\,ds = \infty$$

or

$$\varlimsup_{s\to\infty} \frac{f'(s)}{g'(s)} < \infty \text{ and } \lim_{s\to\infty} g'(s) = \infty.$$

The marginal case $p = 2m-1$ corresponds to the more general condition $f(s)/g(s)g'(s) =$ constant, while the cases $p < 2m-1$ and $p > 2m-1$ correspond to $\lim_{s\to\infty} f(s)/g(s)g'(s) = 0$ or ∞ respectively.

Department of Mathematics, Iowa State University, Ames IA 50011

This research sponsored by the Air Force Office of Scientific Research, Air Force Systems Command, USAF under Grant 84-0252.

Research at MSRI supported in part by NSF Grant DMS 812079-05.

REFERENCES

1. T. F. Chen, H. A. Levine and P. E. Sacks, *Analysis of a convective reaction-diffusion equation*, preprint.
2. J. K. Hale and P. Massat, *Asymptotic behavior of gradient-like systems.*
3. D. Henry, "Geometric Theory of Semilinear Parabolic Equations," Springer-Verlag, 1981.
4. M. Hirsch, *Differential equations and convergence almost everywhere in strongly monotone flows*, Contemp. Math. **17**, 267–285.
5. E. Hopf, *The partial differential equation* $u_t + uu_x = \mu u_{xx}$, Comm. Pure Appl. Math. **3** (1950), 201–230.
6. H. A. Levine, L. E. Payne, P. E. Sacks and B. Straughan, in preparation.
7. H. Matano, *Asymptotic behavior and stability of solutions of semilinear diffusion equations*, Pub. R. I. M. S., Kyoto University **15** (1979), 401–454.
8. H. Matano, *Convergence of solutions of one-dimensional semilinear parabolic equations*, J. Math. Kyoto Univ. (1978), 221–227.
9. H. Matano, *Existence of nontrivial unstable sets for equilibriums of strongly order-preserving systems*, J. Fac. Sci. Univ Tokyo **30** (1983), 645–673.
10. J. Smoller, "Shock Waves and Reaction-Diffusion Equations," Springer-Verlag, 1983.
11. T. I. Zelenyak, *Stabilization of solutions of boundary value problems for a second order parabolic equation in one space variable*, Diff. Equ. **4** (1968), 17–22.

Resonance and Higher Order Quasilinear Ellipticity

VICTOR L. SHAPIRO

1. Introduction.

Let Ω be a bounded domain in $\mathbb{R}^N$. With

$$(1.1) \qquad Qu = \sum_{|\alpha|,|\beta|\le m} (-1)^{\beta} D^{\beta}[a_{\alpha\beta}(x, Du)D^{\alpha}u],$$

we shall study the equation

$$(1.2) \qquad Qu = f(x, Du) + \sum_{|\alpha|\le m} D^{\alpha} g_{\alpha}(x, Du)$$

defined in Ω where $\beta = (\beta_1, \dots, \beta_N)$, $D^{\beta} = D_1^{\beta_1}, \dots, D_N^{\beta_N}$ and $|\beta| = \beta_1 + \dots + \beta_N$. With $M =$ the number of all derivatives $D^{\alpha}u$ for $0 \le |\alpha| \le m-1$, Du will stand for the M-vector whose components are the derivatives $D^{\alpha}u$ for $0 \le |\alpha| \le m-1$. Thus if $m = 1$, then $Du = u$ and $M = 1$. If $m = 2$, $Du = (u, D_1u, \dots, D_Nu)$ and $M = N + 1$. Also, we shall let η stand for a point in $\mathbb{R}^M$. Thus for $m = 2$, $\eta = (\eta_0, \eta_1, \dots, \eta_N)$ where $\eta_0 = \eta_{(0,0,\dots,0)}$ and $\eta_1 = \eta_{(1,0,\dots,0)}$. In particular, we see that η_0 stands for the first component of η and corresponds to u in the M-vector Du.

For the operator Q in (1.1), we shall also assume the following:

(Q-1) The coefficients $a_{\alpha\beta}(x, \eta)$ satisfy the usual Caratheodory conditions for $|\alpha|, |\beta| \le m$, i.e., for each fixed $\eta \in \mathbb{R}^M$, $a_{\alpha\beta}(x, \eta)$ is measurable on Ω; for a.e. $x \in \Omega$, $a_{\alpha\beta}(x, \eta)$ is continuous in $\mathbb{R}^M$.

(Q-2) $a_{\alpha\beta}(x, \eta) \in L^{\infty}(\Omega \times \mathbb{R}^M)$ for $|\alpha|, |\beta| \le m$.

With Q in (1.1) meeting (Q-1) and (Q-2), we associate the form

$$(1.3) \qquad Q(w, u, v) = \sum_{|\alpha|,|\beta|\le m} \int_{\Omega} a_{\alpha\beta}(x, Dw) D^{\alpha}u D^{\beta}v$$

defined for $u, v, w \in H_0^m$ $(= W_0^{m,2}(\Omega))$. We observe that $Q(w, u, v)$ is bilinear in u and v.

Next, we introduce the symmetric uniformly elliptic operator L of order $2m, m \geq 1$.

(1.4)
$$Lu = \sum_{|\alpha|,|\beta| \leq m} (-1)^{|\beta|} D^\beta [b_{\alpha\beta}(x) D^\alpha u]$$

where we assume

(L-1) $b_{\alpha\beta} \in L^\infty(\Omega)$ for $|\alpha|, |\beta| \leq m$. Also if $|\alpha| = |\beta| = m$, $b_{\alpha\beta}$ is uniformly continuous in Ω.

(L-2) L is uniformly elliptic, that is, there is a constant $c_0 > 0$ such that

$$\sum_{|\alpha|=|\beta|=m} b_{\alpha\beta}(x) \xi^\alpha \xi^\beta \geq c_0 |\xi|^{2m}$$

$\forall x \in \Omega$ and $\forall \xi \in \mathbf{R}^N (|\xi|^2 = \xi_1^2 + \cdots + \xi_N^2)$.

To L, we associate the bilinear form

$$\mathcal{L}(u,v) = \sum_{|\alpha|,|\beta| \leq m} \int_\Omega b_{\alpha\beta}(x) D^\alpha u D^\beta v$$

defined for $u, v \in H_0^m$. We shall also assume

(L-3) L is symmetric, that is, $\mathcal{L}(u,v) = \mathcal{L}(v,u)$ $\forall u, v \in H_0^m$.

We shall say Q is quasilinear elliptic of type I with respect to L if under the above assumptions on Q and L,

(1.5)
$$\mathcal{L}(u,u) \leq Q(w,u,u) \qquad \forall u, w \in H_0^m.$$

This definition encompasses the usual definition of quasilinearity in the order 2 case [**2**, p. 259]. To see this, let

$$Qu = \sum_{i,j=1}^N -D_i[a_{ij}(x,u) D_j u]$$

where $Qu = \sum_{i,j=1}^N a_{ij}(x,s) \xi_i \xi_j \geq c_0 |\xi|^2$ $\forall x \in \Omega, s \in \mathbf{R}, \xi \in \mathbf{R}^N$ with $c_0 > 0$.

Take $Lu = -c_0 \Delta u = -c_0 \sum_{i=1}^N D_i[D_i u]$. Then for $u, w \in H_0^m$,

$$Q(w,u,u) = \sum_{i,j=1}^N \int_\Omega a_{ij}(x,w) D_i u D_j u \geq c_0 \sum_{i=1}^N \int_\Omega |D_i u|^2 = \mathcal{L}(u,u).$$

Therefore, with the role of L being played by $-c_0\Delta$, Q is quasilinear elliptic of type I with respect to L and our assertion is established.

Next, we define

$$(1.6)\qquad \lambda_1^* = \inf Q(u,u,u)/\langle u,u\rangle_0 \qquad u \not\equiv 0, u \in H_0^m$$

where $\langle\cdot,\cdot\rangle_0$ is the $L^2(\Omega)$ inner product. If Q were linear and symmetric, then λ_1^* would be the first eigenvalue of Q. Nevertheless, even though Q is quasilinear, we shall show that certain well known theorems which hold for linear symmetric uniformly elliptic operators and their first eigenvalue also hold for Q and λ_1^*. In particular, we shall establish some existence theorems for equation (1.2) of this nature. In order to accomplish this, we shall need some assumptions on f and g. We shall assume

(f-1) $f(x,\eta)$ satisfies the Caratheodory conditions [see (Q-1)].

(f-2) For every $r > 0$, $\exists\varsigma_r \in L^2(\Omega)$ such that if $|\eta_0| \le r$, then $|f(x,\eta)| \le \varsigma_r(x)$ for $\eta \in \mathbf{R}^M$ and a.e. $x \in \Omega$. Here (and in the sequel) η_0 stands for the first component of the vector η which corresponds to u in the vector Du.

(f-3) $\exists\varepsilon_1 > 0$ such that

$$\eta_0 f(x,\eta) \le (\lambda_1^* - \varepsilon_1)\eta_0^2 + h(x)|\eta_0|$$

$\forall\eta \in \mathbf{R}^M$ and a.e. $x \in \Omega$.

(g-1) $g_\alpha(x,\eta)$ satisfies the Caratheodory conditions.

(g-2) $\forall\varepsilon > 0$, $\exists h_{\alpha\varepsilon}(x) \in L^2(\Omega)$ such that $|g_\alpha(x,\eta)| \le \varepsilon|\eta_0| + h_{\alpha\varepsilon}(x)$ $\forall\eta \in \mathbf{R}^M$ and a.e. $x \in \Omega$.

The first theorem we prove is the following.

THEOREM 1. *Suppose that Q in (1.1) is quasilinear elliptic of type I with respect to L and that $f(x,\eta)$ and $g_\alpha(x,\eta)$ satisfy (f-1), (f-2), (f-3), and (g-1), (g-2) respectively for $|\alpha| \le m$. Then $\exists u \in H_0^m$ with $f(x,Du) \in L^1(\Omega)$ such that u is a distribution solution of (1.2), that is*

$$(1.7)\qquad Q(u,u,\varphi) = \int_\Omega f(x,Du)\varphi + \sum_{|\alpha|\le m} (-1)^{|\alpha|} \int_\Omega g_\alpha(x,Du)D^\alpha\varphi$$

$\forall\varphi \in C_0^\infty(\Omega)$ where $Q(u,u,\varphi)$ is defined in (1.3).

The above theorem constitutes a generalization of a well known semilinear elliptic result [**1**, p. 4] to the quasilinear elliptic case (see also [**3**, p. 575]). The $D^\alpha g_\alpha$ part above is new even in the semilinear case.

We shall also establish a resonance result, i.e., where ε_1 in (f-3) is allowed to approach zero, in this paper. But first we need more background. Let L be a symmetric uniformly elliptic operator of order $2m$ given by (1.4) satisfying (L-1), (L-2), and (L-3). Then as is well known, there is associated to L a sequence of number $\lambda_1 \leq \lambda_2 \leq \cdots \leq \lambda_n \leq \cdots \to +\infty$ and a sequence of functions $\{\psi_n\}_{n=1}^{\infty}$ in H_0^m with the following properties:

(1.8) $\quad \mathcal{L}(\psi_n, v) = \lambda_n \langle \psi_n, v\rangle_0 \qquad$ for $v \in H_0^m$ and $n = 1, 2, \dots$;

(1.9) $\quad \{\psi_n\}_{n=1}^{\infty}$ is a complete orthonormal system in $L^2(\Omega)$.

Furthermore,

$$\lambda_1 = \inf \mathcal{L}(u,u)/\langle u,u\rangle_0, \qquad u \not\equiv 0, u \in H_0^m. \tag{1.10}$$

We shall say Ω is a domain of type B with respect to L if in addition to (1.8) and (1.9) the following holds.

$$\begin{aligned} &(a) \quad \psi_n \in L^\infty(\Omega) \cap H_0^m \qquad \text{for } n = 1, 2, \dots ; \\ &(b) \quad \|\varphi - \sum_{k=1}^{n} \hat{\varphi}(k)\psi_k\|_{L^\infty} \to 0 \quad \forall \varphi \in C_0^\infty(\Omega) \\ &\qquad\qquad \text{where } \hat{\varphi}(k) = \langle \varphi, \psi_k\rangle_0. \end{aligned} \tag{1.11}$$

Using [**2**, Thm. 8.15] for the case $m = 1$, it is not difficult to see that every bounded domain in $\mathbb{R}^2$ and $\mathbb{R}^3$ is of type B with respect to L for $m = 1, 2, \dots$.

The resonance result that we shall obtain in this paper will be for f and g of the form $f(x,s)$ and $g(x,\eta)$ where $x \in \Omega$, $s \in \mathbb{R}$, and $\eta \in \mathbb{R}^M$ We shall assume

(f-1)′ $f(x,s)$ satisfies the Caratheodory conditions.

(f-2)′ For every $r > 0$, $\exists \varsigma_r(x) \in L^2(\Omega)$ such that if $|s| \leq r$, then $|f(x,s)| \leq \varsigma_r(x)$ for a.e. $x \in \Omega$.

(f-3)′ $\forall \varepsilon > 0$, $\exists h_\varepsilon \in L^2(\Omega)$ such that

$$s f(x,s) \leq (\lambda_1^* + \varepsilon)s^2 + |s| h_\varepsilon(x) \quad \text{for } s \in \mathbb{R} \text{ and a.e. } x \in \Omega.$$

If Q is quasilinear elliptic of type I with respect to L and if furthermore $\lambda_1 = \lambda_1^*$ where λ_1 is given by (1.10) and λ_1^* by (1.6), we shall say Q is $*$-related to L.

We intend to establish the following resonance result.

THEOREM 2. *Suppose that Q in (1.1) is quasilinear elliptic of type I with respect to L and that $f(x,s)$ and $g_\alpha(x,\eta)$ satisfy (f-1)$'$, (f-2)$'$, (f-3)$'$ and (g-1),(g-2) respectively for $|\alpha| \le m$. Suppose furthermore that Ω is a domain of type B with respect to L and Q is $*$-related to L. Set $\overline{\lim}_{s\to\infty} f(x,s)/s = \mathcal{F}_+(x)$ and $\overline{\lim}_{s\to-\infty} f(x,s)/s = \mathcal{F}_-(x)$. Then if*

$$\int_{\Omega,\psi>0} \mathcal{F}_+\psi^2 + \int_{\Omega,\psi>0} \mathcal{F}_-\psi^2 < \lambda_1 \int_\Omega \psi^2 \tag{1.12}$$

for every λ_1-eigenfunction ψ of L, $\exists u \in H_0^m$ with $f(x,u) \in L^1(\Omega)$ such that u is a distribution solution of

$$Q(u) = f(x,u) + \sum_{|\alpha|\le m} D^\alpha g_\alpha(x, Du). \tag{1.13}$$

The above theorem constitutes a generalization of a well-known semilinear elliptic result [1, p. 10] to the quasilinear elliptic case. The $D^\alpha g_\alpha(x,Du)$ part is also new even in the semilinear case.

It is easy to show that in a certain sense condition (1.12) is best possible, i.e., if we allow equality in (1.12) then no solution of (1.13) may exist. To see this, take $\Omega = \{x : 0 < x_1 < \pi, 0 < x_2 < \pi\}$ and consider the boundary value problem:

$$-\Delta u = 2u - \sin x_1 \sin x_2 \text{ for } x \in \Omega \text{ and } u \in H_0^1.$$

It is clear that this BVP has no solution, but that with $f(x,s) = 2s - \sin x_1 \sin x_2, g_\alpha = 0$ for $|\alpha| \le 1$, and $Q = L = -\Delta$, equality is obtained in (1.12).

Before proceeding with the proof of the above theorems, we give an example of a nontrivial situation where Q is $*$-related to L. Take Ω to be any bounded domain in $\mathbb{R}^3$, take $Lu = \Delta^2 u$ and $Qu = \Delta[1 + (1+u^2)^{-1}]\Delta u$. Then

$$Q(w,u,u) = \mathcal{L}(u,u) + \int_\Omega (1+w^2)^{-1}(\Delta u)^2. \tag{1.14}$$

$\forall w, u \in H_0^2$. Since (1.5) holds, it follows that Q is quasilinear elliptic of type I. It remains to show that $\lambda_1 = \lambda_1^*$. Let ψ_1 be a λ_1-eigenfunction of L with the property that $\langle \psi_1, \psi_1 \rangle_0 = 1$. Then it follows from (1.8), (1.10) and (1.14) on using the sequence $u_n = n\psi_1$ in (1.6) that $\lambda_1 = \lambda_1^*$ provided

$$\lim_{n\to\infty} \int_\Omega (1 + n^2\psi_1^2)^{-1}(\Delta\psi_1)^2 = 0. \tag{1.15}$$

To establish this fact, we observe that ψ_1 is real analytic in Ω. Consequently, the set $\{x \in \Omega : \psi_1(x) = 0\}$ is of μ-measure zero where μ = 3-dimensional Lebesgue measure. From this fact it follows that the set $A_\delta = \{x \in \Omega : |\psi_1(x)| \leq \delta\}$ is such that $\lim_{\delta \to 0} \mu(A_\delta) = 0$.

Next, we observe that

$$\int_\Omega [1 + n^2\psi_1^2]^{-1}(\Delta\psi_1)^2 \leq [n^2\delta^2]^{-1} \int_{\Omega - A_\delta} (\Delta\psi_1)^2 + \int_{A_\delta} (\Delta\psi_1)^2.$$

Since $\psi_1 \in H_0^2, (\Delta\psi_1)^2 \in L^1(\Omega)$. It therefore follows from this last calculation that

$$\overline{\lim_{n\to\infty}} \int_\Omega [1 + n^2\psi_1^2]^{-1}(\Delta\psi_1)^2 \leq \int_{A_\delta} (\Delta\psi_1)^2.$$

But the right hand side of this last inequality is arbitrarily small with δ, and we conclude that (1.15) is indeed true.

2. Fundamental Lemmas.

A consideration of $Q_1 u = Qu + \lambda u$, $L_1 u = Lu + \lambda u$, and $f_1(x,u) = f(x,u) + \lambda u$, shows from the start in proving Theorems 1 and 2, we can assume that

$$\lambda_1 > 0 \qquad \text{and} \qquad \lambda_1^* > 0. \tag{2.1}$$

Assumption (2.1) will hold throughout this paper. We shall also assume henceforth that (Q-1), (Q-2), (L-1), (L-2), and (L-3) hold. Also, we shall assume that Q is quasilinear elliptic of type I with respect to L, i.e., (1.5) holds. The first lemma we prove is the following.

LEMMA 1. *Let $F(x)$ be a nonnegative function in $L^2(\Omega)$ and suppose $f(x,\eta)$ and $g_\alpha(x,\eta)$ satisfy (f-1) and (g-1), respectively for $|\alpha| \leq m$. Suppose also that $|f(x,\eta)| \leq F(x)$ and $|g_\alpha(x,\eta)| \leq F(x)$ $\forall \eta \in \mathbb{R}^M, |\alpha| \leq m$, and a.e. $x \in \Omega$. Then if n is a given positive integer, there are constants $\{\gamma_k^n\}_{k=1}^n$ and a function $u_n = \gamma_1^n\psi_1 + \cdots + \gamma_n^n\psi_n$ such that*

$$Q(u_n, u_n, \psi_k) = \langle f(\cdot, Du_n), \psi_k\rangle_0 + \sum_{|\alpha|\leq m} (-1)^{|\alpha|}\langle g_\alpha(\cdot, Du_n), D^\alpha\psi_k\rangle_0 \tag{2.2}$$

for $k = 1, \ldots, n$.

To establish the lemma, set for each $\xi \in \mathbb{R}^n$,

$$A_{k\ell}(\xi) = Q(\xi_q \psi_q, \psi_\ell, \psi_k) \qquad k, \ell = 1, \ldots, n$$

where the summation convention is used for $q = 1, \ldots, n$. We observe from (Q-1) and (Q-2) that $A_{k\ell}(\xi) \in C(\mathbb{R}^n)$ and from (1.6) that for $\rho = (\rho_1, \ldots, \rho_n) \in \mathbb{R}^n$, $\lambda_1^* |\rho|^2 \leq \rho_k A_{k\ell}(\xi) \rho_\ell$. Consequently it follows from (2.1) that the matrix $A(\xi)$ with components $A_{k\ell}(\xi)$ is nonsingular for every $\xi \in \mathbb{R}^n$. Furthermore, it follows that $\|[A(\xi)]^{-1}\|_{\mathcal{M}} \leq (\lambda_1^*)^{-1} \, \forall \xi \in \mathbb{R}^n$, where $\|\cdot\|_{\mathcal{M}}$ designates the usual $n \times n$ matrix norm. Next for $k = 1, \ldots, n$, we set

$$S_k(\xi) = \langle f(\cdot, D\xi_q \psi_q), \psi_k \rangle_0 + \sum_{|\alpha| \leq m} (-1)^{|\alpha|} \langle g_\alpha(\cdot, D\xi_q \psi_q), D^\alpha \psi_k \rangle_0.$$

The conclusion to the lemma, namely (2.2), follows from the above facts in a manner similar to [**5**, pp. 572–3].

LEMMA 2. *Under the assumptions of Lemma 1, for each positive integer n, let u_n be the solution obtained which satisfies (2.2). Then $\{\|u_n\|_m\}_{n=1}^\infty$ is uniformly bounded.*

Here $\|\cdot\|_m$ designates the usual H^m-norm. To prove the lemma, we put u_n in (2.2), multiply both sides by γ_k^n and sum on k. Consequently, we have that (2.2) holds with u_n replacing ψ_k. Next, we apply (1.5), (1.10), and the hypothesis of the lemma to obtain

$$\lambda_1 \|u_n\|_0^2 \leq \langle F, |u_n| \rangle_0 + \sum_{|\alpha| \leq m} \langle F, |D^\alpha u_n| \rangle_0.$$

Applying Schwartz's inequality and (2.1) to this last inequality, we see that there is a constant K such that $\|u_n\|_0^2 \leq K \|u_n\|_m \; \forall n$.

From Garding's inequality, we have $\exists c$ and c' positive constants such that

$$c\|u_n\|_m^2 - c'\|u_n\|_0^2 \leq \mathcal{L}(u_n, u_n) \leq Q(u_n, u_n, u_n). \tag{2.3}$$

Applying (2.2) once again as above, we obtain

$$\begin{aligned} c\|u_n\|_m^2 &\leq \langle F, |u_n| \rangle_0 + \sum_{|\alpha| \leq m} \langle F, |D^\alpha u_n| \rangle_0 + c'\|u_n\|_0^2 \\ &\leq (M' + 1)\|F\|_0 \|u_n\|_m + c'K\|u_n\|_m \end{aligned}$$

where M' is the number of α such that $|\alpha| \leq m$. Therefore, $\|u_n\|_m \leq c^{-1}[(M'+1)\|F\|_0 + c'K]$. □

LEMMA 3. *Under the assumptions of Lemma 1,* $\exists u \in H_0^m$ *such that*

$$Q(u,u,v) = \langle f(\cdot,u),v\rangle_0 + \sum_{|\alpha|\le m} (-1)^{|\alpha|}\langle g_\alpha(\cdot,Du),D^\alpha v\rangle_0. \tag{2.4}$$

$\forall v \in H_0^m$.

With u_n designating the function in Lemma 1, it follows from Lemma 2 that the sequence $\{\|u_n\|_m\}_{n=1}^\infty$ is uniformly bounded. It consequently follows that there is a function $u \in H_0^m$ and a subsequence, which for ease of notation we take to be the full sequence, such that (i) $D^\alpha u_n \to D^\alpha u$ weakly in $L^2(\Omega)$ for $|\alpha| = m$; (ii) $u_n \to u$ strongly in H_0^{m-1}; (iii) $D^\alpha u_n \to D^\alpha u$ a.e. in Ω for $|\alpha| \le m-1$. It is easy to see from the hypothesis of Lemma 1, (Q-1), (Q-2), these three facts and the uniform boundedness of $\{\|u_n\|_m\}_{n=1}^\infty$ that $\langle a_{\alpha\beta}(\cdot,Du_n)D^\beta\psi_k, D^\alpha u_n\rangle_0 \to \langle a_{\alpha\beta}(\cdot,Du)D^\beta\psi_k, D^\alpha u\rangle_0$, $\langle f(\cdot,Du_n),\psi_k\rangle_0 \to \langle f(\cdot,Du),\psi_k\rangle_0$, $\langle g_\alpha(\cdot,Du_n),D^\alpha\psi_k\rangle_0 \to \langle g_\alpha(\cdot,Du), D^\alpha\psi_k\rangle_0$ as $n\to\infty$ for $|\alpha|,|\beta| \le m$ and $\forall k$. We conclude from (1.3) and Lemma 1 that (2.2) holds with u_n replaced by u and $k = 1,2,\dots$. Next using the notation of (1.11)(b), given $v \in H_0^m$, we set $v_n = \sum_{k=1}^n \hat{v}(k)\psi_k$. Under the assumption of (2.1), it is well known that $\|v_n - v\|_m \to 0$ as $n\to\infty$. Since we have that

$$Q(u,u,v_n) = \langle f(\cdot,Du),v_n\rangle + \sum_{|\alpha|\le m} (-1)^{|\alpha|}\langle g_\alpha(\cdot,Du),D^\alpha v_n\rangle_0,$$

it follows immediately from the convergence of $v_n \to v$ in H_0^m that (2.4) holds. □

Proof of Theorem 1.

To establish Theorem 1, we let $\eta = (\eta_0,\eta^*)$ where η^* represents the components of η other than η_0 and set

$$f^J(x,\eta) = \begin{cases} f(x,J,\eta^*) & \eta_0 \ge J \\ f(x,\eta) & -J \le \eta_0 \le J \\ f(x,-J,\eta^*) & \eta_0 \le -J. \end{cases}$$

We define $g_\alpha^J(x,\eta)$ in a similar manner. It then follows from (f-2) and (g-2) that $f^J(x,\eta)$ and $g_\alpha^J(x,\eta)$ meet the assumptions of Lemma 1. Consequently it follows from Lemma 3 that $\exists$ a sequence $\{u^J\}_{J=1}^{\infty}$ in H_0^m such that

$$(3.1)\quad Q(u^J,u^J,v)=\langle f^J(\cdot,Du^J),v\rangle+\sum_{|\alpha|\le m}(-1)^{|\alpha|}\langle g_\alpha^J(\cdot,Du^J),D^\alpha v\rangle_0$$

$\forall v\in H_0^m$. We claim that

$$(3.2)\qquad \text{the sequence } \{\|u^J\|_m\}_{J=1}^{\infty} \text{ is uniformly bounded.}$$

To show that (3.2) is true, we set $v=u^J$ in (3.1) and observe from (1.6) and (f-3) that

$$\lambda_1^*\|u^J\|_0^2\le(\lambda_1^*-\varepsilon_1)\|u^J\|_0^2+\langle|h|,|u^J|\rangle_0+\sum_{|\alpha|\le m}|\langle g_\alpha^J(\cdot,Du^J),D^\alpha u^J\rangle_0|.$$

So, in particular,

$$(3.2')\quad \|u^J\|_0^2\le\varepsilon_1^{-1}\left[\|h\|_0\|u^J\|_0+\sum_{|\alpha|\le m}|\langle g_\alpha^J(\cdot,Du^J),D^\alpha u^J\rangle_0|\right]\quad\forall J.$$

Next, using Garding's inequality as in (2.3) with u_n replaced by u^J and using (3.1) with v replaced by u^J, we obtain

$$c\|u^J\|_m^2\le c'\|u^J\|_0^2+\langle f^J(\cdot,Du^J),u^J\rangle_0+\sum_{|\alpha|\le m}|\langle g_\alpha^J(\cdot,Du^J),D^\alpha u^J\rangle_0|$$

where c and c' are positive constants. Using $(3.2)'$ and (f-3) in conjunction with this last inequality gives

$$c\|u^J\|_m^2\le c_2\|h\|_0\|u^J\|_0+c_2\sum_{|\alpha|\le m}|\langle g_\alpha^J(\cdot,Du^J),Du^J\rangle_0|$$

where $c_2=(c'+\lambda_1^*-\varepsilon_1)\varepsilon_1^{-1}+1$.

Now the last term in the inequality above, on choosing $\varepsilon=c(2c_2M')^{-1}$ in (g-2), is dominated by $c2^{-1}\|u^J\|_0\|u^J\|_m+\sum_{|\alpha|\le m}\|h_{\alpha\varepsilon}\|_0\|u^J\|_m$ where M' is the number of α such that $|\alpha|\le m$. Using the fact that $\|u^J\|_0\le\|u^J\|_m$, we conclude from the above inequality that

$$c2^{-1}\|u^J\|_m\le c_2\|h\|_0+c_2\sum_{|\alpha|\le m}\|h_{\alpha\varepsilon}\|_0$$

for $J = 1, 2, \ldots$. Since c is a positive constant, (3.2) is established.

From (3.2), it follows that $\exists u \in H_0^m$ such that (using the full sequence for ease of notation)

$$(3.3) \qquad \begin{aligned} &(a) \quad D^\alpha u^J \to D^\alpha u \text{ weakly in } L^2(\Omega) \text{ for } |\alpha| = m \\ &(b) \quad \|u^J - u\|_{m-1} \to 0; \\ &(c) \quad D^\alpha u^J \to D^\alpha u \qquad \text{a.e. in } \Omega \text{ for } |\alpha| \le m - 1. \end{aligned}$$

From (Q-1) and (3.3)(c), it follows that $a_{\alpha\beta}(x, Du^J) \to a_{\alpha\beta}(x, Du)$ a.e. in Ω for $|\alpha|, |\beta| \le m$. Consequently, we see from (Q-2), (1.3), and (3.2) that $Q(u^J, u^J, \varphi) - Q(u, u^J, \varphi) \to 0$ for $\varphi \in C_0^\infty(\Omega)$. On the other hand, we see from (Q-2) and (3.3)(a) that $Q(u, u^J, \varphi) \to Q(u, u, \varphi)$ for $\varphi \in C_0^\infty(\Omega)$. We conclude

$$(3.4) \qquad \lim_{J \to \infty} Q(u^J, u^J, \varphi) = Q(u, u, \varphi) \qquad \forall \varphi \in C_0^\infty(\Omega).$$

We next set $\varepsilon = 1$ in (g-2) and observe from (3.2) that the sequence $\{\langle g_\alpha^J(\cdot, Du^J), D^\alpha u^J \rangle_0\}_{J=1}^\infty$ is uniformly bounded for $|\alpha| \le m$. Consequently, it follows from (3.1) with u^J replacing v, and (1.5), (2.1), and (3.2) that

$$\{\langle f^J(\cdot, Du^J), u^J \rangle_0\}_{J=1}^\infty \text{ is uniformly bounded from below.}$$

Next, setting $A_J = \{x \in \Omega : u^J(x) f(x, Du^J(x)) \ge 0\}$ and proceeding as in [**5**, Lemmas 3 and 4] using (f-3), (3.2), and this last established fact concerning uniform boundedness from below, we obtain that the sequence

$$(3.5) \qquad \{f^J(x, Du^J)\}_{J=1}^\infty \text{ is absolutely equiintegrable on } \Omega,$$

i.e., given $\varepsilon > 0$, $\exists \delta$ such that if $\mu(E) < \delta$, then $\int_E |f^J(x, Du^J)| < \varepsilon \ \forall J$ where μ is N-dimensional Lebesgue measure. From (3.3)(c), (f-2), and the definition of f^J, it follows that

$$(3.6) \qquad \lim_{J \to \infty} f^J(x, Du^J(x)) = f(x, Du(x)) \quad \text{a.e. in } \Omega.$$

Since the closure of Ω is compact, it follows from (3.5) that there is a constant K such that $\int_\Omega |f^J(x, Du^J)| \le K \ \forall J$ and consequently from Fatou's lemma [4, p. 24] that

$$(3.7) \qquad f(x, Du) \in L^1(\Omega).$$

From (3.5), (3.6), (3.7), and Egoroff's theorem [**4**, p. 75], we obtain that

$$(3.8)\qquad \lim_{J\to\infty}\int_\Omega f^J(x,Du^J)\varphi = \int_\Omega f(x,Du)\varphi \qquad \forall\varphi\in C_0^\infty(\Omega).$$

Next, we see from (3.3)(c), (g-2), and the definition of g_α^J that

$$(3.9)\qquad \lim_{J\to\infty} g_\alpha^J(x,Du^J(x)) = g_\alpha(x,Du(x)) \quad \text{a.e. in } \Omega.$$

for $|\alpha|\le m$. Also, it follows from (3.3)(b) that the sequence $\{|u^J|^2\}_{J=1}^\infty$ is absolutely equiintegrable on Ω. Consequently, on choosing $\varepsilon=1$ in (g-1), it follows that the sequence $\{|g_\alpha^J(x,Du^J)|^2\}_{J=1}^\infty$ is absolutely equiintegrable on Ω. This fact in conjunction with (3.9) and Egoroff's theorem shows that $\lim_{J\to\infty}\int_\Omega |g_\alpha^J(x,Du^J)-g_\alpha(x,Du)|^2 = 0$ for $|\alpha|\le m$. But then it follows that

$$(3.10)\qquad \lim_{J\to\infty}\int_\Omega g_\alpha^J(x,Du^J)D^\alpha\varphi = \int_\Omega g_\alpha(x,Du)D^\alpha\varphi$$

for $|\alpha|\le m$ and $\forall\varphi\in C_0^\infty(\Omega)$.

Noting from (3.7) that $f(x,Du)\in L^1(\Omega)$, the conclusion to the theorem, namely (1.7), follows on setting $v=\varphi$ in (3.1) and using (3.4), (3.8), and (3.10). □

Proof of Theorem 2.

We define $f^J(x,s)$ and $g_\alpha^J(x,\eta)$ as in the proof of Theorem 1. Then the assumptions of Lemma 1 apply and for each fixed positive integer n we obtain a sequence of functions $u_n^J = \sum_{k=1}^n \gamma_{kJ}^n\psi_k$ such that

$$(4.1)\qquad Q(u_n^J,u_n^J,\psi_k) = \langle f^J(\cdot,u_n^J),\psi_k\rangle_0 - \langle u_n^J,\psi_k\rangle_0 n^{-1} + \sum_{|\alpha|\le m}(-1)^{|\alpha|}\langle g_\alpha^J(\cdot,Du_n^J),D^\alpha\psi_k\rangle_0$$

$\forall J, |\alpha|\le m, 1\le k\le n$. It is easy to see using the definition of u_n^J that we can replace ψ_k by u_n^J in (4.1). Doing this, choosing $\varepsilon=(2n)^{-1}$ in (f-3)$'$, and using (1.6) and (2.1) shows that

$$(4.2)\qquad \|u_n^J\|_0^2(2n)^{-1} \le \|h_{(2n)^{-1}}\|_0\|u_n^J\|_0 + \sum_{|\alpha|\le m}|\langle g_\alpha^J(\cdot,Du_n^J),D^\alpha u_n^J\rangle_0|.$$

Now from (1.8) and (1.9) we have that

$$\mathcal{L}(u_n^J, u_n^J) = \sum_{k=1}^{n} \lambda_k |\gamma_{kJ}^n|^2 \leq \lambda_n \sum_{k=1}^{n} |\gamma_{kJ}^n|^2 = \lambda_n \|u_n^J\|_0^2.$$

So it follows from Garding's inequality (as in (2.3)) that

$$\text{(4.3)} \qquad \|u_n^J\|_m^2 \leq (\lambda_n + c')c^{-1}\|u_n^J\|_0^2 \qquad \forall J,$$

where c and c' are positive constants. Calling the constant on the right hand side of the inequality in (4.3) K_n^2, and taking $\varepsilon = (K_n 4nM')^{-1}$ in (g-2) where M' = the number of α such that $|\alpha| \leq m$, we see from (4.2) and (4.3)) that

$$\|u_n^J\|_0(4n)^{-1} \leq \|h_{(2n)^{-1}}\|_0 + K_n \sum_{|\alpha|\leq m} \|h_{\alpha\varepsilon}\|_0 \qquad \forall J.$$

We conclude that for fixed n, the sequence $\{\|u_n^J\|_0\}_{J=1}^{\infty}$ is uniformly bounded. Consequently $\sum_{k=1}^{n} |\gamma_{kJ}^n|^2$ is uniformly bounded. Therefore (using the full sequence for ease of notation), we see that $\lim_{J\to\infty} \gamma_{kJ}^n = \gamma_k^n$ for $k = 1, \ldots, n$. Setting $u_n = \sum_{k=1}^{n} \gamma_k^n \psi_k$, it follows from (1.11)(a) that $\lim_{J\to\infty} u_n^J = u_n$ in $L^\infty(\Omega) \cap H_0^m$. Using the techniques that we used in establishing Theorem 1, it then follows from (4.1) that

$$\text{(4.4)} \quad Q(u_n, u_n, \psi_k) = \langle f(\cdot, u_n), \psi_k\rangle_0 - \langle u_n, \psi_k\rangle_0 n^{-1} + \sum_{|\alpha|\leq m} (-1)^\alpha \langle g_\alpha(\cdot, Du_n), D^\alpha \psi_k\rangle_0$$

for $1 \leq k \leq n$, and $n = 1, 2, \ldots$ where $u_n = \sum_{k=1}^{n} \gamma_k^n \psi_k$.

We claim

$$\text{(4.5)} \qquad \text{the sequence } \{\mathcal{L}(u_n, u_n)\}_{n=1}^{\infty} \text{ is uniformly bounded.}$$

Suppose to the contrary that this sequence is unbounded. Then it follows (using the full sequence for ease of notation) that

$$\text{(4.6)} \qquad \lim_{n\to\infty} \mathcal{L}_n = +\infty \text{ where } \mathcal{L}_n = \{\mathcal{L}(u_n, u_n)\}^{\frac{1}{2}}.$$

We shall show (4.6) contradicts (1.12). In order to accomplish this, we set

$$(4.7) \qquad v_n = u_n \mathcal{L}_n^{-1} \text{ where } u_n = \sum_{k=1}^{n} \gamma_k^n \psi_k.$$

From (1.8) and (1.9), it follows that

$$(4.8) \qquad 1 = \mathcal{L}(v_n, v_n) = \sum_{k=1}^{n} \lambda_k |\gamma_k^n|^2 \mathcal{L}_n^{-2} \geq \lambda_1 \|v_n\|_0^2$$

for $n = 1, 2, \ldots$. Also, as is well known, $\mathcal{L}(w, w) = \sum_{k=1}^{\infty} \lambda_k |\hat{w}(k)|^2 \geq \lambda_1 \|w\|_0^2 \ \forall w \in H_0^m$ where $\hat{w}(k) = \langle w, \psi_k \rangle_0$. By assumption (2.1), $\lambda_1 > 0$. Consequently, $\mathcal{L}(\cdot, \cdot)$ is an inner product on H_0^m. From (4.8), we have that both $\{\mathcal{L}(v_n, v_n)\}_{n=1}^{\infty}$ and $\{\|v_n\|_0^2\}_{n=1}^{\infty}$ are uniformly bounded. Furthermore from Garding's inequality (see (2.3)), we have that $\{\|v_n\|_m^2\}_{n=1}^{\infty}$ is uniformly bounded. We conclude that $\exists v \in H_0^m$ (using the full sequence for ease of notation) such that

$$(4.9) \qquad \lim_{n \to \infty} \|v_n - v\|_{m-1} = 0;$$

$$(4.10) \qquad \lim_{n \to \infty} v_n(x) = v(x) \qquad \text{a.e. in } \Omega.$$

Now, as is well known, λ_1 is of finite multiplicity. Suppose it is of multiplicity k_1, i.e., $\lambda_1 = \cdots = \lambda_{k_1} < \lambda_{k_1+1}$. Then we claim

$$(4.11) \qquad v = \gamma_1 \psi_1 + \cdots + \gamma_{k_1} \psi_{k_1}.$$

To establish this claim, we observe first that (4.4) holds with ψ_k replaced by u_n. Next, we use this fact in conjunction with (1.5) and (4.7) to obtain

$$(4.12) \qquad \mathcal{L}(v_n, v_n) + \|v_n\|_0^2 n^{-1} \leq \mathcal{L}_n^{-2} \Big\{ \langle f(\cdot, u_n), u_n \rangle_0 + \sum_{|\alpha| \leq m} |\langle g_\alpha(\cdot, Du_n), D^\alpha u_n \rangle_0| \Big\} \quad \forall n.$$

Now by the hypothesis of the theorem, $\lambda_1 = \lambda_1^*$. Consequently it follows from (f-3)′ that

$$\langle f(\cdot, u_n), u_n \rangle_0 \leq (\lambda_1 + \varepsilon) \sum_{k=1}^{n} |\hat{u}_n(k)|^2 + \|h_\varepsilon\|_0 \|u_n\|_0$$

where $\hat{u}_n(k) = \gamma_k^n$. Also, we observe from (g-2) and Garding's inequality (2.3) that

$$(4.13)\quad |\langle g_\alpha(\cdot, Du_n), D^\alpha u_n\rangle_0| \le \varepsilon c^{-1}[\mathcal{L}_n^2 + c'\|u_n\|_0^2] + \|h_{\alpha\varepsilon}\|_0 c^{-\frac{1}{2}}\left[\mathcal{L}_n^2 + c'\|u_n\|_0^2\right]^{\frac{1}{2}}.$$

We conclude from these last two inequalities in conjunction with (4.8) and (4.12) that for $n > k_1$

$$\sum_{k=k_1+1}^{n} (\lambda_k - \lambda_1)|\hat{v}_n(k)|^2 + n^{-1}\sum_{k=1}^{n}|\hat{v}_n(k)|^2 \le \varepsilon\lambda_1^{-1} + \|h_\varepsilon\|_0\lambda_1^{-\frac{1}{2}}\mathcal{L}_n^{-1} + \varepsilon c^{-1}[1 + c'\lambda_1^{-1}]M' + \mathcal{L}_n^{-1}\sum_{|\alpha|\le m}\|h_{\alpha\varepsilon}\|_0 c^{-\frac{1}{2}}[1 + c'\lambda_1^{-1}]^{\frac{1}{2}}.$$

where M' is the number of α such that $|\alpha| \le m$.

From this last inequality, we obtain from (4.6) and (4.8) that

$$\overline{\lim_{n\to\infty}}\sum_{k=k_1+1}^{n}(\lambda_k - \lambda_1)|\hat{v}_n(k)|^2 \le \varepsilon\lambda_1^{-1} + \varepsilon c^{-1}[1 + c'\lambda_1^{-1}]M'.$$

Since ε is an arbitrary positive number and λ_1, c, and c' are positive constants, it follows in particular that

$$(4.14)\qquad \lim_{n\to\infty}\sum_{k=k_1+1}^{n}(\lambda_k - \lambda_1)|\hat{v}_n(k)|^2 = 0.$$

From the fact that $\lambda_1 > 0$, we have that $\{\psi_k\lambda_k^{-\frac{1}{2}}\}_{k=1}^{\infty}$ is a complete orthonormal system in H_0^m with respect to the inner product $\mathcal{L}(\cdot,\cdot)$. We let H_{02}^m be the infinite dimensional subspace spanned by $\{\psi_k\lambda_k^{-\frac{1}{2}}\}_{k=k_1+1}^{\infty}$ and H_{01}^m be the finite dimensional subspace spanned by $\{\psi_k\lambda_k^{-\frac{1}{2}}\}_{k=1}^{k_1}$. Then $H_0^m = H_{01}^m \oplus H_{02}^m$ and every $w \in H_0^m$ can be written uniquely in the form $w = w' + w''$ where $w' \in H_{01}^m, w'' \in H_{02}^m$ and $\mathcal{L}(w', w'') = 0$. So (4.11) will follow, if we can show $v'' = 0$, i.e., $\hat{v}(k) = 0$ for $k \ge k_1 + 1$. But from (4.14), since $(\lambda_k - \lambda_1) > 0$ for $k \ge k_1 + 1$, it follows that $\lim_{n\to\infty}|\hat{v}_n(k)| = 0$ for $k \ge k_1 + 1$. However, $\hat{v}_n(k) = \langle v_n, \psi_k\rangle_0$, and it follows from (4.9) that $\lim_{n\to\infty}\langle v_n, \psi_k\rangle_0 = \langle v, \psi_k\rangle_0$. Therefore, $\hat{v}(k) = 0$ for $k \ge k_1 + 1$ and (4.11) is established.

Next, since $\lambda_1 = \cdots = \lambda_{k_1} < \lambda_{k_1+1} \leq \cdots \leq \lambda_n \to \infty$, it follows that there is a $\delta > 0$, such that $\delta\lambda_k \leq \lambda_k - \lambda_1$ for $k \geq k_1 + 1$. Consequently, we obtain from (4.14) that

$$\lim_{n\to\infty} \sum_{k=k_1+1}^{n} \lambda_k |\hat{v}_n(k)|^2 = 0. \tag{4.15}$$

On the other hand, we have from (4.7) and (4.8) that for $n \geq k_1 + 1$

$$1 = \sum_{k=1}^{k_1} \lambda_k |\hat{v}_n(k)|^2 + \sum_{k=k_1+1}^{n} \lambda_k |\hat{v}_n(k)|^2 = \mathcal{L}(v_n, v_n).$$

Since by (4.9) $\lim_{n\to\infty} \hat{v}_n(k) = \hat{v}(k)$, we obtain from this last equality and (4.15) that $\lambda_1 \sum_{k=1}^{k_1} |\hat{v}(k)|^2 = 1$. We conclude in particular from this fact and (4.11) that

$$\lim_{n\to\infty} \mathcal{L}(v_n, v_n) = \lambda_1 \|v\|_0^2 = 1. \tag{4.16}$$

Next, we return to (4.4), replace ψ_k by u_n and use (1.5) to obtain

$$n^{-1}\|v_n\|_0^2 + \mathcal{L}(v_n, v_n) \leq \mathcal{L}_n^{-2}\Big\{\langle f(\cdot, u_n), u_n\rangle_0 + \sum_{|\alpha|\leq m} |\langle g_\alpha(\cdot, Du_n), D^\alpha u_n\rangle_0|\Big\}.$$

Observing that ε in (4.13) is an arbitrary positive number, we see that

$$\lim_{n\to\infty} \mathcal{L}_n^{-2} \sum_{|\alpha|\leq m} |\langle g_\alpha(\cdot, Du_n), D^\alpha u_n\rangle_0| = 0.$$

Also, from (4.8), it is clear that $\lim_{n\to\infty} n^{-1}\|v_n\|_0^2 = 0$. We consequently conclude from the above inequality and (4.16) that

$$\lambda_1 \|v\|_0^2 \leq \overline{\lim_{n\to\infty}} \int_\Omega f(x, u_n)\mathcal{L}_n^{-1} v_n. \tag{4.17}$$

From (4.11) and (4.16), we have that v is a nonzero λ_1-eigenfunction of L. Furthermore, we see that if $u_n(x) \neq 0$, $f(x, u_n)\mathcal{L}_n^{-1} v_n = f(x, u_n) u_n^{-1} v_n^2$. Thus, we see that (4.17) is exactly analogous to [1, (21) on p. 10]. Proceeding in a similar manner to that which is done in [1, pp. 10–11] we arrive at the fact that

$$\lambda_1 \|v\|_0^2 \leq \int_{v>0} \mathcal{F}_+ v^2 + \int_{v>0} \mathcal{F}_- v^2.$$

Since v is a nonzero λ_1-eigenfunction of L, we have a contradiction to (1.12). So, (4.6) is false and (4.5) is true.

Now from (4.4), we have that $u_n = \sum_{k=1}^{n} \gamma_k^n \psi_k$. Consequently, from (1.8) and (1.9) we have

$$\mathcal{L}(u_n, u_n) = \sum_{k=1}^{n} \lambda_k |\gamma_k^n|^2 \geq \lambda_1 \sum_{k=1}^{n} |\gamma_k^n|^2 = \lambda_1 \|u_n\|_0^2.$$

Therefore from (2.1) and (4.5), we have that $\{\|u_n\|_0^2\}_{n=1}^{\infty}$ is uniformly bounded, and we conclude from Garding's inequality (see (2.3)) that

$$\text{(4.18)} \qquad \{\|u_n\|_m^2\}_{n=1}^{\infty} \text{ is uniformly bounded.}$$

Next, we observe that (4.4) holds with ψ_k replaced by u_n. Also it follows from (g-2) with $\varepsilon = 1$ and (4.18) that $\Big\{ \sum_{|\alpha| \leq m} |\langle g(\cdot, Du_n), D^\alpha u_n \rangle_0| \Big\}_{n=1}^{\infty}$ is uniformly bounded. By (4.4), (4.5), (1.5) and (4.18), $\{\int_\Omega f(x, u_n) u_n\}_{n=1}^{\infty}$ is uniformly bounded from below. Proceeding as in [5, Lemmas 3 and 4], using (f-3)$'$ and (4.18), we conclude that

$$\text{(4.19)} \qquad \{f(x, u_n)\}_{n=1}^{\infty} \text{ is absolutely equiintegrable.}$$

(See (3.5).)

Next, it follows from (4.18) that $\exists u \in H_0^m$ such that (using the full sequence for ease of notation)

$$\text{(4.20)} \qquad \begin{aligned} &(a) \quad D^\alpha u_n \to D^\alpha u \text{ weakly in } L^2(\Omega) \text{ for } |\alpha| = m \\ &(b) \quad \|u_n - u\|_{m-1} \to 0; \\ &(c) \quad D^\alpha u_n \to D^\alpha u \qquad \text{a.e. in } \Omega \text{ for } |\alpha| \leq m-1. \end{aligned}$$

Using (4.18), (4.20)(a) and (c), and (Q-2) in the exact manner that was used to establish (3.4) in Theorem 1, we obtain

$$\text{(4.21)} \qquad \lim_{n \to \infty} Q(u_n, u_n, \psi_k) = Q(u, u, \psi_k) \qquad \text{for } k = 1, 2, \dots .$$

From (4.20)(c) and (f-1)$'$, we next observe that

$$\text{(4.22)} \qquad \lim_{n \to \infty} f(x, u_n) = f(x, u) \qquad \text{a.e. in } \Omega.$$

Also, using the fact that $\overline{\Omega}$ is compact, we obtain from Fatou's lemma and (4.19), as we did in (3.7), that

$$f(x,u) \in L^1(\Omega). \tag{4.23}$$

From (1.11)(a), (4.19), (4.22), (4.23), and Egoroff's theorem, we conclude that

$$\lim_{n\to\infty} \int_\Omega f(x,u_n)\psi_k = \int_\Omega f(x,u)\psi_k \qquad \text{for } k = 1,2,\dots . \tag{4.24}$$

Next, from (g-1), (g-2), (1.11)(a), (4.20)(b) and (c), proceeding as in the establishment of (3.10) in Theorem 1, we obtain

$$\lim_{n\to\infty} \int_\Omega g_\alpha(x,Du_n)D^\alpha\psi_k = \int_\Omega g_\alpha(x,Du)D^\alpha\psi_k \tag{4.25}$$

for $|\alpha| \le m$ and $k = 1,2,\dots$. Observing from (4.20)(b) that $\lim_{n\to\infty}\langle u_n, \psi_k\rangle_0 n^{-1} = 0\ \forall\psi_k$, we conclude from (4.4), (4.21), (4.24), and (4.25) that

$$Q(u,u,\psi_k) = \int_\Omega f(x,u)\psi_k + \sum_{|\alpha|\le m}(-1)^\alpha \int_\Omega g_\alpha(x,Du)D^\alpha\psi_k. \tag{4.26}$$

Next, using the fact that Ω is a domain of type B, we see from (1.11)(b) that if $\varphi \in C_0^\infty(\Omega)$, $\|\varphi_n - \varphi\|_{L^\infty} \to 0$ where $\varphi_n = \sum_{k=1}^n \hat{\varphi}(k)\psi_k$. Also, as is well known, since $\varphi \in H_0^m$, $\|\varphi_n - \varphi\|_m \to 0$. Consequently, we obtain from (4.23) that

$$\lim_{n\to\infty} \int_\Omega f(x,u)\varphi_n = \int_\Omega f(x,u)\varphi \tag{4.27}$$

and from (g-2) and the fact that $u \in H_0^m$ that

$$\lim_{n\to\infty} \sum_{|\alpha|\le m} \int_\Omega g_\alpha(x,Du)D^\alpha\varphi_n = \sum_{|\alpha|\le m}\int_\Omega g_\alpha(x,Du)D^\alpha\varphi. \tag{4.28}$$

From (Q-2) and (1.3), it is also clear that

$$\lim_{n\to\infty} Q(u,u,\varphi_n) = Q(u,u,\varphi). \tag{4.29}$$

Since $\varphi_n = \sum\limits_{k=1}^n \hat{\varphi}(k)\psi_k$, it is clear that (4.2) holds with ψ_k replaced by φ_n. But then it follows from (4.27)–(4.29) that

$$Q(u,u,\varphi) = \int_\Omega f(x,u)\varphi + \sum_{|\alpha|\le m}(-1)^\alpha \int_\Omega g_\alpha(x,Du)D^\alpha\varphi$$

$\forall\varphi \in C_0^\infty(\Omega)$. Hence u is a distribution solution of (1.13). Since $f(x,u) \in L^1(\Omega)$ by (4.23), the proof of Theorem 2 is complete.

Department of Mathematics and Computer Science, University of California, Riverside CA 92521

REFERENCES

1. D. G. de Figueiredo and J. P. Gossez, *Nonlinear perturbations of a linear elliptic problem near its first eigenvalue*, J. Diff. Eqns. **30** (1978), 1–19.
2. D. Gilbarg and N. S. Trudinger, "Elliptic Partial Differential Equations of Second Order," 2d ed., Springer-Verlag, Berlin, 1983.
3. J. L. Kazdan and F. W. Warner, *Remarks on some quasilinear elliptic equations*, Comm. Pure Appl. Math **28** (1975), 567–597.
4. W. Rudin, "Real and Complex Analysis," 2d ed., McGraw-Hill, New York, 1974.
5. V. L. Shapiro, *Resonance and quasilinear ellipticity*, Trans. Amer Math Soc **294** (1986), 567–582.

Bifurcation from Symmetry

J. SMOLLER AND A. G. WASSERMAN

§1. Introduction.

In this report, we announce several new results concerning the existence of non-symmetric solutions of symmetric equations (complete details of the proofs can be found in [**7**]). Our approach to this question is through bifurcation theory whereby we shall obtain the asymmetric solutions from symmetric ones via a bifurcation procedure.

Our concern is with semilinear elliptic equations of the form

$$\Delta u(x) + f(u(x)) = 0, \quad x \in D_R^n, \tag{1}$$

satisfying the general linear boundary conditions

$$\alpha u(x) - \beta du(x)/dn = 0, \quad x \in \partial D_R^n. \tag{2}$$

Here D_R^n denotes an n-ball of radius R and f is a smooth real-valued function. Now equation (1) is invariant under the full orthogonal group $O(n)$ in the sense that for any $T \in O(n)$, and $x \in D_R^n$,

$$(\Delta u)(Tx) + f(u(Tx)) = \Delta(u \circ T)(x) + f(u(Tx)).$$

This then motivates us to consider radial solutions of (1) (i.e., solutions u of the form $u = u(r)$, $r = |x|$), as the "symmetric" solutions, and any non-radial solution will be termed an "asymmetric" solution. Thus, we shall find non-radial solutions bifurcating off solution branches of radial ones.

In order to discuss the bifurcation of radial solutions, we must find an appropriate bifurcation parameter. To do this, we first note that radial solutions of (1), (2) satisfy the ordinary differential equation

$$u'' + \frac{n-1}{r}u' + f(u) = 0, \quad 0 < r < R, \tag{3}$$

together with the boundary conditions

$$u'(0) = 0 = \alpha u(R) - \beta u'(R). \tag{4}$$

Here "prime" denotes $\frac{d}{dr}$, and the condition $u'(0) = 0$ is a necessary one in order that u be differentiable at $r = 0$. We write (3) as the first-order system

$$u' = v, \quad v' = -\frac{n-1}{r}v - f(u), \tag{5}$$

and we consider orbits of (5) satisfying the initial conditions $u(0) = p > 0$, $v(0) = 0$. This solution will be written as $(u(r,p), v(r,p))$ $r \geq 0$, and p will be taken to be the bifurcation parameter throughout this paper.

We shall consider solutions of (3), (4) which are not necessarily positive. Thus, if $\theta_0 = \tan^{-1}(\alpha/\beta)$, $-\pi \leq \theta_0 < 0$, and k is a fixed nonnegative integer, we define the function $p \mapsto T(p)$ by

$$\begin{aligned} &\text{(i) } \alpha u(T(p),p) - \beta u'(T(p),p) = 0 \\ &\text{(ii) } \theta(T(p),p) = \theta_0 - k\pi, \end{aligned} \tag{6}$$

where $\theta(r,p) \equiv \tan^{-1}(u'(r,p)/u(r,p))$. If (i) and (ii) hold, we say that "$u(\cdot,p)$ lies in the k-th nodal class." Note that since p is allowed to vary, this (generally speaking) forces the radii $T(p)$ of the balls to vary.

We shall assume that $f(u)$ satisfies the following conditions (In [**7**] more general f's are considered, but this entails certain technical modifications in the statements of our results, which we wish to avoid here):

There exist points $b < 0 < c$ such that

$$(H) \quad \begin{cases} \text{(A) } f(c) = 0, \quad f'(c) < 0; \\ \text{(B) } uf(u) > 0 \text{ if } b < u < 0 \text{ or } 0 < u < c; \\ \text{(C) } F(b) = F(c), \text{ (where } F' = f \text{ and } F(0) = 0) \\ \text{(D) } f'(0) > 0. \end{cases}$$

Under these conditions, we have the following result (see [**7**]).

PROPOSITION 1. *If f satisfies (A)–(D), then*

(a) $\mathrm{dom}(T) = (0,c)$;
(b) $\lim_{p\searrow 0} T(p)$ *is finite;*
(c) $\lim_{p\nearrow 0} T(p) = +\infty$.

Linearizing (1) about a given radial solution $u(\cdot, p)$ gives the equation

$$\Delta w + f'(u(\cdot, p))w = 0, \quad |x| < T(p), \tag{7}$$

where w satisfies the boundary conditions (2). Since any function on an n-ball has a spherical harmonic decomposition, we may write w in the form

$$w = \sum_{N=0}^{\infty} a_N(r)\Phi_N(\theta), \quad r = |x|, \quad \theta \in S^{n-1},$$

where Φ_N lies in the N-th eigenspace of the Laplacian on S^{n-1}, corresponding to the N-th eigenvalue λ_N, and a_N satisfies the equation

$$a_N'' + \frac{n-1}{r}a_N' + \left(f'(u(\cdot, p)) + \frac{\lambda_N}{r^2}\right)a_N = 0, \quad 0 < r < T(p), \tag{8}$$

together with the following boundary conditions

$$\begin{aligned} \alpha a_N(T(p)) - \beta a_N'(T(p)) &= 0, \quad N \geq 0 \\ a_N(0) &= 0, \quad N \geq 1 \\ a_0'(0) &= 0. \end{aligned} \tag{9}$$

Now if $u(\cdot, p)$ is a "bifurcation point," then (7) must admit a non-trivial solution. Moreover, if the symmetry breaks on $u(\cdot, p)$, then necessarily $a_N \not\equiv 0$ for some $N > 0$ (see [**2**]). We thus say that: (i) the ***symmetry breaks infinitesimally*** on $u(\cdot, p)$ provided that $a_N \not\equiv 0$ for some $N > 0$, and (ii) the ***symmetry breaks*** on $u(\cdot, p)$ provided that $u(\cdot, p)$ bifurcates into an asymmetric solution. Thus there are two aspects to the symmetry-breaking problem; namely the "linear" part, whereby we must prove the existence of infinitesimal symmetry-breaking, and the "nonlinear" part, whereby we show that actual symmetry-breaking occurs. With regard to the problem of infinitesimal symmetry-breaking, we show that this must occur, and in a universal way. That is, we shall show that there is an integer N_0, *independent of the particular function* f, such that if $N \geq N_0$, there are k points $p_1^N, p_2^N, \ldots, p_k^N$ in the open interval $0 < u < c$ with the property that the symmetry breaks infinitesimally on each radial solution $u(\cdot, p_j^N)$, $j = 1, \ldots, k$, in the sense that the corresponding a_N's are different from zero. Thus the symmetry breaks infinitesimally in a k-fold way for all sufficiently high modes; i.e., for all $N \geq N_0$. As for the actual symmetry-breaking, we shall show that subject to certain additional conditions (needed to rule out some very degenerate situations), e.g. if f is analytic, it actually occurs. This will be done with the aid of the Conley index. These matters will be discussed in the next two sections. In the appendix, we have included a result of C. Pospiech which is related to the bifurcation of positive radial solutions.

§2. Infinitesimal Symmetry-Breaking.

Our objective here is to explain how one shows that infinitesimal symmetry-breaking must occur for all sufficiently high modes. Thus, we want to find points $p \in (0,c)$ and integers $N \geq 1$ such that the symmetry breaks infinitesimally in the N-th mode, on the radial solution $u(\cdot,p)$, in the sense that the corresponding $a_N \not\equiv 0$, where a_N satisfies (8) and (9). To this end, for each $p \in (0,c)$, we define a space of functions Φ_p by

$$\Phi_p = \left\{\phi \in C^2(0,T(p)) : \phi(0) = 0 = \alpha\phi(T(p)) - \beta\phi'(T(p))\right\},$$

and for each integer $N \geq 1$ we define operators $L_N^p : \Phi_p \to C(0,T(p))$ by

$$L_N^p\phi = \phi'' + \frac{n-1}{r}\phi' + \left(f'(u(\cdot,p)) + \frac{\lambda_N}{r^2}\right)\phi.$$

Note that Proposition 1 implies that the operators L_N^p are all well-defined for each $p \in (0,c)$, and each $N \in \mathbb{Z}_+$, $N \geq 1$. Our goal is to show that 0 lies in the spectrum of L_N^p, for certain p and N, since this obviously means that (8), (9) admits a non-trivial solution. In order to state our theorem, we must first define certain quantities.

We consider solutions of the Bessel-type equation

$$u'' + \frac{n-1}{r}u' + u = 0, \quad 0 < r < \rho,$$

lying in the $(k+1)$-th nodal class, which satisfy the boundary conditions $u'(0) = 0 = \alpha u(\rho) - \beta u'(\rho)$. Since this equation is linear, ρ is independent of $u(0)$, and depends only on n, k (and of course, on the given fixed boundary conditions; i.e. on α and β). Suppressing the dependence on α and β, we write $\rho = c_{k,n}$. Next, we define μ to be the principal eigenvalue of the radial Laplacian Δ_R on $(0,1)$ satisfying the given boundary conditions; i.e. $\Delta_R = \frac{d^2}{dr^2} + (n-1)r^{-1}\frac{d}{dr}$. Now it is well-known that λ_N, the N-th eigenvalue of the Laplacian on S^{n-1} is given explicitly by

$$\lambda_N = -N(N+n-2).$$

In these terms, we define an integer $N_0 \in \mathbb{Z}_+$ by

$$N_0 = \min\left\{N \in \mathbb{Z}_+ : -\lambda_N > C_{k,n}^2 + \mu\right\}. \tag{10}$$

Observe that N_0 *is independent of* f. We can now state the main theorem in this section.

THEOREM 1. *Assume that f satisfies hypotheses (H). Let k be a given positive integer defining a fixed nodal class of radial solutions of (1) and (2), and let N_0 be defined by (10). Then for every integer $N \geq N_0$, there are k-distinct points $p_1^N, p_2^N, \ldots, p_k^N$ in the interval $0 < u < c$, for which the equations $L_N^{p_i}\phi = 0$ admit non-trivial solutions.*

We thus see that for every $N \geq N_0$, the symmetry breaks infinitesimally on the k-distinct radial solutions $u(\cdot, p_i^N)$, $i = 1, 2, \ldots, k$. We remark that the corresponding eigenspace of the associated linearized operators have dimension $\ell_N = \binom{N+n-2}{N}\left(\frac{2N+n-2}{N+n-2}\right) = O(N^{n-2})$; see [**6**]. Furthermore, the integer N_0 is "universal" in the sense that it is independent of the particular function f.

We shall now explain the main steps in the proof of this result; as we remarked earlier, complete details can be found in [**7**]. The "philosophy" is to use (3), the equation for u, in order to obtain information about u', and to notice that u' satisfies the a_1 equation; equation (8) where $N = 1$. Since the a_1 and a_N equations differ only by the term $(\lambda_1 - \lambda_N)/r^2$, their corresponding solutions should be close, provided that they are initially close and p is near c (so that r is large). We next outline the procedure in order to carry out this program.

There are three steps in the program:

Step 1. We prove that for p near c, the operator L_1^p has k-positive eigenvalues. This is done by showing that $v = u'$ has at least k (but no more than $(k+1)$) zeros on $0 \leq r \leq T(p)$, and that the orbit (v, v') cuts the boundary line $\alpha v - \beta v' = 0$ in k-distinct points. We do this by obtaining two-sided bounds on $\phi_1(T(p), p) = \tan^{-1}\left[v'(T(p), p)/v(T(p), p)\right]$. Next, we consider the eigenvalue equation $L_1^p a_1^\mu = \mu a_1^\mu$, $0 \leq r \leq T(p)$, $\mu > 0$, and observe that from standard results, if μ is sufficiently large, μ cannot be an eigenvalue of L_N^p. Thus if $\phi_1^\mu(T(p),p) = \tan^{-1}\left[a'_\mu(T(p),p)/a_\mu(T(p),p)\right]$, then for large $\mu, 0 > \phi_1^\mu(T(p),p) > \theta_0 - \pi$. Since the total change in ϕ_1^μ is a continuous function of μ, it follows that there exist k-positive numbers $\mu_1, \mu_2, \ldots, \mu_k$ for which $\phi_1^{\mu_j}(T(p),p) = \theta_0 - j\pi$, $j = 1,2,\ldots,k$. It is easy to see that each such μ_j is an eigenvalue of L_1^p.

Step 2. If $N > 1$ and p is near c, L_N^p has k-positive eigenvalues. This is the most difficult part in the proof. The idea is to use the fact that the operators L_1^p and L_N^p differ by a small term if p is near to c. Thus if $A < c$ is any number for which $f'(u) < 0$ on $A \leq u \leq c$, and $T_1(p)$ is such that $u(T_1(p),p) = A$, then T_1 is defined on $A < p < c$, and $T_1(p) \to \infty$ as $p \to c$. We show that $0 < \phi_N(T_1(p),p) - \phi_1(T_1(p),p) < \frac{c}{T_1(p)^2}$, where c is independent of p.

Next, if $u(T_2(p),p) = A$, and $T_2(p)$ is the "second" time the orbit $(u(\cdot,p),u'(\cdot,p))$ meets $u = A$, we prove that $T_2(p) - T_1(p)$ is bounded independently of p for p near c. This shows that for each $N > 1$, (a_N, a'_N) has the same "rotation number" as (v,v'), provided that (v,v') has rotation number $< 2\pi$. That is, it shows that Step 2 holds if $k \leq 2$. The difficulty to extend this to arbitrary k stems from the fact that for p near c, the orbit (u,u') comes near c on its "second-time-around," and hence stays near c for a long time. This fact makes it much more difficult to obtain a uniform estimate on $\phi_N(r,p) - \phi_1(r,p)$ for $r > T_2(p)$. However, the difficulty can be overcome by obtaining some delicate estimates on the rate at which the "times get large." These facts enable us to show that Step 2 is indeed valid.

Step 3. For $N \geq N_0$, and p near 0, L_N^p has negative spectrum. We shall illustrate this in the easiest case; namely if $0 \geq \theta_0 \geq -\pi/2$. (In particular, our argument will be valid for both Dirichlet and Neumann boundary conditions.) Thus let $L_N^p\phi = \lambda\phi$; then if $\langle,\rangle$ denotes the weighted L_2-inner product on $0 \leq r \leq T(p)$, with "weight" r^{n-1}, we have, if $\|\phi\|^2 = 1$,

$$\begin{aligned}
\lambda = \lambda\langle\phi,\phi\rangle &= \langle L_N^p\phi_1,\phi\rangle \\
&\leq \int_0^{T(p)} \left\{-(\phi')^2 + \left(f'(u) + \frac{\lambda_N}{r^2}\right)\phi^2\right\} r^{n-1}\,dr \\
&\leq \int_0^{T(p)} \left\{\left(f'(u) + \frac{\lambda_N}{r^2}\right)\phi^2\right\} r^{n-1}\,dr.
\end{aligned}$$

Now for p near 0, this last integral is well-approximated by

$$\int_0^{\rho_n^k} \left(f'(0) + \frac{\lambda_N}{r^2}\right) \phi^2 r^{n-1}\, dr,$$

which is negative if $-\lambda_N > (\rho_n^k)^2 f'(0)$. (Here $\rho_n^k = \lim_{p\to 0} T(p)$.) But by a scaling argument, it is easy to show that $\rho_n^k = c_{n,k}/\sqrt{f'(0)}$, and this is where the condition $N > N_0$ comes from.

Now we can put all of these steps together in order to show how to complete the proof of Theorem 1. Thus, let $N \in \mathbb{Z}_+$, $N \geq N_0$. Then from Step 2, there is a point p near c for which $\phi_N(T(p), p) < \theta_0 - k\pi$, and Step 3 shows that we can find a point q near 0 for which $0 > \phi_N(T(q), q) > \theta_0 - \pi$. Since Proposition 1 implies that the domain of T is connected, the intermediate-value theorem applies and we can find points $p_j^N, j = 1, 2, \ldots, k$, $0 < p_j^N < p_{j+1}^N < c$, for which $\phi_N(T(p_j^N), p_j^N) = \theta_0 - j\pi$. That is, 0 is in the spectrum of each operator $L_N^{p_j^N}, j = 1, 2, \ldots, k$.

3. Symmetry-breaking.

In order to show that the symmetry actually breaks, it is necessary to show that bifurcation occurs. This is usually done by appealing to one of the standard bifurcation theorems; e.g., the Crandall-Rabinowitz Theorem [**4**]. This method requires that one verifies the so-called "transversality" condition; i.e., one must show that the eigenvalues cross zero with "non-zero speed." Such a condition is usually very difficult to check (But as we shall show in the appendix, it is always satisfied for positive solutions of the Dirichlet problem; this result is due to C. Pospiech.); and indeed, we have been unable to verify it in general. Instead, we shall approach the problem from a more qualitative, topological point of view, and our main tool will be the Conley index (see [**4**] for a discussion of this notion). Thus we show that there is an increasing sequence of points $\{p_k : k \geq N_0\}$ in the interval $0 < p < c$, $p_k \to c$, such that $h[u(\cdot, p_k)] \neq h[u(\cdot, p_{k+1})]$, where $h[u(\cdot, p)]$ denotes the Conley index of the solution $u(\cdot, p)$. It follows from Conley's Continuation Theorem [**4**, Theorem 23.31] that for some $q_k \in (p_k, p_{k+1})$, the radial solution $u(\cdot, q_k)$ is not an isolated invariant set. This implies that $u(\cdot, q_k)$ is a bifurcation point. However, it does not necessarily follow from this that the symmetry actually breaks on $u(\cdot, q_k)$. Indeed, $u(\cdot, q_k)$ could be a "radial" bifurcation point in the sense that radial solutions bifurcate off of $u(\cdot, q_k)$. We shall however formulate some additional conditions on f which guarantee that the symmetry actually breaks.

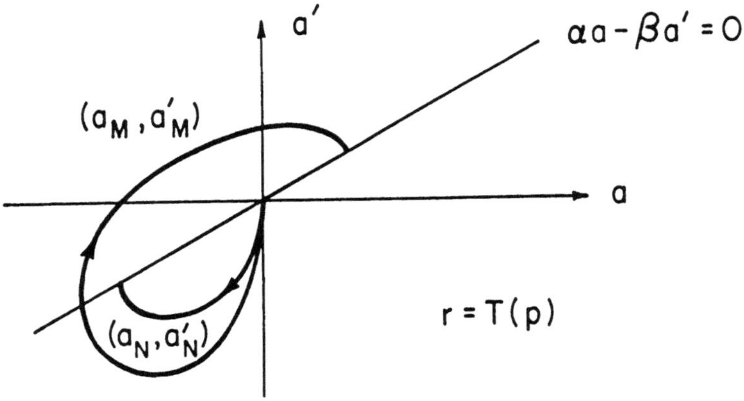

There is a major difficulty in the actual computation of the Conley index. This is due to the fact that it is a priori possible that zero can lie in the spectrum of the linearized operators (about $u(\cdot, p)$), for an entire open interval of p's. As we have shown in [6], such a situation can only occur if $u'(\cdot, p)$ is allowed to change sign, since only in this case can the kernel of the linearized operator about a fixed radial solution contain distinct elements of the form $a_N(r)\Phi_N(\theta)$, and $a_M(r)\phi_M(\theta)$, $N \neq M$; cf., the figure above. We can show, however, that this difficulty can be avoided if the function f is assumed to be analytic. (If u' is of one sign, then we need not make this assumption.) Then using the theorem in the last section, we can compute the Conley indices $h[u(\cdot, p_k)]$ and $h[u(\cdot, p_{k+1})]$, and prove that they are different. Then, as mentioned above, there is a point, $q_k, p_k < q_k < p_{k+1}$ for which $u(\cdot, q_k)$ is a bifurcation point. We further show that the symmetry breaks on this radial solution provided that $T'(p) \neq 0$. This is a rather difficult condition to verify, but we can prove that T' is never zero whenever f satisfies the condition $f(u)/u > f'(u)$ on $(b, c)\backslash\{0\}$. Note that our additional hypotheses on f are needed only to rule out very degenerate situations, which are highly unlikely to actually occur in any particular situation.

Appendix (after C. Pospeich).

We consider positive solutions of the Dirichlet problem

$$(A1)\qquad \begin{aligned} \Delta_y u(y) + f(u(y)) &= 0, \quad y \in D_R^n \\ u(y) &= 0, \quad y \in \partial D_R^n \end{aligned}$$

and we shall show that every degenerate radial solution of (A1) is a bifurcation point. (Recall that a radial solution $u(\cdot, p)$ is called degenerate provided that the problem

$$\begin{aligned} \Delta w(x) + f'(u(|x|, p))w(x) &= 0, \quad |x| < T(p) \\ w(x) &= 0, \quad |x| = T(p), \end{aligned}$$

admits a non-trivial solution.)

In order to prove this result we rely (as in [**5**]) on the Crandall-Rabinowitz Theorem [**4**]. To apply this theorem to our problem, we make a change of scale in (A1) as follows. Let $v(x) = u(Rx)$; then $u(y)$ solves (A1) if and only if v solves

$$(A2)\qquad \begin{aligned} \Delta_x v(x) + R^2 f(v(x)) &= 0, \quad x \in D_1^n \\ v(x) &= 0, \quad x \in \partial D_1^n, \end{aligned}$$

since $\Delta_x v(x) = \Delta_x u(Rx) = R^2 \Delta_y u(Rx) = -R^2 f(u(Rx)) = -R^2 f(v(x))$. Now define the operator

$$M : \{u \in C^2(D_1^n) : u = 0 \text{ on } \partial D_1^n\} \times I \to C^0(D_1^n),$$

$(I = [0,1])$, by

$$(A3)\quad M(z,R)(x) = \Delta_x(z(x) + u(|x|R, p(R)) + R^2(f(z(x) + u(|x|R, p(R))),$$

where $u(|x|R, p(R))$ is a radial solution of (A2) such that (cf. [**5**]):

(a) for every $R \in I$, $u(rR, p(R))$ solves

$$(A4)\qquad u_{rr}(rR, p(R)) + \frac{n-1}{R} u_r(rR, p(R)) + R^2 f(u(rR, p(R)) = 0,$$

for $0 < r \le 1$,

$$(A5)\qquad u(R, p(R)) = 0,$$

$$(A6)\qquad u(0, p(R)) = p(R), \quad u_r(0, p(R)) = 0;$$

(b) for $R \leq \bar{R}$,

$$u(rR, p(R)) > 0, \quad 0 \leq r < 1$$
$$u_r(rR, p(R)) < 0, \quad 0 < r < 1;$$

(c) $u_r(\bar{R}, p(\bar{R})) = 0$.

As shown in [**5**], condition (c) is equivalent to the fact that the symmetry breaks infinitesimally on the solution $u(r\bar{R}, p(\bar{R}))$.

Now if prime denotes differentiation with respect to the first coordinate, we have $u_r(rR,p) = u'(rR,p)R$, $u_R(rR,p) = u'(rR,p)r$, so

$$(A7) \qquad u_R(rR,p) = \frac{r}{R} u_r(rR,p).$$

As was shown in [**6**], conditions (b) and (c) imply that the equation

$$(A8) \qquad 0 = M_z(0,\bar{R})v = \Delta v + \bar{R}^2 f'(u)v$$

has an n-dimensional solution set spanned by

$$\frac{\partial}{\partial x_i} u(\bar{R}x) = \frac{\partial r}{\partial x_i} u_r(r\bar{R}, p(\bar{R})), \quad 1 \leq i \leq n.$$

Now the transversality condition is

$$(A9) \qquad \int_{D_1^n} M_{ZR}(0,\bar{R}) v^2(x) \neq 0,$$

where $v(x)$ is any solution of (A8). Moreover (cf. [**6**]),

$$\begin{aligned} M_{ZR}(0,\bar{R}) &= 2\bar{R} f'(u(r\bar{R}, p(\bar{R}))) \bar{R}^2 f''(u(r\bar{R}, p(\bar{R}))) \big[u_R(r\bar{R}, p(\bar{R})) \\ &\qquad + p'(\bar{R}) u_p(r\bar{R}, p(\bar{R}))\big] \\ &= 2\bar{R} f'(u(r\bar{R}, p(\bar{R}))) + \bar{R} f''(u(r\bar{R}, p(\bar{R}))) u_r(r\bar{R}, p(\bar{R})) r, \end{aligned}$$

in view of (A7), and

$$p'(\bar{R}) = \frac{u_R(\bar{R}, p(\bar{R}))}{u_p(\bar{R}, p(\bar{R}))} = \frac{1}{\bar{R}} \frac{u_r(\bar{R}, p(\bar{R}))}{u_p(\bar{R}, p(\bar{R}))} = 0,$$

because of (c). We can now state the following theorem.

THEOREM. *Assume that hypotheses (a), (b), and (c) hold. Then (A9) holds for every non-zero solution of (A8).*

Note that no assumptions are imposed on f except those necessary to guarantee that hypotheses (a), (b) and (c) are valid (e.g. $f(0) < 0$; see [**5**]). This result generalizes a result in [**5**], where (A9) was shown to hold for a certain class of f; see also [**1**].

We now turn to the proof of the theorem. Let $v(x) = \frac{\partial u(Rx)}{\partial x_n}$; if $\bar{v}$ is any other non-zero solution of (A8), then there is an $a > 0$ and an $A \in O(n)$ such that $\bar{v}(x) = av(A^{-1}x)$. Thus

$$\begin{aligned}\int_{D_1^n} M_{ZR}(0,\bar{R})\bar{v}^2(x)\,dx &= a^2\int_{D_1^n} M_{ZR}(0,\bar{R})v^2(A^{-1}x)\,dx\\ &= a^2\int_{D_1^n} M_{ZR}(0,\bar{R})v^2(x)\,dx.\end{aligned}$$

it suffices to show (A9) where v is as above; namely $v(x) = \frac{x_n}{r}u_r(r\bar{R},p(\bar{R}))$. For this choice,

$$\int_{D_1^n} M_{ZR}(0,\bar{R})v^2(x)\,dx = \int_{D_1^n} M_{ZR}(0,\bar{R})\frac{x_n^2}{r^2}u_r(r\bar{R},p(\bar{R}))^2\,dx.$$

Now use polar coordinates

$$\begin{aligned}x_n &= r\cos\theta_{n-1}\\ x_{n-1} &= r\sin\theta_{n-1}\cos\theta_{n-2},\\ x_{n-2} &= r\sin\theta_{n-1}\sin\theta_{n-2}\cos\theta_{n-3},\ \text{etc.},\end{aligned}$$

where $0 \le r \le 1$, $0 \le \theta_1 \le \pi$, $2 \le i \le n-1$, $0 \le \theta_1 \le 2\pi$. Substituting gives

$$\int_{D_1^n} M_{ZR}(0,\bar{R})v^2(x)\,dx$$

$$\int_0^1\int_0^\pi\cdots\int_0^\pi\int_0^{2\pi} M_{ZR}(0,\bar{R})u_r^2(r\bar{R},p(\bar{R}))\cos^2\theta_{n-1}r^{n-1}$$

$$\prod_{i=1}^{n-1}\sin^{i-1}\theta_i\,d\theta_1\ldots d\theta_{n-1}\,dr$$

$$\int_0^1 M_{ZR}(0,\bar{R})u_r^2(r\bar{R},p(\bar{R})r^{n-1}\,dr\int_0^\pi \cos^2\theta_{n-1}\sin^{n-2}\theta_n\,d\theta_n\cdot$$

$$2\prod_{i=0}^{n-2}\int_0^\pi \sin^{i-1}\theta_i\,d\theta_i$$

$$= k\int_0^1 M_{ZR}(0,\bar{R})u_r^2(r\bar{R},p(\bar{R}))r^{n-1}\,dr,$$

where $k = 2\int_0^\pi \cos^2\theta\sin^{n-2}\theta\,d\theta\prod_{i=1}^{n-2}\int_0^\pi \sin^{i-1}\theta\,d\theta = \frac{\pi^{n/2}}{\Gamma(\frac{n+2}{2})} > 0$, in view of the following formula for the gamma function [**3**]:

$$\frac{\Gamma(x)\Gamma(y)}{\Gamma(x+y)} = 2\int_0^{\pi/2}\cos^{(2x-1)}\theta\sin^{(2y-1)}\theta\,d\theta.$$

It thus suffices to show that

$$(A10)\qquad \int_0^1 M_{ZR}(0,\bar{R})u_r^2(r\bar{R},p(\bar{R}))r^{n-1}\,dr \neq 0.$$

For this, we need the following lemma.

LEMMA. *Let $g\in C^0(I)$, $h\in C^1(I)$, $v\in C^2(I)$ be real-valued functions satisfying*

(i) $rg(r) = \frac{d}{dr}(r^2h(r)$, $\quad r\in I$
(ii) $v''(r) + \frac{n-1}{r}v'(r) + kh(r)v(r) + \frac{\lambda}{r^2}v(r) = 0$, $\quad 0 < r \le 1$, *for some real constants* $\lambda\neq 0$, $k\neq 0$.

Then, for $n\ge 1$,

$$\int_0^1 g(r)v^2(r)r^{n-1}\,dr = \big(h(1)+\frac{\lambda}{k}\big)v^2(1) + \frac{1}{k}v'(1)^2 + \frac{n-1}{k}v'(1)v(1).$$

We postpone the proof of the lemma and prove (A10). Let $v(r) = u_r$, $k=\bar{R}$, $h(r) = \bar{R}f'(u(r\bar{R},p(\bar{R}))$, $g(r) = M_{ZR}(0,\bar{R})$, $\lambda = \lambda_k$; then

$$\begin{aligned} rg(r) = rM_{ZR}(0,\bar{R}) &= 2r\bar{R}f'(u(r\bar{R},p(\bar{R})))\\ &\quad + r^2\bar{R}f''(u(r\bar{R},p(\bar{R})))u_r(r\bar{R},p(\bar{R}))\\ &= \frac{d}{dr}\big[\bar{R}r^2f'(u(r\bar{R},p(\bar{R})))\big] = \frac{d}{dr}(r^2h(r)),\end{aligned}$$

and since

$$(u_r)_{rr} + \frac{n-1}{r}(u_r)_r + \bar{R}^2 f'(u)u_r - \frac{n-1}{r^2}u_r = 0,$$

we see that (ii) also holds. Thus

$$\begin{aligned}\int_0^1 M_{ZR}(0,\bar{R})u_r^2 r^{n-1}\,dr &= \frac{1}{\bar{R}}u_{rr}^2(\bar{R},p(\bar{R}))\\ &= \bar{R}f(u(\bar{R},p(\bar{R})))^2\\ &= \bar{R}f(0)^2 > 0,\end{aligned}$$

in view of (A4), (A5), (c) and the fact that $f(0) < 0$ (see [**5**]).

It remains to prove the lemma. Now from [**7**], $v(r) = rw(r)$, where $w(0) \neq 0$. Thus $v(r)/r = w(r)$ so $v(r)r^{n-3}$, and $v'(r)r^{n-2}$ are bounded as $r \to 0$, for $n \geq 2$, and $\lim_{r\to 0} \frac{v(r)}{r} = w(0)$. Thus $v(r)r^{n-3}$ and $v'(r)r^{n-2}$ are continuous functions on $0 \leq r \leq 1$ for $n \geq 2$; this justifies the integration by parts given below.

$$\begin{aligned}\int_0^1 g(r)v^2(r)r^{n-1} &= \int_0^1 \frac{d}{dr}(r^2h)v^2r^{n-2}\,dr\\ &= h(1)v^2(1)\int_0^1 hvr^{n-1}[2rv' + (n-2)v]\,dr\\ &= h(1)v^2(1) + \int_0^1 (r^{n-1}v') + \lambda r^{n-3}v)[2rv' + (n-2)v]\,dr\\ &= h(1)v^2(1) + 2(v'(1))^2 + (n-2)v(1)v'(1)\\ &\quad - \int_0^1 r^{n-1}v(2rv'' + nv')\,dr\\ &\quad + \lambda\int_0^1 r^{n-3}v[2rv' + (n-2)v]\,dr\\ &= h(1)v^2(1) + 2(v'(1))^2 + (n-2)v(1)v'(1)\\ &\quad - \int_0^1 [2v'v''r^n + (v')^2nr^{n-1}]\,dr\\ &\quad + \lambda\int_0^1 [2vv'r^{n-2} + v^2(n-2)r^{n-3}]\,dr\\ &= (h(1)+\lambda)v^2(1) + (v'(1))^2 + (n-2)v(1)v'(1),\end{aligned}$$

and this completes the proof of the lemma.

We remark that we can modify this proof to apply to general boundary conditions (2), as long as the symmetry breaks infinitesimally in the first (i.e. a_1), mode; see [**6**].

Department of Mathematics, University of Michigan, Ann Arbor MI 48109

Research supported in part by NSF Grant MCS 800-2337.

Research at MSRI supported in part by NSF Grant DMS 812079-05.

REFERENCES

1. G. Cerami, *Symmetry breaking for a class of semilinear elliptic equations*, Nonlinear Anal. **10** (1986), 1–14.
2. E. N. Dancer, *On non-radially symmetric bifurcation*, J. London Math. Soc. **20** (1979), 287–292.
3. K. T. Smith, "Primer of Modern Analysis," UTM (Springer), 1983.
4. J. A. Smoller, "Shock Waves and Reaction-Diffusion Equations," Grundlehren (Springer), 1983.
5. J. A. Smoller and A. Wasserman, *Symmetry-breaking for positive solutions of semilinear elliptic equations*, Arch. Rat. Mech. Anal. **95** (1986), 217–225.
6. J. A. Smoller and A. Wasserman, *Symmetry-breaking for semilinear elliptic equations with general boundary conditions*, Comm. Math. Phys. **105** (1986), 415–441.
7. J. A. Smoller and A. Wasserman, *Symmetry, degeneracy and universality in semilinear elliptic equations*, to appear.

Positive Solutions of Semilinear Elliptic Equations on General Domains

J. SMOLLER AND A. G. WASSERMAN

The existence problem for positive solutions to the equation

$$\Delta U(x) + \lambda^2 f(U(x)) = 0 \tag{1}$$

with homogeneous Dirichlet boundary conditions on a smooth, bounded domain, Ω, has recently been of much interest; cf. the survey article by Lions [**L**]. All the theorems cited therein have required $f(0) > 0$ or $f(0) = 0$ and $f'(0) > 0$. However, it was shown in [**SW1**] that a necessary condition for symmetry breaking of positive solutions of (1) on a disk is that $f(0) < 0$. In this note we outline a general procedure for constructing positive solutions to (1) on Ω which do not require any assumptions on the behavior of f near 0, nor do we restrict ourselves to superlinear $((f(U)/U)' > 0)$ or sublinear $((f(U)/U)' < 0)$ functions; furthermore, we make no assumptions about the sectional curvature of boundary Ω; (cf. [**L**]).

We first introduce the notion of eccentricity of a domain, $e(\Omega)$, which has the properties that $1 \leq e(\Omega) < \infty$ and $e(\Omega) = 1$ if and only if Ω is an n-disk. We then define for any function f on $\mathbb{R}$ the notion of the "nonlinearity" of f, $N(f)$, which has the property that $N(f) = 1$ if f is a linear function, and $N(f) = \infty$ if $f(u) = u^k$, $0 < k < 1$. Our main theorem states that (1) has a positive solution for some λ^2 if $e(\Omega) < N(f)$. We give several applications of this theorem and we give more precise information about the set of λ^2s for which (1) admits a solution.

§1. The Main Theorem.

Let Ω be a smooth bounded region in $\mathbb{R}^n$. We define the "outer radius" of Ω, $R(\Omega)$, by $R(\Omega) = \inf\{r \mid \Omega \subset B_r(p) \text{ for some } p \in \mathbb{R}^n\}$. Note that $\operatorname{diam}(\Omega)/2 \leq R(\Omega) < \operatorname{diam}(\Omega)$.

Since the boundary of Ω is a smooth compact submanifold of $\mathbb{R}^n$ there is an $\eta > 0$, such that, for any $x \in \Omega$ there is a $y \in \Omega$ with $x \in B_\eta(y) \subset \Omega$; the proof uses the tubular neighborhood theorem. We define the inner radius of Ω, $\rho(\Omega)$, to be the sup of all such η.

Finally, we define the eccentricity of Ω, $e(\Omega)$, by $e(\Omega) = R(\Omega)/\rho(\Omega)$. Note that $1 \leq e(\Omega) < \infty$.

If Ω is the elliptical region in $\mathbb{R}^2$ given by $a^2x^2 + b^2y^2 \leq 1$ with $a \geq b$ we have $R(\Omega) = b$, $\rho(\Omega) = \frac{a^2}{b}$, and $e(\Omega) = (b/a)^2$.

Next consider "admissible" 4-tuples (U_1, R_1, U_2, R_2) where

(i) $\Delta U_1(x) + f(U_1(x)) > 0 \quad x \in D^n_{R_1}$ i.e. U_1 is a lower solution of (1) on $D^n_{R_1}$.

(ii) $\Delta U_2(x) + f(U_2(x)) < 0 \quad x \in D^n_{R_2}$ i.e. U_2 is an upper solution of (1) on D^n_R.

(iii) $U_1(x) = 0$, $\frac{\partial U_1}{\partial r}(x) < 0$ if $\|x\| = R_1$.

(iv) $U_2(x) \geq U_1(y) \geq 0$ for all $x \in D^n_{R_2}$, $y \in D^n_{R_1}$.

Define $N(f)$, the "nonlinearity of f", by $N(f) = \sup R_2/R_1$ where the sup is taken over all admissible 4-tuples.

REMARKS:

(1) Note that equations i, ii, for U_1 and U_2 admit radial solutions and thus $N(f)$ can be computed using ODE techniques.

(2) There exists an admissible 4-tuple for f if and only if $\int_o^p f(s)\,ds > 0$ for some $p > 0$; c.f. [**SW3**].

(3) By a simple scaling argument one shows that $N(\mu f) = N(f)$ for $\mu > 0$.

(4) In the applications to follow we shall have $N(f) = \infty$ in each case. However, it is easy to show that the function

$$f(U) = \begin{cases} \mu_1 U & U \leq 1 \\ \mu_2(U-1) + \mu_1 & U \geq 1 \end{cases}$$

has $N(f) = \sqrt{\frac{\mu_1}{\mu_2}}$, if $\mu_1 \geq \mu_2$.

THEOREM. *If f is a C^1 function, Ω a smooth bounded domain in $\mathbb{R}^n$ and $N(f) > e(\Omega)$, then there exists λ^2 such that (1) has a positive solution.*

SKETCH OF PROOF: Choose an admissible 4-tuple (U_1, R_1, U_2, R_2) with $\frac{R_2}{R_1} > \frac{R(\Omega)}{\rho(\Omega)}$; then $\frac{R_2}{R(\Omega)} > \frac{R_1}{\rho(\Omega)}$. Choose λ such that $\frac{R_2}{R(\Omega)} > \lambda > \frac{R_1}{\rho(\Omega)}$. We show that (1) admits a positive solution for this λ.

Let $\Omega \subset B_{R(\Omega)}(p)$; then $V(x) = U_2(\lambda(x-p))$ is an upper solution of (1) on Ω since $\|x-p\| \leq R(\Omega)$ and $\lambda R(\Omega) < R_2$.

We then alter f for $U < 0$ to get a new function $\tilde{f}$ that admits a lower solution, U_0 say, to the equation

$$\Delta U + \lambda^2 \tilde{f}(U) = 0. \tag{2}$$

By the method of upper and lower solutions there exist solutions of (2). We must show that the maximal solution, $\bar{U}$, with $U_0(x) \leq \bar{U}(x) < V(x)$ is, in fact, a positive solution and hence a solution to (1). To that end let $x_0 \in \Omega$ with $x_0 \in B_{R_1/\lambda}(y) \subset B_{\rho(\Omega)}(y) \subset \Omega$. We will show that $\bar{U}(x_0) \geq 0$. Let $V_1(x) = U_1(\lambda(x-y))$. Then V_1 satisfies $U_0(x) \leq V_1(x) \leq V(x)$ for $x \in \Omega$ by condition (iv) of an admissible 4-tuple. Condition (iii) and the fact that $\lambda\rho(\Omega) \geq R_1$ guarantee that V_1 is negative on the boundary of Ω. So V_1 is a lower solution of (2) and thus there is a solution W of (2) with $V_1 \leq W \leq V$. But we then have $\bar{U}(x) \geq W(x)$ by the maximality of $\bar{U}$ and $W(x_0) \geq V_1(x_0) = U_1(\lambda(x_0-y)) \geq 0$ since $\lambda\|x_0-y\| \leq R_1$. Thus $\bar{U}(x_0) \geq 0$ as desired.

REMARK: We can see that the set of λ for which we can prove the existence of positive solutions to (1) is $\cup(R_1/\rho(\Omega), R_2/R(\Omega))$ where the union is taken over all admissible 4-tuples with $R_2/R_1 > e(\Omega)$.

§2. Applications.

First Application. If $f(0) > 0$ then (1) admits positive solutions if $\lambda^2 < \bar{R}_2/R(\Omega)$ where $\bar{R}_2 = \max R_2$ over all admissible 4-tuples.

We need only notice that there exist lower solutions U_1 with R_1 and $\sup\|U_1(x)\|$ as small as we choose, and thus $N(f) = \infty$; see also [**L**].

Second Application. We suppose $f(\gamma) = 0$, $\int_0^\gamma f(s)\,ds > 0$ for some $\gamma > 0$. Then (1) admits positive solutions for $\lambda^2 > \bar{R}_1/\rho(\Omega)$ where $\bar{R}_1 = \inf R_1$, the inf being taken over all admissible 4-tuples with $\sup U_1(x) < \gamma$. It follows from [**SW2**] that there is some lower solution with $\sup U_1(x) < \gamma$. Now we need only note the existence of the upper solutions $U_2(x) \equiv \gamma$ with $R_2 = \infty$. Thus $N(f) = \infty$ and the result follows. A version of this application was also obtained by Clément and Sweers [**CS**].

Third Application. If $\int_0^p f(s)\,ds > 0$ for some $p > 0$ and $\lim_{u\to\infty} \frac{f(u)}{u} = 0$ then $N(f) = \infty$, and hence there exist solutions to (1) for $\lambda^2 > \bar{R}_1/\rho(\Omega)$ where $\bar{R}_1 = \inf R_1$ over all admissible 4-tuples.

The condition $\int_0^p f(s)\,ds > 0$ guarantees the existence of a lower solution [**SW2**] and is also necessary [**SW3**]. It is shown in [**SW3**] that there exists upper solutions to (1) in this case with R_2 and $U_2(R_2)$ arbitrarily large. Thus again $N(f) = \infty$.

Fourth Application. Let $f(u) = \sqrt{u} + C^2u^p$ where $p > 0$. Then there exist positive solutions to (1) for $\lambda^2 < \bar{R}_2/R(\Omega)$ where $\bar{R}_2 = \sup R_2$ over all admissible 4-tuples.

There exist lower solutions, U_1, with R_1 and $\sup \|U_1(x)\|$ arbitrarily small in this case; thus $N(f) = \infty$ as before. However, we cannot apply the theorem directly because f is not C^1. But it is easy first to approximate f by smooth f_K to get solutions U_K to (1) and then show $\lim_{K\to\infty} U_K = U$ exists and satisfies (1).

Department of Mathematics, University of Michigan, Ann Arbor MI 48109

Research supported in part by NSF Grant MCS 800-2337.

REFERENCES

[**L**] P. L. Lions, *On the existence of positive solutions of semilinear elliptic equations*, SIAM Rev. **24** (1982), 441–467.

[**CS**] P. Clément and G. Sweers, *Existence and multiplicity results for a semilinear elliptic eigenvalue problem*, preprint.

[**SW1**] J. Smoller and A. G. Wasserman, *Symmetry-breaking for positive solutions of semilinear elliptic equations*, Arch. Rat. Mech. Anal. **95**, 217–225.

[**SW2**] J. Smoller and A. G. Wasserman, *An existence theorem for positive solutions of semilinear elliptic equations*, Arch. Rat. Mech. Anal. **95**, 211–216.

[**SW3**] J. Smoller and A. G. Wasserman, *Existence of positive solutions for semilinear elliptic equations on general domains*, Arch. Rat. Mech. Anal., to appear.

The Mathematics of Porous Medium Combustion

A. M. STUART

Abstract. Two partial differential equations arising from the theory of porous medium combustion are examined. While both equations possess a trivial steady solution, the form of the reaction rate, which is discontinuous as a function of the dependent variable, precludes bifurcation of non-trivial steady solutions from the branch of trivial solutions. A constructive approach to the existence theory for non-trivial global solution branches is developed. The method relies on finding an appropriate set of solution dependent transformations which render the problems in a form to which local bifurcation theory is directly applicable. Specifically, by taking a singular limit of the (solution dependent) transformation, an artificial trivial solution (or set of solutions) of the transformed problem is created. The (solution dependent) mapping is not invertible when evaluated at the trivial solution(s) of the transformed problem; however, for non-trivial solutions which exist arbitrarily close to the artificial trivial solution, the mapping is invertible. By applying local bifurcation theory to the transformed problem and mapping back to the original problem, a series expansion for the non-trivial solution branch is obtained.

1. Introduction.

In this paper we analyze two time-dependent partial differential equation problems arising from the theory of porous medium combustion [**3**]. The first problem, (P1), defined in section 2, represents a simplified one-dimensional model of porous medium combustion when the depletion of oxygen during the reaction is insignificant. The model comprises a pair of reaction-diffusion equations coupled to a hyperbolic equation. The second problem, (P2), defined in section 3, is a single reaction-diffusion equation; it may be derived from (P1) as the first term in an asymptotic expansion of the governing equations in terms of two small parameters and as such it represents the evolution of temperature in highly exothermic combustion problems before the depletion of reactants becomes significant.

The novel mathematical aspect of the two problems is that the reaction rates are discontinuous as a function of the independent variables. This fact is reflected in the appearance of Heaviside step functions in the reaction terms. We present a unified approach to the existence theory for steady solutions of problems (P1) and (P2).

For problem (P1) we pose the equations on the whole real line and seek steady travelling wave solutions whereas for (P2) we examine steady solutions on a bounded interval. Both problems possess a trivial constant solution corresponding to an ambient state of no chemical reaction. However, the form of the reaction term in both (P1) and (P2) precludes the possibility of bifurcating non-trivial solutions in the neighborhood of the trivial solutions. Thus it is important to develop a constructive approach to the existence problem for both (P1) and (P2) in order that a starting point may be found for numerical studies of the two problems.

In sections 2 and 3 we define the two problems (P1) and (P2), respectively. In section 4 we describe an approach to the existence theory for (P1) and in section 5 we analyze this approach in more detail in the context of problem (P2).

2. Problem (P1).

The simplified model of porous medium combustion described here is derived in [**3**]. The unknowns σ, u and w represent the solid heat capacity and the solid and gas tempreatures respectively. The concentration of combustible solid may be determined as a linear function of the solid heat capacity. The governing equations are

$$\frac{\partial \sigma}{\partial t} = -\lambda r, \tag{2.1}$$

$$\sigma \frac{\partial u}{\partial t} = \frac{\partial^2 u}{\partial z^2} + w - u + r \tag{2.2}$$

$$\text{and } \delta \frac{\partial w}{\partial t} + \mu \frac{\partial w}{\partial z} = u - w, \tag{2.3}$$

where

$$r = \mu^{1/2} H(u - u_c) H(\sigma - \tau) f(w). \tag{2.4}$$

Here $H(X)$ is the Heaviside unit step function defined by

$$H(X) = \begin{cases} 1 & X > 0 \\ 0 & X \leq 0. \end{cases}$$

$f(w)$ is a strictly-positive $C^2(u_a, \infty)$ function where u_a is defined by (2.5). We take $z \in \mathbb{R}$ and impose the boundary conditions that

$$\lim_{z \to \infty} u(\pm z, t) = \lim_{z \to \infty} w(-z, t) = u_a, \tag{2.5}$$

where $u_a < u_c$.

We assume that the discontinuity induced by the reaction rate r in equation (2.2) is taken up entirely by the second spatial derivative of u so that $u(z,t)$ is a piecewise C^2 function of z and a C^1 function of time. Equation (2.1) implies that $\sigma(z,t)$ is a piecewise C^1 function of both z and t.

Taken together with initial conditions, equations (2.1–2.5) define the time-dependent problem (P1). We make no attempt to prove global existence of solutions for all time; we observe, however, that sharp rises in either of the temperatures u and w are naturally limited by the function σ which is non-increasing in time. Further aspects of the time-dependent behavior of problem (P1) are discussed in Norbury and Stuart [**6**].

3. Problem (P2).

Consider the partial differential equations (2.1–2.3). The reaction rate (2.4) may be rewritten as

$$r = \mu^{1/2} d H(u - u_c) H(\sigma - \tau) \bar{f}(w) \tag{3.1}$$

where we have extracted a factor d from the function $f(w)$ and redefined the reaction rate accordingly. We now analyze the partial differential equations defined by (2.1–2.3) and (3.1) in the distinguished limit $\mu \to 0$, $\lambda/\mu^{1/2} \to$ constant, $\delta/\mu^{1/2} \to$ constant, and $d\mu^{1/2} \to 1$. If we expand the solution in the form

$$\begin{aligned} \sigma &= \sigma_0 + O(\mu^{1/2}), \\ u &= u_0 + O(\mu^{1/2}) \\ \text{and } w &= w_0 + O(\mu^{1/2}) \end{aligned} \tag{3.2}$$

then, to first order in $\mu^{1/2}$ equations (2.1) and (2.3) give us

$$\frac{\partial \sigma_0}{\partial t} = 0 \text{ and } w_0 = u_0.$$

Assuming that initially $\sigma_0(z,0) =$ const $> \tau$ equation (2.2) gives, to first order

$$\sigma_0 \frac{\partial u_0}{\partial t} = \frac{\partial^2 u_0}{\partial z^2} + H(u_0 - u_c)\bar{f}(u_0). \tag{3.3}$$

We consider this equation posed on a finite domain and the appropriate boundary conditions are

$$u_0(\pm s, t) = u_a. \tag{3.4}$$

Rescaling the variables by setting

$$v = \frac{u_0 - u_a}{u_c - u_a}, \quad t = s^2\sigma_0\tau \text{ and } y = z/s,$$

equations (3.3) and (3.4) may be reformulated, for some appropriately defined $g(v)$, as the Dirichlet problem

$$\frac{\partial v}{\partial \tau} = \frac{\partial^2 v}{\partial y^2} + H(v-1)g(v), \tag{3.5}$$

with boundary conditions

$$v(\pm 1, t) = 0. \tag{3.6}$$

Together with initial conditions, equations (3.5) and (3.6) form problem (P2). As for problem (P1) we make no attempt to prove global existence of solutions of (P2). Indeed for certain classes of function $g(v)$ (which are likely to arise in practice) finite or infinite time blow-up can occur. These cases are of particular interest since they determine the temporal and spatial scales on which the expansion procedure defined by (3.2) becomes invalid. These and other time-dependent questions are addressed in Norbury and Stuart [**4**].

4. Existence Theory for (P1).

We seek travelling wave solutions of the time-dependent problem defined by equations (2.1–2.5). We will assume that $\delta = 0$. However, this is merely done to clarify the exposition; the Theorems in section 4 are easily modified to cope with the case $\delta > 0$. Defining

$$x = z - ct, \quad Q(x) \equiv c\sigma(z,t), \quad U(x) \equiv u(z,t) \text{ and } W(x) \equiv w(z,t)$$

we obtain

$$\frac{dQ}{dx} = \lambda r \tag{4.1}$$

$$\frac{d^2U}{dx^2} + Q\frac{dU}{dx} + W - U + r = 0 \tag{4.2}$$

$$\text{and } \mu\frac{dW}{dx} = U - W, \tag{4.3}$$

where

$$r = \mu^{1/2} H(U - u_c) H(Q - c\tau) f(W). \tag{4.4}$$

Here $0 < \tau < 1$. Conditions (2.5) transform to

$$\lim_{x \to \infty} U(\pm x) = \lim_{x \to \infty} W(-x) = u_a. \tag{4.5}$$

Assuming that no chemical reaction has taken place at $z = +\infty$ we deduce that σ will be determined by its initial concentration there, so that

$$\lim_{z \to \infty} \sigma(z, t) = 1.$$

Hence the appropriate boundary condition for Q is

$$\lim_{x \to \infty} Q(x) = c. \tag{4.6}$$

The system of equations (4.1–4.6) forms a nonlinear eigenvalue problem for the four-dimensional vector-valued function (Q, U, U', W) and the parameter c. Henceforth we denote this eigenvalue problem by (EVP). If we define the space PC^n to be the set of functions which are piecewise C^n on the whole real line, then we seek solutions (Q, U, W, c) of (EVP) which lie in the space $PC^1 \times PC^2 \times PC^1 \times \mathbb{R}$.

It is clear that (EVP) possesses the family of trivial solutions

$$Q \equiv c \text{ and } U \equiv W \equiv u_a, \text{ for any } c \in \mathbb{R}.$$

It may be demonstrated that all other solutions satisfy $\|U\|_\infty \geq u_c > u_a$. Consequently local bifurcation theory is not directly applicable to the problem since non-trivial solution branches do not exist arbitrarily close to the trivial solution branches. In this section we develop a constructive approach to the existence theory for nontrivial solution branches of (EVP). The first theorem, however, is a non-existence result.

THEOREM 4.1. *Solutions of (EVP) can exist only for* $0 < \lambda < (u_c - u_a)^{-1}$.

PROOF: Integrating equation (4.2) with respect to x over the whole real line and employing equation (4.3) we obtain

$$\int_{-\infty}^{+\infty} Q\frac{dU}{dx}\,dx + \int_{-\infty}^{+\infty} r\,dx = 0,$$

since bounded travelling wave solutions have $w(\infty) = u_a$. Integrating by parts and applying the boundary conditions (4.5) and (4.6) gives us

$$u_a\big[Q(+\infty) - Q(-\infty)\big] + \int_{-\infty}^{+\infty}\left(r - \frac{dQ}{dx}U\right)dx = 0.$$

However, combining this with equation (4.1) implies that

$$\int_{-\infty}^{+\infty}\big[1 + \lambda(u_a - U)\big]r = 0.$$

Since r is strictly positive on some non-zero interval of x, and since $u_c \leq U \leq \infty$ on this interval, we deduce that

$$0 < \lambda < (u_c - u_a)^{-1}. \tag{4.8}$$

This completes the proof.

□

We now aim to determine where solution branches exist within the allowable parameter regime defined by Theorem 4.1; in particular we shall attempt to determine the ends of solution branches. Since the reaction rate is zero outside a finite interval of x-space the equations defining (EVP) may be integrated explicitly in these regions. By so doing it is possible to convert (EVP) into a two-point free boundary problem posed on a finite domain. There are two distinct cases to consider dependent upon whether the reaction rate r, defined by (4.4), becomes zero at its left-hand end because U attains its critical value u_c, or because Q attains its critical value rc. We make the following definitions.

DEFINITION 4.2. *We define the reaction zone for EVP to be the region of x-space $(0, L)$ in which $r \neq 0$.*

DEFINITION 4.3. *We define a (U, U) switch solution to be a solution of EVP which satisfies $U(0) = U(L) = u_c$.*

DEFINITION 4.4. *We define a (Q, U) switch solution to be a solution of EVP which satisfies $Q(0) = \tau c$ and $U(L) = u_c$.*

There are two distinct free boundary problems governing the two distinct types of solutions defined above. The details of their derivation are described in Norbury and Stuart [**5**]. We now treat the two cases separately.

Free Boundary Problem 1; (U, U) solutions.

We denote the following free boundary problem by (FBP1). Find $(Q, U, W, \alpha, L) \in C^1(0, L) \times C^2(0, L) \times C^1(0, L) \times R \times R$, where

$$\begin{aligned} \frac{dQ}{dx} &= \lambda\mu^{1/2} f(W), \\ \frac{d^2U}{dx^2} &+ Q\frac{dU}{dx} + W - U + \mu^{1/2} f(W) = 0 \\ \text{and } \mu\frac{dW}{dx} &= U - W, \end{aligned}$$

together with the initial conditions

$$\begin{aligned} Q(0) &= \frac{\mu - \alpha - \mu\alpha^2}{1 + \mu\alpha}, \\ U(0) &= u_c, \\ \frac{dU}{dx}(0) &= (u_c - u_a)\alpha \\ \text{and } W(0) &= u_a + \frac{(u_c - u_a)}{1 + \mu\alpha}, \end{aligned}$$

and matching conditions

$$\begin{aligned} U(L) &= u_c \\ \text{and } \frac{dU}{dx}(L) &+ (u_c - u_a)Q(L) - \mu[W(L) - u_a] = 0. \end{aligned}$$

THEOREM 4.5. *Solutions of (FBP1) are in a one-to-one correspondence with solutions of (EVP) iff the solutions of (FBP1) satisfy $L > 0$ and*

$$\tau Q(L) \le Q(0) < \mu < Q(L). \tag{4.9}$$

PROOF: See Norbury and Stuart [**5**].

□

The formulation of EVP as a free boundary problem on a finite domain has a decided advantage: by rescaling the independent variable we derive a problem to which local bifurcation theory is directly applicable. Under the transformation $s = x/L$ (FBP1) becomes

$$\frac{dQ}{ds} = \lambda\mu^{1/2}Lf(w), \tag{4.10}$$

$$\frac{d^2U}{ds^2} + LQ\frac{dU}{ds} + L^2(W-U) + \mu^{1/2}L^2 f(W) = 0 \tag{4.11}$$

$$\text{and } \mu\frac{dW}{ds} = L(U-W). \tag{4.12}$$

The initial conditions become

$$Q(0) = \frac{\mu - \alpha - \mu\alpha^2}{1+\mu\alpha}, \tag{4.13}$$

$$U(0) = u_c \text{ and } \frac{dU}{ds}(0) = (u_c - u_a)\alpha L \tag{4.14}$$

$$\text{and } W(0) = u_a + \left(\frac{u_c - u_a}{1+\mu\alpha}\right). \tag{4.15}$$

The matching conditions are

$$U(1) = u_c \tag{4.16}$$

$$\text{and } \frac{dU}{ds}(1) + L(u_c - u_a)Q(1) - \mu L[W(1) - u_a] = 0. \tag{4.17}$$

We denote the problem defined by equations (4.10–4.17) by (FBP1*).

THEOREM 4.6. *Solutions of (FBP1*) which satisfy $L > 0$ are in a one-to-one correspondence with solutions of (FBP1).*

PROOF: If $L > 0$ then the mapping between the two problems is a bijection.

□

Examination of (FBP1*) shows that it possesses the family of trivial solutions

$$Q \equiv \frac{\mu - \bar\alpha - \mu\bar\alpha^2}{1+\mu\bar\alpha}, \quad U \equiv u_c, \quad W \equiv u_a + \frac{(u_c - u_a)}{1+\mu\bar\alpha}, \quad \alpha = \bar\alpha$$

$$\text{and } L = 0,$$

for all $\bar{\alpha} \in \mathrm{R}$. None of these trivial solutions correspond to a true travelling wave solutions since they do not satisfy $L > 0$ and (4.9). However, we may now apply local bifurcation theory to determine the possible location of bifurcation points from the family of trivial solutions. The most important of these trivial solutions is the one corresponding to $\bar{\alpha} = 0$, namely

$$Q \equiv \mu, \quad U \equiv W \equiv u_c, \quad \alpha = 0 \text{ and } L = 0, \tag{4.18}$$

since it is the only solution which, when perturbed by an arbitrarily small amount, can satisfy (4.9).

THEOREM 4.7. *Consider the trivial solution (4.18). Then a necessary condition for bifurcation from this solution into non-trivial solutions of (FBP1) is* $\lambda = \lambda_c = (u_c - u_a)^{-1}$.

PROOF: Let $\underline{\psi} = (\psi, \theta, \phi, \xi, \eta)$ represent the linearization of (Q, U, W, L, α) about the trivial solution (4.18). We consider the solution and its linearization as elements of the Banach space $C^1(0,1) \times C^2(0,1) \times C^1(0,1) \times \mathrm{R} \times \mathrm{R}$. Then $\underline{\psi}$ satisfies

$$\frac{d\psi}{ds} = \xi\lambda\mu^{1/2} f(u_c) \text{ and } \frac{d^2\theta}{ds^2} = \frac{d\phi}{ds} = 0$$

where the initial conditions are

$$\begin{aligned} \psi(0) &= -(1+\mu^2)\eta, \\ \theta(0) &= \frac{d\theta}{ds}(0) = 0 \\ \text{and } \phi(0) &= -(u_c - u_a)\mu\eta. \end{aligned}$$

The matching conditions are

$$\begin{aligned} &\theta(1) = 0 \\ &\text{and } \xi \int_0^1 \left[1 + \lambda(u_a - u_c)\right] \mu^{1/2} f(u_c)\, ds = 0. \end{aligned} \tag{4.19}$$

Here we have replaced (4.17) by the equivalent condition [**5**]

$$L \int_0^1 [1 + \lambda(u_a - u(s))] \mu^{1/2} f(w) ds$$

which may be derived by applying the method of proof employed in Theorem 4.1 to (FBP1); see [**5**].

This linear problem for $\underline{\psi}$ defines the Frechet derivative of (FBP1*) with respect to the trivial solution (4.18). In the case $\lambda \neq \lambda_c = (u_c - u_a)^{-1}$ the null-space of the Frechet derivative is one-dimensional and spanned by the eigenfunction

$$(\psi, \theta, \phi, \xi, \eta) = \left(-(1+\mu^2), 0, -(u_c - u_a)\mu, 0, 1\right). \tag{4.20}$$

This eigenfunction corresponds to bifurcation into the trivial branch of solutions parametrized by $\bar{\alpha}$. However, in the case $\lambda = \lambda_c$, the null-space of the Frechet derivative is two-dimensional and spanned by (4.20) and a second eigenfunction

$$(\psi, \theta, \phi, \xi, \eta) = (\lambda_c \mu^{1/2} f(u_c)s, 0, 0, 1, 0).$$

This extra eigenfunction derives from the non-invertibility of the expression (4.19) for ξ at $\lambda = \lambda_c$. Consequently $\lambda = \lambda_c$ is the only point at which bifurcation into non-trivial solutions can occur.

□

That bifurcation actually does occur at $\lambda = \lambda_c$ may be shown by constructing a series expansion for the solution of (FBP1) in powers of $L \ll 1$. The details may be found in [**5**]. Thus the reformulation of (EVP) as a free boundary problem and the rescaling technique employed subsequently has led us to a constructive approach to the existence of non-trivial travelling wave solutions.

Free Boundary Problem 2; (Q, U) solutions.

We denote the following free boundary problem by (FBP2). Find $(Q, U, W, u_b, \alpha, L) \in C^1(0, L) \times C^2(0, L) \times C^1(0, L) \times \mathrm{R} \times \mathrm{R} \times \mathrm{R}$ where

$$\frac{dQ}{dx} = \lambda \mu^{1/2} f(W) \tag{4.21}$$

$$\frac{d^2U}{dx^2} + Q\frac{dU}{dx} + W - U + \mu^{1/2} f(W) = 0 \tag{4.22}$$

$$\text{and } \mu \frac{dW}{dx} = U - W, \tag{4.23}$$

together with initial conditions

$$Q(0) = \frac{\mu - \alpha - \mu\alpha^2}{1 + \mu\alpha}, \tag{4.24}$$

$$U(0) = u_b \text{ and } \frac{dU}{dx}(0) = (u_b - u_a)\alpha \tag{4.25}$$

$$\text{and } W(0) = u_a + \left(\frac{u_b - u_a}{1 + \mu\alpha}\right), \tag{4.26}$$

and matching conditions

$$Q(0) = \tau Q(L) \tag{4.27}$$

$$U(L) = u_c \tag{4.28}$$

$$\text{and } \frac{dU}{dx}(L) + (u_c - u_a)Q(L) - \mu[W(L) - u_a] = 0. \tag{4.29}$$

THEOREM 4.8. *Solutions of (FBP2) are in a one-to-one correspondence with solutions of (EVP) iff the solutions satisfy $\alpha > 0$, $L > 0$, $Q(0) < \mu < Q(L)$ and $U(0) \geq u_c$.*

PROOF: See Norbury and Stuart [**5**].

□

As for (FBP1) the reformulation of (EVP) as a free boundary problem has a number of analytical advantages. These are particularly clear in the case $\tau = 0$. In this case the boundary conditions (4.27) and (4.24) reduce to

$$Q(0) = 0 \tag{4.30}$$

$$\text{and } \alpha = \frac{-1 + [1 + 4\mu^2]^{1/2}}{2\mu}, \tag{4.31}$$

respectively.

Solutions of the simplified version of (FBP2) defined by (4.21–4.23), (4.25), (4.26), and (4.28–4.31) are characterized by a form of bifurcation from infinity as $\mu \to 0$. By this we mean there exists a branch of non-trivial solutions of (FBP2) at least one component of which approaches infinity as the critical value of the bifurcation parameter is approached. As a result the eigenvalue problem determining the existence of the bifurcating branch in the neighborhood of the bifurcation point is *nonlinear* in character. (A similar example of bifurcation from infinity in the case where the value of the bifurcation *parameter* at which bifurcation occurs is infinite may be found in [7]. In that case the leading order eigenvalue problem in the neighborhood of the bifurcation point is also nonlinear.)

For μ small a dimensional analysis shows that

$$Q \sim \mu^{1/4} Q_0, \quad U \sim U_0, \quad W \sim W_0, \quad u_b \sim U_{b0} \text{ and } L \sim \mu^{-1/4} L_0$$

where a subscript zero indicates an order one quantity. If we seek a series expansion of the reduced version of (FBP2) with $\tau = 0$ in powers of $\mu^{1/4} \ll 1$, then the leading order problem for $(Q_0, U_0, W_0, U_{b0}, L_0) \in C^1(0, L_0) \times C^2(0, L_0) \times C^1(0, L_0) \times \mathrm{R} \times \mathrm{R}^+$ is as follows:

$$\frac{dQ_0}{dy} = \lambda f(U_0) \tag{4.32}$$

$$\frac{d^2U_0}{dy^2} + Q_0 \frac{dU_0}{dy} + f(U_0) = 0 \tag{4.33}$$

$$\text{and } W_0 = U_0, \tag{4.34}$$

subject to the initial conditions

$$Q(0) = \frac{dU}{dy}(0) = 0 \text{ and } U_0(0) = U_{b0} \tag{4.35}$$

and the matching conditions

$$U_0(L_0) = u_c \tag{4.36}$$

$$\text{and } \frac{dU_0}{dy}(L_0) + (u_c - u_a) Q_0(L_0) = 0. \tag{4.37}$$

The existence of a solution of the free boundary problem defined by (4.32–4.37) is a necessary condition for the existence of a branch of solutions bifurcating from infinity as $\mu \to 0$. The following theorem establishes this result, for the case $f(U) \propto U^2$ which occurs in practice.

THEOREM 4.8. *For $0 < \lambda < \lambda_c$ and $f(U) \propto U^2$ there exists an odd number, greater than or equal to one, of solutions of the free boundary problem defined by equations (4.32–4.37).*

PROOF: The proof employs a shooting technique. The details may be found in Stuart [8].

□

5. Existence Theory for (P2).

In this section we discuss the existence theory for the steady solutions of problem (P2). It may be shown that all steady solutions $V(y)$ of (3.5) and (3.6) are symmetric and so they satisfy

$$\frac{d^2V}{dy^2} + \lambda H(V-1)g(V) = 0 \tag{5.1}$$

$$\text{and } \frac{dV}{dy}(0) = V(1) = 0, \tag{5.2}$$

where $g(V)$ is a strictly positive $C^2(1,\infty)$. Note that we have redefined g by placing a parameter λ outside it. For the purposes of this section we assume that $g(1) \neq 0$. The trivial solution $V \equiv 0$ satisfies (5.1) and (5.2). However, by virtue of the Heaviside step function appearing in the forcing term, all other solutions satisfy $\|V\|_\infty > 1$. Thus, as for the steady travelling wave solutions of problem (P1), bifurcation from the trivial solution is precluded. Hence we develop an analogous approach to that in section 4 and apply it to the boundary value problem defined by equations (5.1) and (5.2).

If we seek solutions of (5.1) and (5.2) satisfying $\|V\|_\infty > 1$ and define $y = s$ to be the unique point $s \in (0,1)$ such that $V(s) = 1$ then such solutions may be shown to be in one-to-one correspondence with solutions of the following free boundary problem: find $(V(y), s) \in C^2(0,s) \times [0,1)$ with $V(y) \geq 1$ for $y \in (0,s)$ such that

$$\frac{d^2V}{dy^2} + \lambda g(V) = 0, \tag{5.3}$$

$$\text{and } \frac{dV}{dy}(0) = 0, \quad V(s) = 1 \text{ and } \frac{dV}{dy}(s) = -\frac{1}{1-s}. \tag{5.4}$$

Rescaling the independent variable y and the parameter λ by setting

$$z = y/s \text{ and } \bar{\lambda} = \lambda s \tag{5.5}$$

we obtain, from equations (5.3) and (5.4), the associated problem (FBP3), for $(V(z), s) \in C^2(0,1) \times \mathsf{R}$, namely

$$\frac{d^2V}{dz^2} + \bar{\lambda} s g(V) = 0 \tag{5.6}$$

$$\text{and } \frac{dV}{dz}(0) = 0, \quad V(1) = 1 \text{ and } \frac{dV}{dz}(1) = -\frac{s}{1-s}. \tag{5.7}$$

THEOREM 5.1. *Solutions $(V(z), s)$ of (FBP3) which satisfy $s \in (0,1)$ and $V(z) \geq 1$ for $z \in [0,1]$ are in a one-to-one correspondence with solutions $(V(y), s)$ of the original boundary value problem defined by (5.1) and (5.2).*

PROOF: Provided that $s \in (0,1)$ the mappings between the two problems are all bijections.

□

Thus we examine solutions of (FBP3). Clearly the trivial solution $V \equiv 1$ and $s = 0$ satisfies (FBP3). This solution exists for all values of the parameter $\bar{\lambda}$, but does not correspond to a solution of equations (5.1) and (5.2) since it does not satisfy $s > 0$. However, by applying local bifurcation theory to (FBP3) we may determine the possible location of bifurcation points for non-trivial branches of solutions which satisfy $s > 0$.

THEOREM 5.2. *There exists a branch of non-trivial solutions to (FBP3) bifurcating from the trivial solution $V \equiv 1$ and $s = 0$, at $\lambda = 1/g(1)$.*

PROOF: We define the function $W(z)$ by $W(z) = V(z) - 1$. We study solutions of (FBP3) which have the property that s is bounded away from unity by a finite amount. This will necessarily be the case for non-trivial solutions which bifurcate from the trivial solution $(W(z), s) = (0,0)$. In this case (FBP3) may be written in the form

$$M(\lambda, x) = Bx - \bar{\lambda}Ax + N(\bar{\lambda}, x) = 0$$

where

$$B = \begin{pmatrix} \frac{d^2}{dz^2} & 0 \\ \frac{d}{dz}\Big|_{z=1} & 1 \end{pmatrix}$$

$$A = \begin{pmatrix} 0 & -g(1) \\ 0 & 0 \end{pmatrix}$$

$$\text{and } N = \begin{pmatrix} \bar{\lambda}s[g(W+1) - g(1)] \\ \frac{s^2}{1-s} \end{pmatrix}$$

for

$$x \in X = \{(W, s) : W \in C^2[0,1], \frac{dW}{dz}(0) = 0, W(1) = 0, \text{ and } s \in \mathbb{R}\}.$$

M defines a nonlinear mapping between the Banach spaces $\mathsf{R} \times X$ and Z, where $Z = \{(\tilde{W}, \tilde{s}) : \tilde{W} \in C[0,1] \text{ and } \tilde{s} \in \mathsf{R}\}$. We equip $\mathsf{R} \times X$ with the norm

$$\left\|(\bar{\lambda}, W, s)\right\| = |\bar{\lambda}| + \sum_{j=0}^{2} \sup_{z\in[0,1]} |W^j(z)| + |s|$$

(where ^{j} denotes the j-th derivative with respect to z) and Z with the norm

$$\left\|(\tilde{W}, \tilde{s})\right\| = \sup_{z\in[0,1]} |\tilde{W}(z)| + |\tilde{s}|.$$

In order that we may apply the standard theorems of local bifurcation theory [**1**] we require M to be $C^2(\mathsf{R}\times X, Z)$. However, in the neighborhood of $s = 1$ this is not the case since the nonlinear operator N includes the term $\frac{s^2}{1-s}$. Thus, in practice, we define a C^2 extension of the operator N by modifying N for $s \geq 1 - \delta$ where δ is a small strictly positive number. However, since the theorem concerns the local behavior of solutions in the neighborhood of the trivial solution $(W(z), s) = (0,0)$ we do not spell out the details of the extension. With the appropriate extension of N, M defines a mapping $\in C^2(\mathsf{R} \times X, Z)$.

Let $y = (0,0)$, the trivial solution. Then $N(\bar{\lambda}, y) = 0$. Also

$$D_x N(\bar{\lambda}, x) = \begin{pmatrix} \bar{\lambda} s \frac{dg}{dV}(W+1) & \bar{\lambda}[g(W+1) - g(1)] \\ 0 & \frac{2s-s^2}{(1-s)^2}. \end{pmatrix}$$

Thus $D_x N(\bar{\lambda}, y) = 0$. Hence, by Theorem 5.3 in Chapter 5 of [**1**], we deduce that if λ_0 is a simple eigenvalue of the pair (B, A) then (λ_0, y) is a bifurcation point of $M(\bar{\lambda}, x) = 0$. Thus we determine the location of simple eigenvalues of the pair (B, A); if we denote the corresponding eigenvectors by $x_0 = (\phi, \eta)$ then they satisfy

$$\frac{d^2\phi}{dz^2} + \eta\bar{\lambda}g(1) = 0, \tag{5.8}$$

$$\text{and } \frac{d\phi(0)}{dz} = 0, \quad \phi(1) = 0 \text{ and } \frac{d\phi}{dz} = -\eta. \tag{5.9}$$

Integration of the eigenvalue problem defined by (5.8–5.9) shows that $(B - \bar{\lambda}A)x_0 = 0$ has a nontrivial solution

$$x_0 = (\phi_0, \eta_0), \text{ where } \phi_0 = \frac{\eta_0}{2}(1 - z^2), \quad \eta_0 \in \mathsf{R}\backslash\{0\},$$

if and only if

$$\bar{\lambda} = \lambda_0 = 1/g(1).$$

A necessary and sufficient condition for the eigenvalue λ_0 to be simple is that the equation

$$(B - \lambda_0 A)x = Ax_0 \tag{5.10}$$

does not possess a solution. Defining $x = (\Phi, \gamma)$ equation (5.10) may be written as

$$\frac{d^2\Phi}{dz^2} + \gamma = -g(1)\eta \tag{5.11}$$

together with boundary conditions

$$\frac{d\Phi(0)}{dz} = 0, \quad \Phi(1) = 0 \text{ and } \frac{d\Phi(1)}{dz} = -\gamma. \tag{5.12}$$

Integration from $z = 0$ to $z = 1$ shows that equations (5.11–5.12) do not possess a solution unless $g(1) = 0$. However this case is excluded at the beginning of the section and thus we have proved that λ_0 is a simple eigenvalue of the pair (B, A). This completes the proof.

□

In the neighborhood of the bifurcation point we deduce from Theorem 5.3 in Chapter 5 of [**1**] that the solutions of (FBP3) satisfy

$$\bar{\lambda} = 1/g(1) - O(\varepsilon), \tag{5.13}$$

$$V(z) = 1 + \frac{\varepsilon}{2}(1 - z^2) + O(\varepsilon^2) \tag{5.14}$$

$$\text{and } s = \varepsilon + O(\varepsilon^2) \tag{5.15}$$

where $\varepsilon \ll 1$ is a measure of proximity to the bifurcation point. The value $\varepsilon = 0$ corresponds to the bifurcation point itself. Clearly for $0 < \varepsilon \ll 1$ we will have $0 < s \ll 1$ and hence, by Theorem 5.1, these solutions correspond to genuine solutions of the original boundary value problem defined by equations (5.1–5.2). By virtue of the transformation (5.5) we deduce from (5.13) and (5.15) that in the neighborhood of the bifurcation point in the original problem

$$\lambda = 1/\varepsilon g(1) + O(1)$$

and thus that since bifurcation corresponds to $\varepsilon \to 0$, it occurs from $\lambda = \infty$. Furthermore, in terms of the original independent variable y, the solution (5.14) corresponds to a solution

$$\begin{aligned} V(y) &= \frac{1+y}{1-\varepsilon} + O(\varepsilon^2), \text{ for } -1 < y < -s \\ V(y) &= 1 + \frac{\varepsilon}{2}(1 - y^2/\varepsilon^2) + O(\varepsilon^2), \text{ for } -s < y < s \\ \text{and } V(y) &= \frac{1-y}{1+\varepsilon} + O(\varepsilon^2), \text{ for } s < y < 1. \end{aligned}$$

Taking the limit $\varepsilon \to 0$, which corresponds to approaching the bifurcation point at $\lambda = \infty$, we obtain

$$\lim_{\varepsilon \to 0} V(y) = 2g(y; 0)$$

where $g(y; z)$ is the Greens function for the problem, satisfying

$$\frac{d^2 g}{dy^2} + \delta(y - z) = 0$$

and

$$g(\pm 1) = 0.$$

The coefficient 2 appears so that the solution satisfies the condition that

$$\lim_{\lambda \to \infty} V(0) = 1.$$

Similar results on the approach of solutions to Greens functions in certain parameter limits have been found by Keady and Norbury [**4**].

6. Summary.

In this paper we have presented a constructive approach to the existence theory for two steady state problems (P1) and (P2) to which local bifurcation theory is not directly applicable. Although both problems possess a trivial solution the appearance of a Heaviside step function in the reaction terms (which consequently have a switch-like behavior) prevents bifurcation from the trivial solutions.

However, for both (P1) and (P2), the equations are linear outside a finite interval of unknown length, in which the reaction terms are non-zero. Thus, by solving the equations explicitly outside this region the problems may both be reformulated as two-point free boundary problems. A rescaling, which is problem dependent, renders both problems in a form to which local bifurcation theory is directly applicable. In particular we have proved that for problem (P2) bifurcation occurs from a parameter value at infinity and that the solution approaches a multiple of the Greens function for the domain as that parameter approaches infinity.

Acknowledgments.

I am extremely grateful to Dr. John Norbury for helpful advice on various aspects of this work. I would also like to thank the Science and Engineering Research Council, UK, for funding my research.

Department of Mathematics, MIT, Cambridge MA 02139

Research at MSRI supported in part by NSF Grant DMS 812079-05.

REFERENCES

1. S. N. Chow and J. Hale, "Methods of Bifurcation Theory," Springer-Verlag, New York, 1982.
2. G. Keady and J. Norbury, *A semilinear elliptic eigenvalue problem II.*, Proc. Roy. Soc. Edinburgh **87A** (1980), 83–109.
3. J. Norbury and A. M. Stuart, *Models for porous medium combustion*, Submitted to Quart. J. Mech. Appl. Math. (1987).
4. J. Norbury and A. M. Stuart, *Parabolic free boundary problems arising in porous medium combustion*, To appear in IMA J. Appl. Math. (1988).
5. J. Norbury and A. M. Stuart, *Travelling combustion waves in a porous medium. Part I - Existence*, SIAM J. Appl. Math **48** (1988), 155-169.
6. J. Norbury and A. M. Stuart, *Travelling combustion waves in a porous medium. Part II - Stability*, SIAM J. Appl. Math **48, No. 2** (1988).
7. S. Rosenblat and S. H. Davis, *Bifurcation from infinity*, SIAM J. Appl. Math. **37** (1980), 1–19.
8. A. M. Stuart, *The existence of solutions of a two-point free boundary problem arising from the theory of porous medium combustion*, IMA J. Appl. Math. **38** (1987), 23–34.

Connection Problems Arising from Nonlinear Diffusion Equations

D. TERMAN

I. Introduction.

A central feature in the study of many partial differential equations is the existence of special solutions. In this paper we consider two such examples. These are the existence of traveling wave solutions of reaction diffusion systems and the existence of radial solutions of semilinear elliptic equations. Both of these problems can be reduced to finding a particular solution of a system of ordinary differential equations. As we shall see, the particular solution corresponds to a trajectory in phase space which "connects" two equilibrium points. That is, the particular solution satisfies a system of ordinary differential equations of the form

$$U'(z) = F(U), \quad U \in \mathbb{R}^n, \tag{1.1a}$$

together with boundary conditions

$$\lim_{z \to -\infty} U(z) = A \text{ and } \lim_{z \to +\infty} U(z) = B \tag{1.1b}$$

where A and B are equilibria of (1.1a). We refer to a solution of (1.1) as an "$A \to B$ connection" or just "connection".

There are two main parts to this paper. In the first we discuss general methods for proving the existence of connections. A central feature of these methods is the Conley index. The second part of the paper is concerned with specific applications of the general theory.

This paper is organized as follows. In Section 2 we demonstrate how the existence of traveling wave and radial solutions can be reduced to proving the existence of a connecting orbit in phase space. This is not obvious for the radial solution problem. In Section 3 we present some simple examples which are meant to motivate the index arguments which follow in later sections. In particular, the examples help motivate the main theorems which have direct applications to the traveling wave and radial solution problems. In order to state the theorems, we must present the basic features of the Conley index. This is done in Section 4. Finally, in Section 6 we present some applications to combustion. We prove the existence of traveling wave solutions for a one-step combustion process. We also describe a result in [8] where we prove the existence of traveling wave solutions for a two-step reaction.

II. Reduction to a Connection Problem.

II.A. Reaction-Diffusion Equations.

By a reaction-diffusion system we mean a system of partial differential equations of the form

$$u_t = Du_{xx} + f(u), \quad U \in \mathbb{R}^n, \tag{2A.1}$$

where D is a nonnegative, diagonal matrix, and $f : \mathbb{R}^n \to \mathbb{R}^n$ is assumed to be smooth. These systems arise as models in various areas of mathematical biology and chemistry. Examples include Fisher's equation in population genetics, the Hodgkin-Huxley equations as a model for the propagation of electrical impulses in a nerve axon, models for interacting species in ecology, and models for laminar flames in combustion. See Smoller [7] and Fife [5] and the references cited there for a detailed description of these and other models.

A traveling wave solution of (2A.1) is a bounded solution of the form

$$u(x,t) = U(z), z = x + \theta t. \tag{2A.2}$$

That is, a traveling wave solution is a solution of (2A.1) which appears to be traveling with constant shape and velocity. We look for a traveling wave solution which satisfies certain boundary conditions. We assume that

$$\lim_{z \to -\infty} U(z) = \alpha \text{ and } \lim_{z \to +\infty} U(z) = \beta$$

where $f(\alpha) = f(\beta) = \mathcal{O}$, the zero vector in $\mathbb{R}^n$.

Note that if $V(z) = U'(z)$, then $(U(z), V(z))$ satisfies the system of ordinary differential equations

$$\begin{aligned} U' &= V \\ DV' &= \theta V - f(U) \end{aligned} \tag{2A.3}$$

together with the boundary conditions

$$(U,V)(-\infty) = (\alpha, \mathcal{O}) \equiv A \text{ and } (U,V)(+\infty) = (\beta, \mathcal{O}) \equiv B.$$

Hence, a traveling wave solution corresponds to a connecting orbit. Note that θ is an unknown parameter. A traveling wave solution will only exist for special values of θ.

II.B. Radial Solutions of Semilinear Elliptic Equations.

Consider the system

$$\Delta u + f(u) = 0, \quad u \in \mathrm{R}^m, \tag{2B.1}$$

where Δ is the Laplace operator, $u(x)$ is a function of $x \in \mathrm{R}^n$, $n > 1$, and $f : \mathrm{R}^m \to \mathrm{R}^m$. We assume that $f(\mathcal{O}) = \mathcal{O}$. A radial solution of (2B.1) is one of the form

$$u(x) = U(r), r = \|x\|.$$

We always assume that a radial solution satisfies

$$\lim_{r\to\infty} U(r) = \mathcal{O}. \tag{2B.2}$$

If $V(r) = U'(r)$, then $(U(r), V(r))$ satisfies the first order system of ordinary differential equations

$$\begin{aligned} U' &= V \\ V' &= -\frac{n-1}{r} V - f(U), \quad r > 0, \end{aligned} \tag{2.B3}$$

together with the boundary conditions

$$(U,V)(0) = (U_0, \mathcal{O}) \text{ and } (U,V)(\infty) = (\mathcal{O}, \mathcal{O}) \tag{2B.4}$$

for some $U_0 \in \mathbb{R}^m$.

Note that $(U, V)(r)$ is not a connecting orbit. We describe a procedure to transform (2B.3), (2B.4) into a connection problem.

Consider r as a dependent variable by introducing $z = r$ as the new independent variable, and compactify by letting

$$r = \frac{1}{2} \ln \frac{1+p}{1-p}. \tag{2B.5}$$

Then (2B.3) becomes

$$\begin{aligned} U' &= V \\ V' &= -\phi(\rho)V - f(U) \\ \rho' &= 1 - \rho^2 \end{aligned} \tag{2B.6}$$

where differentiation is with respect to z, and

$$\phi(\rho) = (n-1)\left[\frac{1}{2}\ln\frac{1+\rho}{1-\rho}\right]^{-1} \quad \text{for } 0 < \rho \leq 1.$$

A radial solution must satisfy

$$(U, V, \rho)(0) = (U_0, \mathcal{O}, 0) \text{ and } (U, V, \rho)(\infty) = (\mathcal{O}, \mathcal{O}, 1) \tag{2B.7}$$

for some U_0. For $\varepsilon > 0$, let

$$\phi_\varepsilon(\rho) = \begin{cases} \frac{1}{\varepsilon} & \text{for } -1 \leq \rho \leq \rho_\varepsilon \\ \phi(\rho) & \text{for } \rho_\varepsilon < \rho \leq 1 \end{cases}$$

where ρ_ε satisfies $\phi(\rho_\varepsilon) = \frac{1}{\varepsilon}$. Assume that $\beta \neq \mathcal{O}$ satisfies $f(\beta) = \mathcal{O}$, and consider the system for $z \in \mathbb{R}$,

$$\begin{aligned} U' &= V \\ V' &= -\phi_\varepsilon(\rho)V - f(U) \\ \rho' &= 1 - \rho^2 \end{aligned} \tag{2B.8ε}$$

together with the boundary conditions

$$(U, V, \rho)(-\infty) = (\beta, \mathcal{O}, -1) \text{ and } (U, V, \rho)(+\infty) = (\mathcal{O}, \mathcal{O}, 1). \tag{2B.9}$$

This is a connection problem. The idea will be to prove that for each $\varepsilon > 0$, there exists a solution of (2B.8ε), (2B.9). We must then prove that as $\varepsilon \to 0$, some subsequence of the solutions converge to a solution of (2B.6), (2B.7) for $z > 0$. Of course, this may not always be true, or, if it is true, be easy. This technique was, however, used in [**10**] to prove the existence of infinitely many radial solutions for a particular system.

III. The Scalar Equation.

We present simple examples with the hope of motivating things to come. Consider the scalar equation

$$u_t = u_{xx} + f(u), \quad u \in \mathbb{R}^n, \tag{3.1}$$

where, either

$$\begin{aligned} &\text{(a)} \quad f(u) = u(1-u) \text{ or} \\ &\text{(b)} \quad f(u) = u(1-u)(u-a), \quad 0 < a < \tfrac{1}{2}. \end{aligned} \tag{3.2}$$

In case (3.2a) we refer to (3.1) as Fisher's equation. In case (3.2b) we refer to (3.1) as the bistable equation. The basic results concerning the existence of traveling wave solutions of (3.1) are

THEOREM 3.1.

(a) *For Fisher's equation there exists θ_0 such that for each $\theta > \theta_0$ there exists a traveling wave solution of (3.1) which satisfies $0 \leq U(z) \leq 1$ for all z,*

$$\lim_{z \to -\infty} U(z) = 0, \text{ and } \lim_{z \to +\infty} U(z) = 1. \tag{3.3}$$

(b) *For the bistable equation there exists a unique θ^* for which there exists a traveling wave solution of (3.1) which satisfies (3.3).*

For a detailed proof of this result, see Aronson and Weinberger [**1**]. Here we give a few details of the proof. The traveling wave equations associated with (3.1), (3,3) are

$$\begin{aligned} U' &= V \\ V' &= \theta V - f(U), \end{aligned} \tag{3.4a}$$

$$\lim_{z \to -\infty} (U,V)(z) = (0,0), \text{ and } \lim_{z \to +\infty} (U,V)(z) = (1,0). \tag{3.4b}$$

Consider Fisher's equation. The phase space associated with (3.4a), for large θ, is as shown in Figure 1.

The equilibrium points (0,0) and (1,0) are, respectively, a node and a saddle. Let W_A^S and W_A^U be, respectively, the stable and unstable manifolds at an equilibrium A. An $A \to B$ connection must lie in both W_A^U and W_B^S. For Fisher's equation, $\dim W_{(0,0)}^U = 2$ and $\dim W_{(1,0)}^S = 1$. It immediately follows that $W_{(0,0)}^U$ and $W_{(1,0)}^S$ intersect transversely. In general, we make the following definition:

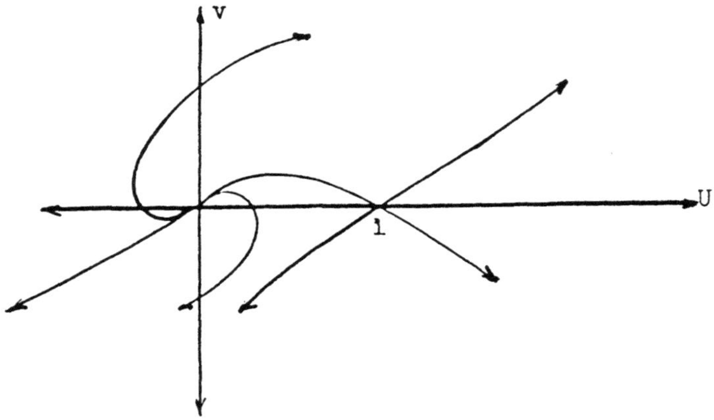

Figure 1.

DEFINITION. *An $A \to B$ connection is transversal if W_A^U and W_B^S intersect transversely, and A and B are nondegenerate equilibria.*

Note that a necessary condition for an $A \to B$ connection to be transveral is that

$$\dim W_A^u + \dim W_B^S \geq n + 1$$

where n is the dimension of the phase space.

Now consider the bistable equation. To prove the existence of one traveling wave solution we use a shooting argument. The phase spaces associated with (3.4a) for $\theta = 0$ and θ very large are shown, respectively, in Figure 2.

The equilibria $(0,0)$ and $(1,0)$ are saddles for all θ. Let $\gamma_\theta(z)$ be the trajectory in $W^u_{(0,0)}$ which points into the first quadrant. When $\theta = 0$, $\gamma_\theta(z)$ eventually crosses the U-axis to the left of $(1,0)$. When θ is very large $\gamma_\theta(z)$ never crosses the U-axis. Because $\gamma_\theta(z)$ changes continuously with respect to θ, it is not hard to see that this implies that there exists a $\theta^\star$ for which $\gamma_\theta(z)$ corresponds to a $(0,0) \to (1,0)$ connection.

Note that $\dim W^u_{(0,0)} = 1 = \dim W^S_{(1,0)}$. Hence, the connection is not transversal. By perturbing a parameter, namely θ, we can perturb away the connection.

The preceeding examples are rather simple because the phase space is two-dimensional. For larger systems, things are much more complicated. The phase space, and stable and unstable manifolds will be of higher dimensions, so it will be impossible to follow individual trajectories explicitly.

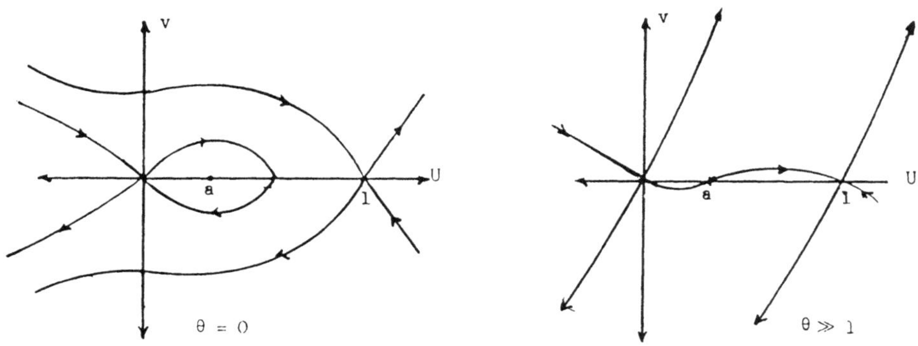

Figure 2

However, to prove the existence of an $A \to B$ connection, it is not necessary to know everything about the flow defined by (1.1a). It may be possible to prove the existence of a connection by knowing very little information about the flow.

The information we seek takes the form of an index. To each connecting orbit and equilibrium we associate a topological object which we call the Conley index of that set. The Conley index of an invariant set depends only on the behavior of the flow on the boundary of a compact neighborhood of the set. The Conley index has the property that it does not change if we perturb the underlying equations. Hence, if the Conley index is used to prove the existence of an $A \to B$ connection for $\theta = \theta_0$, then there must exist an $A \to B$ connection for all values of θ close to θ_0. This is the case for Fisher's equation where the connection is transveral. These remarks imply that the Conley index cannot be used directly to prove the existence of a nontransversal connection. If the connection is not transversal, it will be necessary to somehow alter the flow if we hope to use an index argument. To illustrate what we mean, consider the bistable equation (3.1), (3.2B).

On a solution of (3.4a), θ is constant. To (3.4a) we add the equation

$$\theta' = 0. \tag{3.5}$$

In the now three-dimensional phase space,

$$\ell_A = \{(U,V,\theta) : (U,V) = (0,0)\} \text{ and } \ell_B = \{(U,V,\theta) : (U,V) = (1,0)\}$$

correspond to lines of equilibrium points. Now instead of (3.5), consider the equation

$$\theta' = \varepsilon(\theta^2 - \theta_0^2) \tag{3.6}$$

where $0 < \varepsilon \ll 1$ and $\theta_0 \gg 1$. Instead of the lines of equilibria, ℓ_A and ℓ_B, we now have four equilibria. These are at the following values of (U, V, θ):

$$A_1 = (0,0,\theta_0), \quad A_2 = (0,0,-\theta_0), \quad B_1 = (1,0,\theta_0), \quad B_2 = (1,0,-\theta_0).$$

There are now two main steps in the analysis. These are

Step 1. Prove that for each $\varepsilon > 0$, there exists an $A_1 \to B_2$ connection, $\gamma_\varepsilon(z)$, of (3.4a), (3.6).

Step 2. Let $\varepsilon \to 0$ and prove that $\gamma_\varepsilon(z)$ converges, somehow, to the desired solution of (3.4).

To prove the existence of the $A_1 \to B_2$ connection in Step 1, we use the Conley index. The point of this construction is that by attaching the flow (3.6) we have added a dimension to both $W^u_{(0,0)}$ and $W^s_{(1,0)}$. More precisely, $\dim W^u_{A_1} = 2 = \dim W^S_{B_2}$. Because the dimension of the phase space is three, we can expect that the $A_1 \to B_2$ connection is transversal. Hence, there is hope that an index argument applies.

IV. The Conley Index.

We present some basic ingredients of the Conley index. We only discuss those features needed for the later sections of this paper. More complete presentations may be found in Conley [**2**], Conley and Zehnder [**4**], and Smoller [**7**].

Let Γ be a topological space. A flow on Γ is a continuous map from $\Gamma \times R$ into Γ, $(\gamma, t) \to \gamma \cdot t$, satisfying for all $\gamma \in \Gamma$ and all $s, t \in \mathbb{R}$ the two conditions $\gamma \cdot 0 = \gamma$ and $(\gamma \cdot s) \cdot t = \gamma \cdot (s + t)$. For two subsets $\Gamma' \subset \Gamma$ and $R' \subset R$ we set $\Gamma' \cdot R' = \{\gamma \cdot t \in \Gamma \mid \gamma \in \Gamma' \text{ and } t \in R'\}$. A subset $I \subset R$ is called invariant if $I \cdot R = I$. If $N \subset \Gamma$ is a subset we denote the invariant set contained in N by $I(N)$:

$$I(N) = \{\gamma \in N \mid \gamma \cdot R \subset N\}.$$

In all of the examples of this paper, Γ will be n-dimensional phase space. The flow will be defined by a system of ordinary differential equations. The invariant sets will be either equilibrium points, connecting orbits, or combinations of these.

DEFINITION. *Let N be a compact subset of Γ. If $I(N) \subset \operatorname{int} N$, then N is called an isolating neighborhood and $I(N)$ is called an isolated invariant set.*

If $Z \subset Y \subset \Gamma$ are subsets, we call Z positively invariant relative to Y if $y \in Z$ and $\gamma \cdot [0,t] \subset Y$ together imply that $\gamma \cdot [0,t] \in Z$. By a compact pair (Z_2, Z_1) we mean an ordered pair of compact spaces with $Z_1 \subset Z_2$.

DEFINITION. *Let S be an isolated invariant set. A compact pair (N_1, N_0) in Γ is called an index pair for S if*

(i) *$cl(N_1/N_0)$ is an isolating neighborhood for S,*
(ii) *N_0 is positively invariant relative to N_1,*
(iii) *if $\gamma \in N_1$ and $\gamma \cdot R^+ \not\subset N_1$, then there exists $t \geq 0$ such that $\gamma \cdot [0,t] \subset N_1$ and $\gamma \cdot t \in N_0$.*

Observe that positive orbits can leave N_1 only through the "face" N_0. In many applications N_0 will be the exist set of N_1; that is, those points on the boundary of N_1 through which trajectories leave in forward time.

In order to define the Conley index of an isolated invariant set we need the notion of a pointed space. For any pair (X, A), the pointed space X/A is the pair

$$X/A = ((X \backslash A) \cup [A], [A]).$$

The points of X/A consist of the points $x \in X \backslash A$ and an additional distinguished point $[A]$. The topology of X/A is described in [**2**]. In [**2**] it is proved that the homotopy type N_1/N_0 depends only on S; that is, if (N_1, N_0) and $(\bar{N}_1, \bar{N}_0)$ are two index pairs for the isolated invariant set S, then the pointed spaces N_1/N_0 and $\bar{N}_1/\bar{N}_0$ are homotopically equivalent. We can therefore associate with S a unique equivalence class $h(S)$. We call $h(S)$ the Conley index of S.

A key ingredient in proving the existence of traveling wave solutions is the sum formula. The sum $\vee$ of two pointed spaces (A, a) and (B, b) is defined as $A \cup B/\{a, b\}$, or, in other words, it is the pointed space obtained by taking the union and identifying the distinguished points a and b. The sum is denoted by $(A, a) \vee (B, b)$. The following important result is proved in [**4**].

LEMMA. *If S_1 and S_2 are isolated invariant sets with $S_1 \cap S_2 = \phi$, then $S_1 \cup S_2$ is an isolated invariant set and $h(S_1 \cup S_2) = h(S_1) \vee h(S_2)$.*

The way this lemma will be used is as follows:

COROLLARY. *Assume that N is an isolating neighborhood, and S_1 and S_2 are isolated invariant sets in N with $S_1 \cap S_2 = \phi$. If $h(I(N)) \neq h(S_1) \vee h(S_2)$, then $I(N) \neq S_1 \cup S_2$.*

In the applications, S_1 and S_2 will be equilibria, say A and B. N will be an isolating neighborhood which contains A and B as its only equilibria, and our goal will be to prove that there exists an $A \to B$ connection in N. If we can prove that $h(I(N)) \neq h(S_1) \vee h(S_2)$, then it will follow that there must exist an invariant set besides A and B which lies entirely in N. Often we can prove that there exists a Liapounov function in N; that is, a function which is decreasing on trajectories in N. This will imply that the only other possibility for a bounded solution in N, besides A and B, is an $A \to B$ connection.

In view of these remarks, we introduce a more general definition of a transversal connection. Suppose that $\gamma(z)$ is an $A \to B$ connection. Let

$$S = \{A\} \cup \{B\} \cup \{\gamma(z) : z \in \mathbb{R}\}.$$

We say that $\gamma(z)$ is *transversal* if S is an isolated invariant set and

$$h(S) \neq h(A) \vee h(B).$$

If a connection is transversal in the old sense, then it is certainly transversal in the new sense, if $\dim W_A^U + \dim W_B^S = n+1$. This last condition is needed so that S is isolated.

For a subset $Y \subset \Gamma$ we define the ω-limit sets

$$\omega(Y) = I\big(\mathrm{cl}\{Y \cdot [0,\infty)\}\big) \text{ and } \omega^*(Y) = I\big(\,\mathrm{cl}\{Y \cdot (-\infty,0]\}\big).$$

DEFINITION. *Assume that I is a compact, Hausdorff, invariant set in Γ. A* Morse decomposition *of I is a finite collection $\{M_\pi\}_{\pi \in P}$ of subsets $M_\pi \subset I$, which are disjoint, compact, and invariant, and which can be ordered $(M_1, M_2, \ldots, M_n)$ so that for every $\gamma \in I \backslash U_{1 \le j \le n} M_j$ there are indices $i < j$ such that $\omega(\gamma) \subset M_i$ and $\omega^*(\gamma) \subset M_j$.*

Indices (i.e., pointed spaces) can also be multiplied: the (smash) product of two pointed spaces is obtained from their topological product on collapsing to a point the set of pairs either of whose entries is a distinguished point. That is, if (X, A) and (Y, B) are compact pairs, then their smash product is defined as

$$[X/A] \wedge [Y/B] = [X \times Y/(X \times B \cup A \times Y)].$$

V. The Flow for θ Very Large.

V.A. The Main Theorem.

We now present a theorem which was first proved by Conley and Fife [**3**]. This result states that the structure of the set of bounded solutions of (2A.3) is identical to the structure of the set of bounded solutions of

$$U' = f(U), \quad U \in \mathbf{R}^n. \tag{5A.1}$$

We then prove two corollaries which have direct applications to the traveling wave and radial solution problems.

THEOREM 1. *Let M_0 be an isolated invariant set of (5A.1) and $(M_1, \ldots, M_k)$ a Morse decomposition of M_0. Assume that for each i, N_i is an isolating neighborhood of $M_i, \delta > 0$ and*

$$N_i' = \{(U, V) : U \subset N_i, \|V\| \leq \delta\}, i = 0, 1, \ldots, k.$$

There exists $\theta_0 > 0$ such that if $\theta > \theta_0$ and $M_i' = I(N_i')$ relative to (2A.3), then for each i,

(a) *N_i' is an isolating neighborhood of M_i',*
(b) *$h(M_i') = h(M_i) \wedge \Sigma^n$,*
(c) *$(M_1', \ldots, M_k')$ is a Morse decomposition of M_0'.*

A similar result holds if $\theta < -\theta_0$. To state this theorem precisely, consider the equation

$$U' = -f(U). \tag{5A.2}$$

THEOREM 2. *Let M_0 be an isolated invariant set of (5A.2) and $(M_1, \ldots, M_k)$ a Morse decomposition of M_0. Assume that for each i, N_i is an isolating neighborhood of M_i, $\delta > 0$, and*

$$N_i' = \{(U, V) : U \subset N_i, \|V\| \leq \delta\}, i = 0, 1, \ldots, k.$$

Then θ_0 can be chosen such that if $\theta < -\theta_0$ and $M_i' = I(N_i')$ relative to (2A.3), then for each i,

(a) *N_i' is an isolating neighborhood of M_i',*
(b) *$h(M_i') = h(M_i)$,*
(c) *$(M_1', \dots, M_k')$ is a Morse decomposition of M_0'.*

The key idea in the proof of these theorems is to make the change in variables $W = \theta V - f(U)$ in (2A.3). We multiply the resulting equations by θ and change variables again to obtain the system

$$\begin{aligned} U' &= f(U) + W \\ DW' &= \theta^2 W - (df)(U)[W + f(U)]. \end{aligned} \tag{5A.3}$$

Because $\|f(U)\|$ and $\|df(U)\|$ are bounded, there exists K such that on any bounded solution $\|W\| \leq K/\theta^2$. It now follows from the first equation in (5A.3) that for $|\theta|$ large, bounded solutions of (2A.3) lie close to bounded solutions of (5A.1). A complete proof can be found in [**3**].

V.B. Two Consequences of the Main Theorem.

In what follows, we assume that α and β are nondegenerate equilibria of (5A.1), and $A = (\alpha, \mathcal{O})$, $B = (\beta, \mathcal{O})$ are the corresponding equilibria of (2A.3).

PROPOSITION 1. *Assume that there exists a transversal $\alpha \to \beta$ connection, $u(z)$, of (5A.1) and N is an isolating neighborhood of this connection. Fix $\delta > 0$. Then there exists θ_0 such that for $\theta > \theta_0$, there exists a transversal $A \to B$ connection, $(U(z), V(z))$, of (2A.3). Moreover, $U(z) \in N$ and $\|V(z)\| < \delta$ for each z.*

PROOF: Let $M = I(N)$. By assumption,

$$h(M) \neq h(\alpha) \vee h(\beta). \tag{5B.4}$$

Clearly, (α, β) forms a Morse decomposition of M. Let $N' = \{(U, V) : U \in N, \|V\| \leq \delta\}$ and $M' = I(N')$. Theorem 1 implies that (A, B) forms a Morse decomposition of M', and, together with (5B.4), $M' \neq A \cup B$. The result now easily follows.

The second proposition is concerned with the case when there does not exist an $\alpha \to \beta$ connection. There may still exist an $A \to B$ connection for some θ, as illustrated by the bistable equation, but one cannot expect the $A \to B$ connection to be transversal. The remarks in Section 3 motivate us to consider the system

$$\begin{aligned} U' &= V \\ V' &= \theta V - f(U) \\ \theta' &= \varepsilon(\theta^2 - \theta_0^2) + \lambda \end{aligned} \tag{5B.5λ}$$

where $0 < \varepsilon \ll 1$, $\lambda \geq 0$, and $\theta_0 \gg 1$. We will be primarily interested in this system when $\lambda = 0$. If $\lambda = 0$, then the following values of (U, V, θ) are equilibria:

$$\begin{aligned} A_1 &= (A, \theta_0), \quad A_2 = (A, -\theta_0), \\ B_1 &= (B, \theta_0), \quad B_2 = (B, -\theta_0). \end{aligned}$$

We make the following assumptions:

(A1) α and β are nondegenerate attractors for (5A.1).

(A2) There exists a simply connected, smooth, compact set N which contains α and β, and is attracting. That is, if $p \in \partial N$ and n is an outward normal to ∂N at p, then $n \cdot f(p) < 0$.

(A3) There exists $L > 0$ such that if (U, V, θ) is a bounded solution of (5B.5) for $\varepsilon, \lambda \geq 0$, then $U \in N$ and $\|V\| \leq L$.

With these assumptions we prove

PROPOSITION 2. *Assume (A1)–(A3), $\varepsilon > 0$, and $\lambda = 0$. Then θ_0 can be chosen so that there exists an $A_1 \to B_2$ connection.*

PROOF: Because α is a nondegenerate attractor, there exists a neighborhood $N_\alpha \subset N$ of α such that $\alpha = I(N_\alpha)$ and trajectories enter N_α along its boundary. There also exists a corresponding neighborhood N_β of β. Let

$$\begin{aligned} N_1 &= \mathrm{cl}(N \backslash N_\alpha), \quad N_2 = \mathrm{cl}(N \backslash N_\beta), \\ \hat{S}_1 &= I(N_1), \quad \hat{S}_2 = I(N_2), \\ M_0 &= \{(U, V, \theta) : U \in N, \quad \|V\| \leq L, \quad |\theta| \leq \theta_0 + 1\}, \\ M_3 &= \{(U, V, \theta) : U \in N_1, \quad \|V\| \leq L, \quad |\theta - \theta_0| \leq 1\}, \\ M_1 &= \{(U, V, \theta) : U \in N_2, \quad \|V\| \leq L, \quad |\theta + \theta_0| \leq 1\}, \\ M_2 &= \mathrm{cl}(M_0 / (M_1 \cup M_3)), \\ S_i &= I(M_i), \quad i = 0, 1, 2, 3, 4. \end{aligned}$$

Clearly, S_1 and S_3 lie in $\{\theta = -\theta_0\}$ and $\{\theta = \theta_0\}$, respectively. The proof is now broken up into a few Lemmas.

LEMMA 1. *(S_1, S_2, S_3) is a Morse decomposition of S_0.*

PROOF: Let $S^+ = S_0 \cap \{\theta = \theta_0\}$ and $S^- = S_0 \cap \{\theta = -\theta_0\}$. Clearly, (S^-, S^+) is a Morse decomposition of S_0. Assumptions (A1) and (A2), together with the theorems, imply that (A_1, S_3), respectively (S_1, B_2), is a Morse decomposition of S^+, respectively S^-. Because (B_2, A_1) is a Morse decomposition of S_2, the result follows.

LEMMA 2. *$h(S_0) = \bar{0}$, the index of the empty set.*

PROOF: Let $S^\lambda = I(M_0)$ for the system (5B.5λ). Because of (A.3), M_0 is an isolating neighborhood for each λ. From a continuation argument (see Robbin and Solomon [6]) it follows that $h_0 = h(S^\lambda)$, does not depend of λ. However, if λ is sufficiently large, then $\theta(z) > 0$ for all z. Hence, $S^\lambda = \emptyset$ for large λ, and the result follows.

LEMMA 3. *$h(S_1) = h(S_3) = \bar{0}$.*

PROOF: Assumptions (A1) and (A2) imply that $(N_1, \partial N_\alpha)$ is an index pair for $\hat{S}_1$, and $(N_2, \partial N_\beta)$ is an index pair for $\hat{S}_2$. A direct computation shows that $h(\hat{S}_1) = h(\hat{S}_2) = \bar{0}$. Because S_1 and S_3 lie, respectively, in $\{\theta = -\theta_0\}$ and $\{\theta = \theta_0\}$, the results follows from Theorem 1 and Theorem 2, if θ_0 is sufficiently large.

LEMMA 4. *$h(S_2) = \bar{0}$.*

PROOF: This is proved in [8, Lemma 5E.4]. The main idea is to use the Morse inequalities, see Conley and Zehnder [4], to relate $h(S_2)$ with $h(S_0)$, $h(S_1)$ and $h(S_3)$.

LEMMA 5. *There exists an $A_1 \to B_2$ connection.*

PROOF: It follows from assumption (A1) and the Theorems that $h(A_1) \neq \bar{0}$ and $h(B_2) \neq \bar{0}$. Hence $h(S_2) \neq h(A_1) \vee h(B_2)$, and $S_2 \neq A_1 \cup B_2$. However, A_1 and B_2 are the only equilibria in S_2, and on any bounded solution in S_2, $\theta(z)$ is decreasing. The result now easily follows.

VI. Applications.

VI.A. The One-Step Combustion Process $A \to B$.

We present an application of Proposition 1 to combustion. This is concerned with the one-step reaction $A \to B$. Let $T(x,t)$ be the temperature and $Y(x,t)$ the mass fraction of A. Then $U = (T,Y)$ satisfies the system

$$U_t = DU_{xx} + F(U) \tag{6A.1}$$

where D is a positive, diagonal matrix, and $F(T,Y) = (QYf(T), -Yf(T))$, $Q > 0$. For $f(T)$ we assume

$$\begin{aligned} &\text{(a)} \quad f(0) = 0, \quad f'(0) = 0 \\ &\text{(b)} \quad f(T) > 0 \text{ for } T > 0 \\ &\text{(c)} \quad f \text{ is continuous.} \end{aligned} \tag{6A.2}$$

For example, f may be the usual Arrhenius term $f(T) = Be^{\frac{E}{-T}}$ with B, $E > 0$.

A traveling wave solution of (6A.1) corresponds to a combustion front in a premixed, reactive gas. We seek a traveling wave solution, $U(x,t) = U(x + \theta t)$, which satisfies boundary conditions of the form

$$U(-\infty) = (0, Y_-) \text{ and } U(+\infty) = (T_+, 0). \tag{6A.3}$$

REMARK: It is not hard to see from the structure of $F(U)$ that $T_+ = QY_-$.

PROPOSITION 6A. *There exists θ^* such that for each $\theta > \theta^*$ there exists a traveling wave solution of (6A.1) which satisfies (6A.3).*

PROOF: The first step in the proof is to consider the reaction equations

$$U' = F(U). \tag{6A.4}$$

Note that there is a $(0, Y_-) \to (QY_-, 0)$ connection , but it is not transversal; this is because the T and Y axes consist of equilibrium points. We perturb the equations (6A.4) so that the $(0, Y_-) \to (QY_-, 0)$ connection is transversal. Consider the system $U' = F_\varepsilon(U)$, which we write out explicitly as

$$\begin{aligned} T' &= QYf(T) + \varepsilon(T + QY - QY_-) \\ Y' &= -Yf(T) + \varepsilon(T + QY - QY_-). \end{aligned} \tag{6A.5}$$

For each $\varepsilon > 0$, there exists a $(0, Y_-) \to (QY_-, 0)$ connection of (6A.5) which is transversal. Proposition 1 now implies that there exists θ^* such that for each $\theta > \theta^*$ there exists a traveling wave solution $U^\varepsilon(z)$ of the system

$$U_t^\varepsilon = DU_{xx}^\varepsilon + F_\varepsilon(U^\varepsilon)$$

which satisfies (6A.3). We must now let $\varepsilon \to 0$ and show that some subsequence of $\{U^\varepsilon(z)\}$ converges to the desired traveling wave solution of (6A.1). This is straightforward, so we do not give the details.

VI.B. An Application of Proposition 2.

We briefly describe the main result in [**8**]. The key step in the proof of this result is Proposition 2.

Consider the sequential, two-step reaction

$$A \to B \to C. \tag{6B.1}$$

In order to understand what sort of phenomenon may arise from this reaction, we first imagine (6B.1) as taking place in two steps, and thus producting two flames.

The first reaction $A \to B$ in (6B.1) will convert the given unburned state to a partially burned state. This will produce a flame, $F1$, with speed, say θ_1. The second reaction, $B \to C$ in (6B.1) will then act on the product, or final state, of $F1$ and convert the partially burned state to a completely burned state. We denote the second flame by $F2$, and its velocity by θ_2.

The above flames will proceed at different velocities. For example, if the speed of $F2$, which is built on the products of $F1$, is slower than $F1$ itself, then one can imagine two flames both existing, but the distance between them ever-increasing. Another, more interesting, phenomenon is if $F2$ is faster than $F1$. In this case, the rear flame approaches the forward one. As it does so, its effect is to heat up the forward one. One expects the speed of the forward wave to then increase. One may then expect that what eventually evolves is a single configuration moving with constant velocity. In [**8**] we prove that such a third wave does indeed exist. A precise statement of our results and a short description of the proof is given in [**11**].

The key ingredient in the proof is Proposition 2. The proof demonstrates the usefulness of introducing the flow (3.6) in the θ direction.

Department of Mathematics, Ohio State University, Columbus OH 43210

This work was supported in part by the National Science Foundation under grant DMS-8401719.

References

1. D. G. Aronson and H. F. Weinberger, *Nonlinear diffusion in population genetics, combustion and nerve propagation*, in "Proceedings of the Tulane Program in Partial Differential Equations and Related Topics," Springer Lecture Notes in Math., vol. 446, 1975, pp. 5-49.
2. C. Conley, *Isolated invariant sets and the Morse index*, in "CBMS Regional Conference Series in Mathematics," AMS Conference Series in Math., vol. 38, 1978.
3. C. Conley and P. Fife, *Critical manifolds, traveling waves, and an example from population genetics*, J. Math. Biology **14** (1982), 159–176.
4. C. Conley and E. Zehnder, *Morse-type index theory and periodic solutions for Hamiltonian equations*, Comm. Pure Appl. Math. **39** (1984), 207–253.
5. P. C. Fife, *Mathematical aspects of reacting and diffusing systems*, Springer Lecture Notes in Biomath. **28** (1979).
6. J. Robbin and D. Salomon, *Dynamical systems, shape theory, and the Conley index*, preprint.
7. J. Smoller, "Shock Waves and Reaction-Diffusion Equations," Springer-Verlag, 1983.
8. D. Terman, *Traveling wave solutions arising from a combustion model*, I.M.A. Preprint # 216, (submitted to SIAM J. on Math. Analysis) (1986).
9. D. Terman, *Infinitely many traveling wave solutions of a gradient system*, Trans. of the AMS **301** (1987), 537–556.
10. D. Terman, *Infinitely many radial solutions of an elliptic system*, accepted by Annales de l'Institut Henri Poincaré, Analyse Nonlinéaire.
11. D. Terman, *An application of the Conley index to combustion*, to appear in Proceedings of the NATO Conference on Infinite Dimensional Dynamical Systems, J. Hale, ed.

Singularities of some Quasilinear Equations

LAURENT VÉRON

Introduction.

Let Ω be an open subset of $\mathbb{R}^N, N \geq 2$, containing 0, $\Omega' = \Omega \backslash \{0\}$ and $p > 1$. A function $u \in C^1(\Omega')$ is a weak singular solution of the p-Laplace equation

$$\operatorname{div}(|Du|^{p-2}Du) = 0 \tag{0.1}$$

in Ω' (or u is p-harmonic in Ω') if

$$\int_{\Omega'} |Du|^{p-2}Du \cdot D\varsigma \, dx = 0 \tag{0.2}$$

holds for any $\varsigma \in C_0^1(\Omega')$. It is clear that (0.1) admits a simple radial solution (with ω_N = volume of $B_1(0)$)

$$\mu(x) = \mu(|x|) = \begin{cases} \frac{p-1}{N-p}(N\omega_N)^{-1/(p-1)}|x|^{(p-N)/(p-1)} & \text{if } p \neq N, \\ (N\omega_N)^{-1/(N-1)} \log(1/|x|) & \text{if } p = N, \end{cases} \tag{0.3}$$

which is singular when $1 < p \leq N$ and satisfies

$$-\operatorname{div}(|Du|^{p-2}Du) = \delta_0 \tag{0.4}$$

in $\mathcal{D}'(\Omega)$. When u is singular and nonnegative a celebrated result from Serrin [**26**] asserts, as a particular case, that there exist $C > 0$ and $K > 0$ such that

$$C^{-1} \leq u/\mu \leq C \tag{0.5}$$

holds near 0 when $1 < p \leq N$. Moreover

$$-\operatorname{div}(|Du|^{p-2}Du) = K\delta_0 \tag{0.6}$$

in $\mathcal{D}'(\Omega)$. In Section 1 we present sharp results obtained from Kichenassamy [**17**] which give the exact behavior of a solution of (0.1) near 0 such that $u/\mu \in L^\infty_{\text{loc}}(\Omega)$.

In Section 2 we study the behavior near 0 of any $u \in C^1(\Omega')$ satisfying

$$-\operatorname{div}(|Du|^{p-2}Du) + u|u|^{q-1} = 0 \tag{0.7}$$

in Ω'. When $0 < q \le p-1$, (0.7) falls within the scope of Serrin's work [**27**] so we shall restrict ourselves to the case $q > p-1$. If we first look for solutions of (0.7) under the radial form $u(x) = \alpha|x|^\beta$ we find $\beta = -p/(q+1-p)$ and

$$\alpha = \gamma_{N,p,q} = \left(\left(\frac{p}{q+1-p}\right)^{p-1}\left(\frac{pq}{q+1-p} - N\right)\right)^{1/(q+1-p)}, \tag{0.8}$$

and it is clear that $\gamma_{N,p,q}$ exists only if $p \ge N$ or if $1 < p < N$ and $p-1 < q < N(p-1)/(N-p)$. In fact, when $1 < p < N$ and $q \ge N(p-1)/(N-p)$, Vazquez and Veron [**36**] proved that all the isolated singularities of solutions of (0.7) were removable and so $u \in C^1(\Omega)$. This is not the case when $\gamma_{N,p,q}$ does exist and we give the classification theorem obtained with Friedman [**13**] which extends Veron's results in the semilinear case $(p=2)$ [**38**].

In Section 3 we present some results obtained with Guedda [**16**] concerning positive solutions of

$$\operatorname{div}(|Du|^{p-2}Du) + u^q = 0. \tag{0.9}$$

It is also clear that if we look for solutions of (0.9) under the form $u(x) = \alpha|x|^\beta$ we find $\beta = -p/(q+1-p)$ and

$$\alpha = \lambda_{N,p,q} = \left(\left(\frac{p}{q+1-p}\right)^{p-1}\left(N - \frac{pq}{q+1-p}\right)\right)^{1/(q+1-p)}, \tag{0.10}$$

and this last quantity only exists when $q > \frac{N(p-1)}{N-p}$ (if $1 < p < N$).

The results we get are still partial (u is supposed to be radial) but they show the existence of only one type of singularity in any case ([**15,16**]) as in [**11,14**].

In Section 4 we look for solutions of (0.1) under the form

$$u(x) = |x|^{-\beta}\omega(x/|x|) \tag{0.11}$$

with $\beta > 0$. We obtain all of them when $N = 2$ and for $N \geq 3$ we prove the existence of such solutions ω as automorphic functions for any finite subgroup of $O(N)$ generated by reflections through hyperplanes containing O. Anisotropic singularities of (0.7) are given under the form

$$u(x) = |x|^{p/(q+1-p)}\omega(x/|x|). \tag{0.12}$$

The contents of this contribution are the following:

1. Singularities of the p-Laplace equation.

The main result of this section is the following [**17**]:

THEOREM 1.1. *Assume $1 < p \leq N$ and $u \in C^1(\Omega')$ is p-harmonic in Ω' such that*

$$u/\mu \in L^\infty(B_r(0)), \tag{1.1}$$

for $r > 0$ small enough, then there exists $\gamma \in \mathbb{R}$ such that

$$u - \gamma\mu \in L^\infty_{\text{loc}}(\Omega). \tag{1.2}$$

Moreover, Du satisfies

$$\lim_{x\to 0} |x|^{(N-1)/(p-1)}D(u - \gamma\mu)(x) = 0, \tag{1.3}$$

and the following equation holds in $\mathcal{D}'(\Omega)$

$$-\operatorname{div}(|Du|^{p-2}Du) = \gamma|\gamma|^{p-2}\delta_0. \tag{1.4}$$

PROOF: Without any loss of generality we assume $\Omega \subset \bar{B}_1(0) = \{x : |x| \leq 1\}$.

Step 1. A priori estimate. For $0 < a < \frac{1}{2}$ and $0 < |x| < 1$ we set $x = ay/(1+a)$ and $v(y) = u(ay/(1+a))/\mu(a)$. The function v is p-harmonic and bounded in $\Gamma = \{y : \frac{1}{2} \leq |y| \leq 3\}$. So we deduce from the a priori estimates of [**30**] (see also [**10**], [**20**], [**33**] and [**34**]) that

$$\|Du\|_{C^\alpha(\bar{B}_2(0)\setminus B_1(0))} \leq C \tag{1.5}$$

for some $C \geq 0$ and $\alpha \in (0,1)$. This clearly implies

$$|Du(x)| \leq C|x|^{-1}\mu(x) \tag{1.6}$$

$$|Du(x) - Du(x')| \leq C|x - x'|^\alpha |x|^{-1-\alpha}\mu(x). \tag{1.7}$$

Step 2. Strict comparison principle. The following easy-to-prove the result is due to Tolksdorff [**31**, Prop. 3.3.2]: if u_1 and u_2 are two p-harmonic functions in a connected open subset G of R^N such that $u_1 \geq u_2$ in G and ∇u_2 never vanishes in G, then either $u_1 > u_2$ in G or $u_1 = u_2$ in G. As a consequence, if G does not contain 0 and φ is p-harmonic in G, and if $\psi = \varphi/\mu$ or $\varphi - \mu$ achieves its maximum in G then ψ is constant.

Step 3. The case $1 < p < N$. We define

$$\gamma = \limsup_{x \to 0} u(x)/\mu(x), \tag{1.8}$$

and we may assume $\gamma > 0$ (or $\gamma < 0$ in the same way). We can also assume $\sup_{|x|=1} u(x) = 0$ and let $\tilde{g}$ be the function defined on $(0,1]$ by

$$\tilde{g}(r) = \sup_{r \leq |x| \leq 1} u(x)/\mu(x). \tag{1.9}$$

$\tilde{g}$ is nonnegative and nonincreasing on $(0,1]$ (from Step 2) and $\lim_{r\downarrow 0} \tilde{g}(r) = \gamma$. Moreover there exists x_r such that $|x_r| = r$ and $\tilde{g}(r) = u(x_r)/\mu(r)$. We now define the function u_r on $\Lambda_r = \{\xi : 0 < |\xi| < \frac{1}{r}\}$ by

$$u_r(\xi) = u(r\xi)/\mu(r). \tag{1.10}$$

From Step 1 we have

$$\begin{cases} |u_r(\xi) \leq C\mu(\xi), \\ |Du_r(\xi) \leq C|\xi|^{(1-N)/(p-1)}, \\ |Du_r(\xi') - Du_r(\xi)| \leq C|\xi - \xi'|^\alpha |\xi|^{(1-N)/(p-1)-\alpha}, \end{cases} \tag{1.11}$$

for $0 < |\xi| \leq |\xi'| \leq \frac{1}{r}$, where C is independent of r. From the Arzela-Ascoli theorem there exists a p-harmonic function v and a sequence $\{r_n\} \to 0$ such that $\{u_{r_n}\}$ converges to v in the $C^1_{\mathrm{loc}}(\mathbb{R}^N \backslash \{0\})$ topology. Moreover we have

$$u_r(\xi)/\mu(\xi) = u(r\xi)/(\mu(r\xi)\mu(1)) \leq \gamma/\mu(1); \tag{1.12}$$

and if we set $\xi_r = x_r/r$, then $u_r(\xi_r)/\mu(\xi_r) = \tilde{g}(r)/\mu(1)$. We can also assume that $\{\xi_{r_n}\}$ converges to some $\xi_0 \in S^{N-1}$, so we get

$$v(\xi)/\mu(\xi) \leq \gamma/\mu(1) \text{ and } v(\xi_0)/\mu(\xi_0) = \gamma/\mu(1). \tag{1.13}$$

Step 2 implies: $v(\xi) = \lim_{r\to 0} u_r(\xi) = (\gamma/\mu(1))\mu(\xi)$. If we take $|\xi| = 1$ in particular, we deduce

$$\lim_{x\to 0} u(x)/\mu(x) = \gamma \tag{1.14}$$

and (1.3).

In order to prove the boundedness of $u - \gamma\mu$ we consider for $\epsilon > 0$ the two following p-harmonic functions

$$v_\epsilon^+(x) = (\gamma + \epsilon)\mu(x) - (\gamma + \epsilon)\mu(1) + \sup_{|x|=1} u(x), \tag{1.15}$$

$$v_\epsilon^-(x) = (\gamma - \epsilon)\mu(x) - (\gamma - \epsilon)\mu(1) + \inf_{|x|=1} u(x). \tag{1.16}$$

From (1.14) we have $v_\epsilon^-(x) \leq u(x) \leq v_\epsilon^+(x)$ for $0 < |x| \leq 1$, which implies (1.2). As for (1.4) it is a straightforward application of (1.3) and Green's formula.

Step 4. The case $p = N$. As in Step 3 we define γ and $\tilde{g}$, $\gamma > 0$ and $\tilde{g} = \sup_{r\leq|x|\leq 1/2} u(x)/\mu(x)$, and $u_r(\xi) = u(r\xi)/\mu(r)$. From Step 1 we have the following estimates:

$$\left\{\begin{aligned} |u_r(\xi)| &\leq C(1 + |\log(|\xi|)/\log(\tfrac{1}{r})), \\ |Du_r(\xi)| &\leq C(1 + |\log(|\xi|)/\log(\tfrac{1}{r}))|\xi|^{-1}, \\ |Du_r(\xi) - Du_r(\xi')| &\leq C|\xi - \xi'|^\alpha(1 + |\log(|\xi|)/\log(\tfrac{1}{r}))|\xi|^{-1-\alpha}, \end{aligned}\right. \tag{1.17}$$

for $0 < |\xi| \leq |\xi'| \leq \frac{1}{(2r)}$. So there exists a sequence $\{r_n\}$ and an N-harmonic function v defined in $\mathbb{R}^N\backslash\{0\}$ such that $\{u_{r_n}\}$ converges to v in the $C^1_{\mathrm{loc}}(\mathbb{R}^N\backslash\{0\})$ topology. From (1.17) v is bounded, hence it can be extended to $\mathbb{R}^N$ as a bounded N-harmonic function [**26**], say $\tilde{v}$ and $\tilde{v}$ is constant from [**25**]. As in Step 3 we have

$$\gamma = \tilde{v}(\xi) = \lim_{x\to 0} u(x)/\mu(x). \tag{1.18}$$

The proof of the boundedness of $u - \gamma\mu$ is as in Step 3. For proving (1.3) and (1.4) we define

$$v_r(\xi) = u(r\xi) - \gamma\mu(r); \tag{1.19}$$

v_r is N-harmonic in Λ_r and satisfies

$$|v_r(\xi) \leq \gamma|\mu(\xi)| + C. \tag{1.20}$$

This implies, from Step 1,

$$\begin{cases} |Dv_r(\xi)| \leq C|\xi|^{-1}, \\ |Dv_r(\xi) - Dv_r(\xi')| \leq C|\xi - \xi'|^{\alpha}|\xi|^{-1-\alpha}, \end{cases} \tag{1.21}$$

for $0 < |\xi| \leq |\xi'| \leq \frac{1}{(2n)}$. Moreover the same estimates hold with v_r replaced by u, ξ by x, and ξ' by x' with $0 < |x| \leq |x'| \leq 1$. From Step 2 the supremum of $u(x) - \gamma\mu(x)$ in $B_{1/2}(0)$ is achieved either at 0 or for $|x| = \frac{1}{2}$ (otherwise $u - \gamma\mu$ is constant and everything is done). If we assume that it is achieved at 0, then the function $\tilde{\lambda}$ defined by

$$\tilde{\lambda}(r) = \sup_{r\leq|x|\leq 1/2} (u(x) - \gamma\mu(x)) \tag{1.22}$$

is nonincreasing and $\lim_{r\to 0} \tilde{\lambda}(r) = \lambda$. From (1.20) and (1.21) there exist an N-harmonic function v and a sequence $\{r_n\} \to 0$ such that $\{v_{r_n}\}$ converges to v in the $C^1_{\mathrm{loc}}(\mathbb{R}^N\backslash\{0\})$ topology and $\xi_0 \in S^{N-1}$ such that

$$v(\xi) \leq \gamma\mu(x) + \lambda, \qquad v(\xi_0) = \gamma\mu(\xi_0) + \lambda. \tag{1.23}$$

Hence $v = \gamma\mu + \lambda$ which implies $\lim_{x\to 0}(u(x) - \gamma\mu(x)) = \lambda$ and (1.3). If the supremum of $u - \gamma\mu$ is achieved for $|x| = \frac{1}{2}$ we can use the conformal invariance of the N-Laplace equation, and, performing the inversion $x \mapsto x/|x|^2$, prove (1.3) as above. The proof of (1.4) is the same as in Step 3.

Thanks to Theorem 1.1 we can study the singular Dirichlet problem

$$\begin{cases} -\operatorname{div}(|Du|^{p-2}Du) = \gamma|\gamma|^{p-2}\delta_0 & \text{in } \mathcal{D}'(\Omega) \\ u = g & \text{on } \partial\Omega. \end{cases} \tag{1.24}$$

THEOREM 1.2. *Assume $1 < p \leq N$, Ω is bounded with a C^1 boundary $\partial\Omega$, $g \in L^\infty(\Omega) \cap W^{1,p}(\Omega)$ and $\gamma \in \mathbb{R}$. Then there exists a unique $u \in C^1(\Omega')$ such that $|Du|^{p-1} \in L^1(\Omega)$, $Du \in L^p(\Omega \backslash B_r(0))$ for $r > 0$ small enough and (1.1) satisfying (1.24). Moreover u satisfies (1.2) and (1.3).*

REMARK 1.1: The proof of the existence of u is rather easy but the uniqueness is explicitly based on (1.2) and (1.3). A similar result has been obtained in [**7**] with a more general quasilinear operator but in a smaller class of uniqueness. The general problem of proving existence and uniqueness for

$$-\operatorname{div}(|Du|^{p-2}Du) = \nu \tag{1.25}$$

where ν is a bounded measure is still unsolved.

As a consequence of Theorem 1.1, we have a global result:

THEOREM 1.3. *Assume $1 < p \leq N$ and u is p-harmonic in $\mathbb{R}^N \backslash \{0\}$ such that $|u(x)| \leq a|\mu(x)| + b$ for any $x \neq 0$. Then there exist two constants γ and λ such that*

$$u(x) = \gamma\mu(x) + \lambda \tag{1.26}$$

for any $x \neq 0$.

Another interesting consequence of Theorem 1.1 is the following:

THEOREM 1.4. *Assume C is a discrete subset of $\mathbb{R}^N$ and u is a nonnegative N-harmonic function in $\mathbb{R}^N \backslash \{C\}$. Then u is constant.*

2. The Classification Theorem.

The main result of this section, which generalizes those of [**38**], is the following ([**13**])

THEOREM 2.1. *Assume $0 < p - 1 < q < N(p-1)/(N-p)$ if $1 < p < N$ or $q > p - 1$ if $p = N$ and $u \in C^1(\Omega')$ is a weak nonnegative solution of (0.7) in Ω'. Then we have the following:*

(i) *either $\lim_{x \to 0} |x|^{p/(q+1-p)} u(x) = \gamma_{N,p,q}$ (defined by (0.8)),*

(ii) *there exists a real number γ which can take any positive value such that $\lim_{x\to 0} u(x)/\mu(x) = \gamma$ and u satisfies*

$$-\operatorname{div}(|Du|^{p-2}Du) + u|u|^{q-1} = \gamma|\gamma|^{p-2}\delta_0 \tag{2.1}$$

in $\mathcal{D}'(\Omega)$, or

(iii) *$u(x)$ admits a finite limit when x tends to 0 and u can be extended as a $C^1(\Omega)$ function satisfying (0.7) in $\mathcal{D}'(\Omega)$.*

The main ingredient for proving Theorem 2.1 is the following extension of Theorem 1.1.

PROPOSITION 2.1. *Assume $0 < p-1 < q < N(p-1)/(N-p)$ if $1 < p < N$ or $q > p-1$ if $p = N$ and $u \in C^1(\Omega')$ is a weak solution of (0.7) in Ω' such that u/μ remains bounded near 0. Then there exists a real number γ such that*

$$\lim_{x\to 0} u(x)/\mu(x) = \gamma. \tag{2.2}$$

Moreover

$$\lim |x|^{(N-1)/(p-1)} D(u-\gamma u)(x) = 0, \tag{2.3}$$

and (2.1) holds in $\mathcal{D}'(\Omega)$.

The proof of Proposition 2.1 is essentially based on the same ingredients as those of Theorem 1.1: estimates on Du and scaling methods. Moreover, an extended strict comparison principle and Theorem 1.3 allow us to end the proof (see [**13**] for details). Moreover, singularities of type (ii) do exist as we have the following result:

PROPOSITION 2.2. *Assume $0 < p-1 < q < N(p-1)/(N-p)$ if $1 < p < N$ or $q > p-1$ if $p = N$ and let R be any positive number. Then for any real numbers α, γ there exists a unique function $\psi = \psi^{\alpha,\gamma} \in C^1((0,R])$ satisfying*

$$\begin{cases} -\left(r^{N-1}|\psi_r|^{p-2}\psi_r\right)_r + r^{N-1}|\psi|^{q-1}\psi = 0 & \text{in } (0,R), \\ \lim_{r\to 0} \psi(r)/\mu(r) = \gamma, \quad \psi(R) = \alpha. \end{cases} \tag{2.4}$$

Furthermore

$$\lim_{r\to 0} \psi_r(r)/\mu_r(r) = \gamma, \tag{2.5}$$

the function $(\alpha,\gamma) \mapsto \psi^{\alpha,\gamma}$ *is separately nondecreasing in* α *or in* γ, *and for any* α' *we have*

$$\left|\psi^{\alpha,\gamma}(r) - \psi^{\alpha',\gamma}(r)\right| \le |\alpha - \alpha'|, \tag{2.6}$$

on $(0, R]$.

The proof is essentially based upon the fact that under the hypotheses on N, p, q, $|u|^q$ is integrable near 0. The equation (2.4) is replaced by

$$\begin{cases} -\left(r^{N-1}|\psi_r^\epsilon|^{p-2}\psi_r^\epsilon\right)_r + r^{N-1}|\psi^\epsilon|^{q-1}\psi^\epsilon = 0 \quad \text{in } (\epsilon, R), \\ \psi^\epsilon(r) = \alpha, \quad \psi_r^\epsilon(\epsilon) = \gamma\mu_r(\epsilon), \end{cases} \tag{2.7}$$

where $0 < \epsilon < R$. It can then be proved that $\lim_{\epsilon\to 0}\psi^\epsilon$ exists and satisfies (2.4) and (2.5). As a consequence of Theorem 2.1 and Proposition 2.2 we can solve the singular Dirichlet problem for (0.7).

THEOREM 2.2. *Assume* $0 < p-1 < q < N(p-1)/(N-p)$ *if* $1 < p < N$ *or* $q > p-1$ *if* $p = N$, *that* Ω *is bounded and* $\partial\Omega$ *is* C^1. *Then for any* $g \in L^\infty(\Omega) \cap W^{1,p}(\Omega)$ *there exists a unique function* $u \in C^1(\Omega')$ *such that* $|Du|^{p-1} \in L^1(\Omega)$, $|u|^q \in L^1(\Omega)$, $Du \in L^p(\Omega \backslash B_r(0))$ ($r > 0$ *small enough) and* $u/\mu \in L^\infty(B_r(0))$ *satisfying*

$$\begin{cases} -\operatorname{div}(|Du|^{p-2}Du) + u|u|^{q-1} = \gamma|\gamma|^{p-2}\delta_0 & \text{in } \mathcal{D}'(\Omega), \\ u = g & \text{on } \partial\Omega. \end{cases} \tag{2.8}$$

Moreover (2.2) and (2.3) are satisfied.

We now give the sketch of the proof of Theorem 2.1.

LEMMA 2.1. *There exists a constant* $C > 0$ *such that for any* $\rho \in (0,1]$ $(B_1(0) \subset \Omega)$ *we have*

$$\max_{|x|=\rho} u(x) \le C \min_{|x|=\rho} u(x). \tag{2.9}$$

PROOF: From Vazquez's a priori estimate [**35**] (see also [**5**] and [**36**]) there exists $A = A(N,p,q) > 0$ such that

$$u(x) \le A|x|^{-p/(q+1-p)} \tag{2.10}$$

for $0 < |x| \le \frac{1}{2}$. Moreover it is clear that $u > 0$ in $B_1(0)$. Let $y \in B_{1/2}(0)$, $\rho = |y|/2$; then $B_\rho(y) \subset B_1(0)\backslash\{0\}$. We write (0.7) in $B_\rho(y)$ under the form

$$-\operatorname{div}(|Du|^{p-2}Du) + b^p u^{q-1} = 0 \tag{2.11}$$

with $b^p = u^{q+1-p}$. By a result of Trudinger [**32**], there exists a constant $C_1 = C_1(N,p,q,\rho\,\|b\|_{L^\infty(B_\rho(y))})$ such that

$$\max_{x\in B_{\rho/3}(y)} u(x) \le \min_{x\in B_{\rho/3}(y)} u(x). \tag{2.12}$$

By (2.9) $\rho\,\|b\|_{L^\infty(B_\rho(y))} \le C_2$, independent of ρ. If x_1 and x_2 are such that $0 < |x_1| = |x_2| \le \frac{1}{2}$, we can connect them by 10 connected balls of radius $|x_1|/6$ with center on $\partial B_{|x_1|}(0)$. Hence (2.9) follows with $C = C_2^{10}$.

LEMMA 2.2. *Assume $0 < p-1 < q < N(p-1)/(N-p)$ if $1 < p < N$ or $q > p-1$ if $p = N$ and let $\psi \in C^1((0,1])$ be a nonnegative solution of*

$$-\left(r^{N-1}|\psi_r|^{p-2}\psi_r\right)_r + r^{N-1}\psi^q = 0 \text{ in } (0,1], \tag{2.13}$$

such that $\psi(r)/\mu(r)$ is unbounded in $(0,\frac{1}{2}]$. Then there exists a constant $C = C(N,p,q)$ such that

$$\left|r^{p/(q+1-p)}\psi(r) - \gamma_{N,p,q}\right| \le Cr^\tau \tag{2.14}$$

for $0 < r \le \frac{1}{2}$ where τ is the positive root of the equation

$$\chi^2 + \left(N - \frac{p(q+1)}{q+1-p}\right)\chi + \frac{p}{p-1}\left(N - \frac{pq}{q+1-p}\right) = 0. \tag{2.15}$$

PROOF: **Step 1.** There exists $\alpha > 0$ such that

$$\alpha^{-1} \le r^{p/(q+1-p)}\psi(r) \le \alpha \tag{2.16}$$

on $(0,\frac{1}{2})$. We give the proof when $1 < p < N$ and set

$$s = r^{(p-N)/(p-1)}, \quad \varphi(s) = \psi(r). \tag{2.17}$$

Then

$$(|\varphi_s|^{p-2}\varphi_s)_s = \left(\frac{p-1}{N-p}\right)^p s^{-p(N-1)/(N-p)}\varphi^q \tag{2.18}$$

in $(1,+\infty)$. Hence φ_s is increasing and $\lim_{s\to+\infty}\varphi_s(s) = +\infty$ from the hypothesis. From convexity $\varphi(s) \le s\varphi_s(s)$ which implies

$$(\varphi_s^{p-1})_s \le \left(\frac{p-1}{N-p}\right)^p s^{q-p(N-1)/(N-p)}\varphi_s^q \tag{2.19}$$

for s large enough. Integrating (2.19) yields

$$\liminf_{s\to+\infty} s^{[q-p(N-1)/(N-p)]/(q+1-p)}\varphi_s(s) > 0, \tag{2.20}$$

which gives the left-hand side of (2.16). The right-hand side is just (2.10).

Step 2.

$$\lim_{r\to 0} r^{p/(q+1-p)}\psi(r) = \gamma_{N,p,q}.$$

We consider $v(r) = r^{p/(q+1-p)}\psi(r)$; then v satisfies

$$\left|rv_r - \frac{p}{q+1-p}v\right|^{p-2}\left\{r^2(p-1)v_{rr} + r\left(N-1-2\frac{p(p-1)}{q+1-p}\right)v_r \right.$$
$$\left. + p\frac{pq-N(q+1-p)}{(q+1-p)^2}v\right\} = v^q \tag{2.21}$$

in $(0,1)$ except at the points where $rv_r - \frac{p}{q+1-p}v = 0$.

But it is clear that v is monotone near 0. If we set $v_k(r) = v(r/k)$ then v_k is bounded and there exists $\{k_n\} \to +\infty$ and $\tilde{v}$ such that $\{v_{k_n}\}$ converges to $\tilde{v}$ in the $C^1_{\text{loc}}((0,+\infty))$ topology and $\tilde{v} = \lim_{r\to 0} v(r) = v_0$. Using the weak form on (2.21) we deduce

$$\left(\frac{p}{q+1-p}v_0\right)^{p-2}\left(p\frac{pq-N(q+1-p)}{(q+1-p)^2}\right)v_0 = v_0^q. \tag{2.22}$$

As $v_0 \geq \alpha^{-1}$ we deduce $v_0 = \gamma_{N,p,q}$.

Step 3. We set $\gamma = \gamma_{N,p,q}$, $\psi(r) = \gamma r^{-p/(q+1-p)}(1+z(r))$. From Step 2 $z(r)$, $rz_r(r)$, and $r^2 z_{rr}(r)$ tend to 0 as $r \to 0$. Moreover

$$\gamma^{p-1}\left|rz_r - \frac{p}{q+1-p}(1+z)\right|^{p-2}\left\{(p-1)r^2 z_{rr} + \right. \tag{2.23}$$
$$\left. r\left(N-1-2\frac{p(p-1)}{q+1-p}\right)z_r + \frac{p^2q-Np(q+1-p)}{(q+1-p)^2}(1+z)\right\} = \gamma^q(1+z)^q$$

which gives

(2.24)
$$r^2(p-1)z_{rr} + r\left[N(p-1)-1+2p-\frac{p^2q}{q+1-p}\right]z_r + p\left(N-\frac{pq}{q+1-p}\right)z$$
$$= 0(z^2+r^2z_r^2+r^4z_{rr}^2).$$

If we introduce $t = \log r$, $\tilde{z}(t) = z(r)$ then the left-hand side of (2.24) is an elliptic linear operator with constant coefficients. By classical elliptic estimates $r^2z_r^2 = 0(z^2)$ and $r^4z_{rr} = 0(z^2)$.

Let us now introduce the linear equation

$$r^2(p-1)Z_{rr}+r\left[N(p-1)-1+2p-\frac{p^2q}{q+1-p}\right]Z_r+p\left(N-\frac{pq}{q+1-p}\right)Z=0$$

whose solutions are $Z_1(r)=r^\tau$ and $Z_2(r)=r^{\tau'}$ where τ and τ' are respectively the positive and negative roots of (2.15). By a classical comparison argument we deduce $|Z(r)|\le CZ_1(r)$ which ends the proof.

□

PROOF OF THEOREM 2.1: We can assume that u/μ is not bounded near 0, otherwise we use Proposition 2.1. So there exists $\{x_n\}\to 0$ such that

$$\lim_{n\to+\infty} u(x_n)/\mu(x_n)=+\infty. \tag{2.26}$$

If we set $r_n=|x_n|$, then Lemma 2.1 implies

$$\lim_{n\to+\infty}\min_{|y|=r_n} u(y)/\mu(y)=+\infty. \tag{2.27}$$

We consider now φ_n (resp. ψ_n) the solution of (2.13) in $(r_n,1)$ such that $\varphi_n(1)=0$ (resp. $\psi_n(1)=\max_{|x|=1}u(x)$) and $\varphi_n(r_n)=\min_{|x|=r_n}u(x)$ (resp. $\psi_n(r_n)=\max_{|x|=r_n}u(x)$). From the maximum principle $\varphi_n\le u\le\psi_n$ in $(r_n,1)$. From estimate (2.10) applied to φ_n and ψ_n there exists φ and ψ satisfying (2.13) in $(0,1)$ and $\{n_k\}\to+\infty$ such that $\{\varphi_{n_k}\}\to\varphi$ and $\{\psi_{n_k}\}\to\psi$ as $\{n_k\}\to+\infty$. Moreover $\varphi\le u\le\psi$. An easy verification shows that $\varphi(r)/\mu(r)$ is not bounded near 0 which implies

$$\left|\gamma_{N,p,q}-|x|^{p/(q+1-p)}u(x)\right|\le C|x|^\tau \tag{2.28}$$

for $0<|x|\le\frac{1}{2}$, from Lemma 2.2, which ends the proof.

□

REMARK 2.1: Under the same restrictions on N, p, and q we can show that if u is positive in $\mathrm{R}^N\backslash K$ (K compact) and satisfies (0.7) in $\mathrm{R}^N\backslash K$ then

$$\limsup_{|x|\to+\infty}|x|^{-\tau'}\left|\gamma_{N,p,q}-|x|^{p/(q+1-p)}u(x)\right|<+\infty, \tag{2.29}$$

where τ' is the negative root of (2.15). As a consequence we have a global result

THEOREM 2.3. *Assume* $0 < p-1 < q < N(p-1)/(N-p)$ *if* $1 < p < N$ *or* $q > p-1$ *if* $p = N$, $u \in C^1(\mathbb{R}^N \backslash \{0\})$ *is a nonnegative solution of (0.7) in* $\mathbb{R}^N \backslash \{0\}$. *Then* u *is radial and we have the following situation:*

(i) *either* $u(x) = \gamma_{N,p,q}|x|^{-p/(q+1-p)}$,

(ii) *or there exists* $\gamma > 0$ *such that* $\tilde{u}(|x|) = u(x)$ *and* $\tilde{u}$ *is the unique* $C^1(0,+\infty)$*-function satisfying*

$$\left\{ \begin{array}{l} -(r^{N-1}|\tilde{u}_r|^{p-2}\tilde{u}_r)_r + r^{N-1}|\tilde{u}|^{q-1}\tilde{u} = 0 \text{ on } (0,+\infty), \\ \lim_{r \to 0} \tilde{u}(r)/\mu(r) = \gamma, \end{array} \right. \tag{2.30}$$

in $\mathcal{D}'(\Omega)$,

(iii) *or* u *is the zero function.*

3. Singularities of $\operatorname{div}(|Du|^{p-2}Du) + u^q = 0$.

In this section we present some results concerning the isolated singularities of the positive radial solutions of (0.9), that is of

$$(r^{N-1}|u_r|^{p-2}u_r)_r + r^{N-1}u^q = 0. \tag{3.1}$$

If we look for solutions of (3.1) under the form $u(r) = \alpha r^\beta$ then we get $\beta = -p/(q+1-p)$ (we always assume $q > p-1$, otherwise (3.1) falls within the scope of Serrin's works) and

$$\alpha = \lambda_{N,p,q} = \left(\left(\frac{p}{q+1-p} \right)^{p-1} \left(N - \frac{pq}{q+1-p} \right) \right)^{1/(q+1-p)} \tag{3.2}$$

and it is clear that $\lambda_{N,p,q}$ only exists when $N > p$ and $q > N(p-1)/(N-p)$; even in that case there is another critical value for q which is $p^* = \frac{Np}{N-p} - 1$ and corresponds to the embedding of $W_0^{1,p}$ into L^{p^*}. The results of this section are mostly due to Guedda and Veron [**15,16**].

THEOREM 3.1. *Assume* $1 < p \leq N$, $p-1 < q < N(p-1)/(N-p)$ *and* u *is a positive radial solution of (0.9) in* Ω'. *Then either* u *is regular in* Ω *or there exists* $\alpha > 0$ *such that*

$$\lim_{|x| \to 0} u(x)/\mu(x) = \alpha. \tag{3.3}$$

Moreover, u satisfies

$$(3.4) \qquad -\operatorname{div}(|Du|^{p-2}Du) - u^q = \alpha^{p-1}\delta_0.$$

in $\mathcal{D}'(\Omega)$.

We first start with the following a priori estimates of the singularity which have been obtained in a different way by Ni and Serrin [**23**].

PROPOSITION 3.1. *Assume* $1 < p \leq N$, $q \geq p-1$ *and* u *is a positive radial solution of (0.9) in* Ω'. *Then we have the following:*

(i) *if* $p - 1 \leq q < N(p-1)/(N-p)$, $u(x)/\mu(x)$ *remains bounded near 0,*

(ii) *if* $q = N(p-1)/(N-p)$, $N > p$,

$$u(x)/\left(\mu(x)\left(\log(1/|x|)^{(p-N)/(p(p-1))}\right)\right)$$

remains bounded near 0,

(iii) *if* $q > N(p-1)/(N-p)$, $N > p$, $|x|^{p/(q+1-p)}u(x)$ *remains bounded near 0.*

PROOF: For simplicity let us assume $1 < p < N$. We make the change of variable (2.17), $s = r^{(p-N)/(p-1)}$, $u(r) = v(s)$; then

$$(3.5) \qquad \left(|v_s|^{p-2}v_s\right)_s + \left(\frac{p-1}{N-p}\right)^p s^{-p(N-1)/(N-p)}v^q = 0$$

for $s \geq 1$. Hence the function v_s is decreasing and v is concave, then either $\lim_{s\to+\infty} v(s)$ is finite and u is regular in Ω or $\lim_{s\to+\infty} v(s) = +\infty$. Moreover $v(s)/s$ is bounded at infinity and $u(x)/\mu(x)$ is bounded near 0, which implies (i). When $q \geq N(p-1)/(N-p)$ we can improve this estimate. From concavity and $\lim_{s\to+\infty} v(s) = +\infty$ we have $v(s) \geq sv_s(s)(1+o(1))$ and (3.5) becomes (as $v_s > 0$)

$$(3.6) \qquad (v_s^{p-1})_s + Cs^{-p(N-1)/(N-p)+q}v_s^q \leq 0$$

for some $c > 0$. If we set $\psi(s) = v_s^{p-1}$ we get

$$(3.7) \qquad (\psi(s))^{-(q+1-p)/(p-1)} \geq C'\begin{cases} s^{q-N(p-1)/(N-p)} & \text{for } q > N(p-1)/(N-p) \\ \log s & \text{for } q = N(p-1)/(N-p) \end{cases}$$

for some C' and $s > 2$, or equivalently

$$0 \le v_s(s) \le C'' \begin{cases} s^{[N(p-1)/(N-p)-q]/(q+1-p)} & \text{for } q > N(p-1)/(N-p) \\ (\log s)^{-(p-1)/p(N-p)} & \text{for } q = N(p-1)/(N-p). \end{cases}$$

Integrating this last relation yields

$$v(s) \le C''' \begin{cases} s^{p(p-1)/((N-p)/(q+1-p))} & \text{for } q > N(p-1)/(N-p) \\ s(\log s)^{-(p-1)/p(N-p)} & \text{for } q = N(p-1)/(N-p) \end{cases} \tag{3.8}$$

which is the desired estimate.

In the case $p = N$ the previous change of variable is replaced by $s = \log \frac{1}{r}$, $v(s) = u(r)$.

PROOF OF THEOREM 3.1: We still assume $1 < p < N$. From the concavity of v either u is regular in Ω or $\lim_{x\to 0} u(x) = +\infty$. If we write (0.9) under the form

$$\operatorname{div}(|Du|^{p-2}Du) + d(x)u^{p-1} = 0 \tag{3.9}$$

with $d = u^{q+1-p}$, then for $\epsilon > 0$ small enough $d \in L^{N/(p-\epsilon)}_{\text{loc}}(\Omega)$ as $q < N(p-1)/(N-p)$ and $|u(x)| \le C|x|^{(N-p)/(p-1)}$. We then deduce from [**27**] that if u is singular at 0 there exists $\beta > 0$ such that

$$\beta|x|^{(p-N)/(p-1)} \le u(x) \le \frac{1}{\beta}|x|^{(p-N)/(p-1)} \tag{3.10}$$

for $0 < |x| \le 1$. If we return to v we have

$$\beta \le v(s)/s \le \frac{1}{\beta}. \tag{3.11}$$

As $(v(s) - v(1))/(s-1)$ is decreasing and bounded below by $\beta/2$ for s large it admits a positive limit α as $s \to +\infty$ which gives (3.3). The proof of (3.4) is the same as in Theorem 2.1 (ii).

REMARK 3.1: In the case $q \ge N(p-1)/(N-p)$ (3.3) and (3.4) still hold with $\alpha = 0$. *In the case $p - 1 < q < N(p-1)/(N-p)$ there truly exist functions u satisfying (3.4) with $\alpha > 0$ or, equivalently, solutions v of (3.5) such that* $\lim_{s\to+\infty} v(s) = +\infty$. They can be obtained in the following way: for $\gamma > 0$ let v_γ be the solution of (3.5) such that $v_\gamma(1) = 0$, $v_{\gamma s}(1) = \gamma$; v_γ is defined on some maximal interval $[1, T)$ where it stays positive and we have three possibilities:

(i) either $T < +\infty$ and $v_\gamma(T) = 0$,
(ii) $T = +\infty$ and v_γ admits a finite limit at infinity,
(iii) or $T = +\infty$ and $\lim_{s\to+\infty} v_\gamma(s) = +\infty$.

In cases (i) and (ii) there exists $\tau \in (1,T]$ such that $v_{\gamma s}(\tau) = 0$. As $v(s) \leq \gamma s$ (concavity) we get

$$(v_{\gamma s}^{p-1})_s + c_{N,p}\gamma^q s^{q-p(N-1)/(N-p)} \geq 0, \tag{3.12}$$

on $[1,\tau)$, which implies

$$c'_{N,p}\gamma^q(1-\tau^{q-p(N-1)/(N-p)}) \geq \gamma^{p-1}, \tag{3.13}$$

as $q < N(p-1)/(N-p)$. As $q > p-1$ (3.13) is impossible for $\gamma \ll 1$ and we are left with the singular case (iii).

The methods of Proposition 3.1 can be used for studying the behavior of solutions of (0.9) at infinity.

PROPOSITION 3.2. *Assume $1 < p \leq N$, $q > p-1$ and $G = \{x \in \mathbb{R}^N : |x| \geq 1\}$. Then we have the following:*

(i) *if $p-1 < q \leq N(p-1)/(N-p)$ when $1 < p < N$ or $q > p-1$ when $p = N$ there exist no positive radial solutions of (0.9) in G.*
(ii) *if $1 < p < N$ and $q > N(p-1)/(N-p)$ and if u is a positive radial solution of (0.9) in G there exist two positive constants C and C' such that*

$$C\mu(x) \leq u(x) \leq C'|x|^{-p/(q+1-p)}. \tag{3.14}$$

PROOF: **Step 1.** Assume $1 < p \leq N$, $q > p-1$ and u is a positive radial solution of (0.9) in G, then $\lim_{|x|\to+\infty} u(x) = 0$. For that we shall restrict ourself to $1 < p < N$ and make the change of variable (2.17). As v is concave in (0,1] it admits a nonnegative limit α at 0. Let us assume that α is nonzero, then we deduce from (3.5) that the following relation

$$-C_1 s^{-p(N-1)/(N-p)} \leq \left(|v_s|^{p-2}v_s\right)_s \leq -\frac{1}{C_1}s^{-p(N-1)/(N-p)} \tag{3.15}$$

holds for s small enough and $C_1 > 0$, which implies

$$C'_1 s^{-N(p-1)/(N-p)} - D \leq |v_s|^{p-2}v_s \leq \frac{1}{C'_1}s^{-N(p-1)/(N-p)} + D \tag{3.16}$$

for $0 < s \leq s_0 \ll 1$. Hence $v_s > 0$ on $(0, s_0)$. Taking the $(p-1)$-th root of (3.16) and integrating it on $(0, s_0)$ yields $v(0) = -\infty$, a contradiction.

Step 2. Assume $1 < p \leq N$ and $p-1 < q \leq N(p-1)/(N-p)$; then (i) holds. As $v(0) = 0$ and v is concave $v(s) \geq s v_s(s)$. Hence (3.5) becomes

$$(3.17) \qquad \left(v_s^{p-1}\right)_s + C_{N,p} s^{q-p(N-1)/(N-p)} v_s^q \leq 0.$$

If we set $\psi(s) = v_s^{p-1}(s)$ we get

(3.18)
$$(\psi(s))^{-(q+1-p)/(p-1)} \geq C' \begin{cases} s^{q-N(p-1)/(N-p)} & \text{for } q < N(p-1)/(N-p) \\ \log(1/s) & \text{for } q = N(p-1)/(N-p) \end{cases}$$

for $s \ll 1$ and $C' > 0$. As $\psi(0) > 0$ we get a contradiction. When $p = N$ we just set $\tau = \log(1/r)$, $w(\tau) = u(r)$.

Step 3. Assume $1 < p < N$ and $q > N(p-1)/(N-p)$; then (ii) holds. We shall make a distinction based on whether v_0^{p-1} is finite. In the first case it is clear that $u(r) \backsim \rho\mu(r)$ for some $\rho > 0$. In the second case we deduce from (3.17)

$$(3.19) \qquad v_s \leq C' s^{-[q-N(p-1)/(N-p)]/(q+1-p)}$$

near 0 for some $C' > 0$ which implies the right-hand side of (3.14). As for the left-hand side it is just the consequence of $v(s) \geq Cs$ near 0.

REMARK 3.2: Proposition 3.2 has been obtained by Ni and Serrin [**23**] under a slightly stronger hypothesis with a completely different method. Moreover, when $1 < p < N$ and $q > N(p-1)/(N-p)$ there exist solutions of (0.9) in G such that $u(x) \backsim \gamma\mu(x)$ or $u(x) \backsim \lambda_{N,p,q}|x|^{-p/(q+1-p)}$ at infinity: the first type of asymptotic behavior can be obtained through the following fixed-point relation

$$(3.20) \qquad v(s) = \int_0^s \left[\alpha - \int_0^\sigma \tau^{-p(N-1)/(N-p)} v^q(\tau)\, d\tau\right]^{1/(p-1)} d\sigma.$$

As a consequence of Proposition 3.2 we have

PROPOSITION 3.3. *Assume $1 < p < N$, then*

(i) *if $p-1 < q < Np/(N-p)-1$ there exist no positive radial solutions of (0.9) in $\mathbb{R}^N$.*

(ii) *if $q = Np/(N-p)-1$ the only positive radial solutions of (0.9) in $\mathbb{R}^N$ are the functions*

$$(3.21) \qquad y_a(x) = \left[Na\left(\frac{N-p}{p-1}\right)^{p-1}\right]^{1/(q+1-p)} \left(a + |x|^{p/(p-1)}\right)^{(p-N)/p}$$

where a is any positive real number.

PROOF: In the case $N(p-1)/(N-p) < q < Np/(N-p)-1$ (i) is the consequence of (3.14), and of the following Pohozaev-type relation [**24,16**]

$$(3.22)\qquad -\int_{|x|\le R} u^q(x\cdot Du)\,dx + \frac{p-N}{p}\int_{|x|\le R}|Du|^p\,dx = \int_{|x|=R}|Du|^{p-2}(x\cdot Du)u_\nu\,d\sigma - \frac{1}{p}\int_{|x|=R}|Du|^p(x\cdot\nu)\,d\sigma,$$

and the fact that $|Du(x)| \le C|x|^{-(q+1)/(q+1-p)}$ (from (3.14)). In fact, the right-hand side of (3.22) converges to 0 as $R\to+\infty$; as for the left-hand side, it converges to

$$\left(\frac{1}{q+1}-\frac{N-p}{Np}\right)\int_{\mathbb{R}^N} u^{q+1}\,dx,$$

and $u\equiv 0$. For (ii) it is first clear that y_a is a regular solution of (0.9) in $\mathbb{R}^N$ such that $Dy_a(0)=0$. The uniqueness comes from the fact that for $b>0$ there exists a unique local solution of

$$(3.23)\qquad \begin{cases}(r^{N-1}|u_r|^{p-2}u_r)_r + r^{N-1}u^q = 0 \text{ in } [0,\rho_{\max}),\\ u(0)=b, u_r(0)=0,\end{cases}$$

(this was proved by McLeod and Serrin [**22**] for $1<p\le 2$ and by Serrin and Veron [**28**] for $p>2$) and from the fact that $u_r(r)<0$ for $r>0$ which implies that there is no bifurcation along the trajectory of u.

Finally, we can improve Theorem 3.1 when $q > N(p-1)/(N-p)$ and we get

THEOREM 3.2. *Assume $1<p<N$ and $N(p-1)/(N-p) < q < Np/(N-p)-1$ and u is a positive radial solution of (0.9) in Ω'. Then we have the following:*

(i) *either* $\lim_{x\to 0}|x|^{p/(q+1-p)}u(x) = \lambda_{N,p,q}$,
(ii) *or u is regular in Ω.*

PROOF: Let us define $\psi(r) = r^{p/(q+1-p)}u(r)$; ψ is bounded and satisfies

(3.24)
$$\left|r\psi_r - \frac{p}{q+1-p}\psi\right|^{p-2}\left\{(p-1)r^2\psi_{rr} + \left(N-1-2\frac{p(p-1)}{q+1-p}\right)r\psi_r + \frac{p^2q-Np(q+1-p)}{(q+1-p)}\psi\right\} + \psi^q = 0.$$

Let $t = \log(1/r)$, $\delta = p/(q+1-p)$ and $w(t) = \psi(r)$. Then

(3.25)
$$|w_t + \delta w|^{p-2}\{(p-1)w_{tt} - (N - \delta(q+p-1))w_t - \delta(N-\delta q)w\} + w^q = 0$$

holds for $t \geq 0$.

Step 1. We claim that $w_t + \delta w > 0$. If we write (3.25) as

$$(p-1)|w_t + \delta w|^{p-2}(w_t + \delta w_t) - (N - \delta q)|w_t + \delta w|^{p-2}(w_t + \delta w) + w^q = 0,$$

we get as w_t is bounded (see the proof of Theorem 2.1),

$$|w_t + \delta w|^{p-2}(w_t + \delta w) = e^{N-\delta q)t}\int_t^{+\infty} w^q(s)e^{(\delta q - N)s}\,ds > 0. \tag{3.26}$$

As a consequence (3.25) can be written as

$$(p-1)w_{tt} - (N - \delta(q+p-1))w_t - \\ \delta(N-\delta q)w + w^q(w_t + \delta w)^{2-p} = 0, \tag{3.27}$$

and we consider (3.27) as a nonlinear system in $\mathbb{R}^2$ in the unknowns (w, w_t).

Step 2. The point $(\lambda_{N,p,q}, 0)$ is asymptotically stable. For simplicity we set $\lambda = \lambda_{N,p,q}$ and we get by linearization around $(\lambda, 0)$

$$(p-1)\chi_{tt} - \left[N - \delta(q+p-1) + (p-2)\lambda^{q+1-p}\delta^{1-p}\right]\chi_t \\ - \left[\delta(N-\delta q) - q\lambda^{q+1-p}\delta^{2-p} + (p-2)\lambda^{q+1-p}\delta^{2-p}\right]\chi = 0 \tag{3.28}$$

and the characteristic equation is

$$(p-1)\rho^2 - (p-1)[N - \delta(q+q)]\rho + \delta(N-\delta q)(q+1-p) = 0. \tag{3.29}$$

Hence (3.29) admits two roots with negative real part as $p - 1 < q < Np/(N-p) - 1$, which proves the asymptotic stability.

Step 3. Assume $w(t)$ decreases to 0 as t tends to $+\infty$; then

$$w(t) \leq Ce^{-\delta t} \tag{3.30}$$

for some $C > 0$ and $t \geq 0$. To see that, we have

$$(w_t + \delta w)^{p-1}(t) \leq \frac{w^q(t)}{N - \delta q} \tag{3.31}$$

from (3.26), which implies

$$w_t + \delta w - C' w^{q/(p-1)} \le 0. \tag{3.32}$$

If we integrate (3.32) we get (3.30). *Moreover, it must be noticed that if w satisfies (3.30), then u can be extended to Ω as a regular solution of (0.9) in Ω.*

Let $\mathcal{E}$ be the trajectory of (w, w_t) in the phase plane $\mathbb{R}^+ \times \mathbb{R}$ and let Γ^+ be the ω-limit set of $\mathcal{E}$ at $+\infty$. We have two possibilities:

(i) either $(\lambda_{N,p,q}, 0) \in \Gamma^+$,
(ii) or $(\lambda_{N,p,q}, 0) \notin \Gamma^+$.

In case (i) we deduce from Step 2 that $\lim_{t\to+\infty} w(t) = \lambda_{N,p,q}$. So let us assume that (ii) holds. Then there exists a neighborhood $\mathcal{O}$ of $(\lambda_{N,p,q}, 0)$ in $\mathbb{R}^+ \times \mathbb{R}$ such that $\mathcal{E} \cap \mathcal{O} = \{\emptyset\}$. We are left with two possibilities:

(iii) either $w(t)$ is monotone for t large enough,
(iv) or $w(t)$ is not asymptotically monotone.

In case (iii) it is clear that $w(t)$ converges to an equilibrium as t tends to $+\infty$. This equilibrium is necessarily 0 and (3.30) holds.

We now assume (iv): there exists a sequence $\{t_n\}$ converging to $+\infty$ such that $w_t(t_n) = 0$ and

$$w(t_{2n}) \text{ is a local maximum}, w(t_{2n+1}) \text{ is a local minimum.} \tag{3.33}$$

As $\text{sign}(w_{tt}(t_n)) = (-1)^{n-1}$ we deduce from (3.27) that

$$0 < w(t_{2n+1}) < \lambda_{N,p,q} < w(t_{2n}) \tag{3.34}$$

for any $n \ge 0$. We again have two possibilities:

(v) either $(0,0) \notin \Gamma^+$,
(vi) or $(0,0) \in \Gamma^+$.

In case (v) there would exist a neighborhood $\mathcal{O}'$ of $(0,0)$ such that $\mathcal{E} \cap \mathcal{O}' = \{\emptyset\}$ (it is clear that $\{w(t_{2n+1})\}$ and $\{w(t_{2n})\}$ are respectively decreasing and increasing (or the converse) from the uniqueness theorem. We then deduce from the Poincaré-Bendixon theorem that Γ^+ *is a cycle*, surrounding $(\lambda_{N,p,q}, 0)$. in order to prove the nonexistence of such a cycle we set $y = (w_t + \delta w)^{p-1}$, now (3.27) can be written as

$$\begin{cases} w_t = y^{1/(p-1)} - \delta w = P(w,y) \\ y_t = (N - \delta q) y - w^q = Q(w,y). \end{cases} \tag{3.35}$$

From the Green-Riemann formula we have

$$\int_d\int (Q_y + P_w)\,dy\,dw = \int_{\Gamma^+} (Q\,dw - P\,dy) = 0 \tag{3.36}$$

where D is the bounded domain of the (w,y)-plane with boundary Γ^+. But $Q_y + P_w = N - \delta(q+1) > 0$ and we get a contradiction. So *there exists no cycle* and we are left with the case (vi). It is clear in that case that $\{w(t_{2n+1})\}$ decreases to 0 and $\{w(t_{2n})\}$ increases and converges to some $\Lambda > \lambda_{N,p,q}$. If $\tilde{\mathcal{O}}$ is the trajectory of (3.27) passing through $(\Lambda, 0)$ in the (w, w_t) plane we deduce from the Jordan curve theorem that the ω-limit set of $\tilde{\mathcal{O}}$ at $+\infty$ is $(0,0)$ and its α-limit set at $-\infty$ is also $(0,0)$. If $(\tilde{\omega}(0), \tilde{\omega}_t(0)) = (\Lambda, 0)$ then $\tilde{\omega}(t)$ decreases to 0 as t tends to $+\infty$ and $\tilde{\omega}(t)$ satisfies (3.30). If we set

$$\tilde{u}(x) = |x|^{-p/(q+1-p)}\tilde{\omega}(\log(1/|x|)); \tag{3.37}$$

then $\tilde{u}$ is a nonzero regular solution of (0.9) in $\mathbb{R}^N$, which is impossible, and (vi) is not true, which ends the proof.

□

In the critical case we get the following extension of Fowler's results [**16**]:

THEOREM 3.3. *Assume* $1 < p < N$, $q = N(p-1)/(N-p)$ *and* u *is a positive radial solution of (0.9) in* Ω'. *Then we have the following:*

(i) *either*

$$\lim_{x\to 0} |x|^{\frac{N-p}{p-1}} (\log(1/|x|))^{\frac{p-N}{p(p-1)}} u(x) = \left(\left(\frac{N-p}{p}\right)\left(\frac{N-p}{p-1}\right)^{p-1}\right)^{\frac{N-p}{p(p-1)}}$$

(ii) *or* u *can be extended to* Ω *as a solution of (0.9) in* Ω.

Finally, we can propose the following question: *Are the results of this section still true if we no longer assume that* u *is radial?* When $p = 2$ the answer is yes [**1,14**]! When $p \neq 2$ Theorem 3.1 still holds if we assume that u is a positive solution of (0.9) in Ω' such that $u(x)/\mu(x)$ remains bounded near 0.

4. Anisotropic singularities.

If we look for singular p-harmonic functions u in the form

$$u(x) = |x|^{-\beta}\omega(x/|x|) = r^{-\beta}\omega(\sigma), \quad \beta > 0, \tag{4.1}$$

where $(r, \sigma) \in (0, +\infty) \times S^{n-1}$, a straightforward calculation shows that $\omega \in C^1(S^{N-1})$ and must satisfy

$$\begin{aligned} -\operatorname{div}_{S^{N-1}} &\left((\beta^2\omega^2 + |\nabla_\tau\omega|^2)^{(p-2)/2}\nabla_\tau\omega\right) \\ &+ \beta\gamma(\beta^2\omega^2 + |\nabla_\tau\omega|^2)^{(p-2)/2}\omega = 0, \end{aligned} \tag{4.2}$$

where $\gamma = N - 1 - (\beta + 1)(p - 1)$, ∇_τ is the tangential gradient on S^{N-1} identified with the covariant derivative on S^{N-1} and $\operatorname{div}_{S^{N-1}}$ is the divergence operator acting on C^1 vector fields on S^{N-1}. Using a shooting method introduced by Tolksdorf and the symmetry invariance of the p-Laplace equation we construct many spherical p-harmonic functions for $p > 1$.

PROPOSITION 4.1. *Assume S is a connected open subset of S^{N-1} with a C^∞ boundary ∂S. Then there exists only one (β, ω) with $\beta > 0$, $\omega \in C^1(S)$, ω positive in S with maximum value 1, $\omega = 0$ on ∂S such that the function u defined by (4.1) is p-harmonic in $\mathbb{R}^N \setminus \{0\}$. Moreover $\omega \in C^\infty(S)$.*

SKETCH OF THE PROOF: **Step 1: Existence.** As in [**31**] we define a conical domain $K_S(R, R') = \{r\sigma : \sigma \in S, R < r < R'\}$ and its conical boundary $B_S(R, R') = \{r\sigma : \sigma \in \partial S, R < r < R'\}$ and let g be the function defined by

$$g(x) = \begin{cases} 2 - |x| & \text{if } |x| \le 2, \\ 0 & \text{if } |x| \ge 2. \end{cases} \tag{4.3}$$

For $n \ge 2$ set u_n to the unique solution of the Dirichlet problem

$$\begin{cases} -\operatorname{div}(|Du_n|^{p-2}Du_n) = 0 & \text{in } K_S(1, n), \\ u_n = g & \text{on } \partial K_S(1, n). \end{cases} \tag{4.4}$$

From the Hopf maximum principle [**31**], $u_n > 0$ in $K_S(1, n)$. When n tends to $+\infty$ the sequence $\{u_n\}$ increases and converges in $W^{1,p}_{\text{loc}}(\overline{K_S(1, \infty)})$ to some u which is the unique function in $L^p_{\mathrm{loc}}(\overline{K_S(1,\infty)})$ such that $\nabla u \in L^p(K_S(1,\infty))$ satisfying

$$
(4.5) \qquad \begin{cases} -\operatorname{div}(|Du|^{p-2}Du) = 0 & \text{in } K_S(1,\infty), \\ u = g & \text{on } \partial K_S(1,\infty) = \bar{S} \cup B_S(1,\infty), \\ \lim_{|x|\to+\infty} u(x) = 0. \end{cases}
$$

Moreover $u > 0$ in $K(1,\infty)$ and $u \in C^{1,\alpha}(\overline{K(2+\epsilon,\infty)})$. From the strong maximum principle [**31**] the function

$$
(4.6) \qquad R \mapsto C(R) = \sup_{x\in K(R,\infty)} u(x)
$$

is decreasing and the supremum is achieved on $|x| = R$. Using Tolksdorf's equivalence principle [**31**, Lemma 2.1] we have

$$
(4.7) \qquad u(Rx) \leq (1 - \epsilon(R-1))u(x)
$$

for some $\epsilon > 0$ and any $R \in (1,2)$, and

$$
(4.8) \qquad C(R) \leq kC(2R)
$$

for some positive constant k and $R > 3$. We now define

$$
(4.9) \qquad u_R(x) = U(Rx)/C(R).
$$

From (4.7) and (4.8) we have as in Theorem 1.1

$$
(4.10) \qquad |Du(x)| \leq MC(|x|)|x|^{-1}
$$

$$
(4.11) \qquad |Du(x) - Du(x')| \leq MC(|x|)|x|^{-1-\alpha}|x - x'|,
$$

for $1 \leq |x| \leq |x'|$, which implies that for any compact subset K in $K_S(0,\infty) \cup B_S(0,\infty)$ there exists $C(k) > 0$ such that

$$
(4.12) \qquad |u_R|_{C^{1,\alpha}(K)} \leq C(K)
$$

and there exists a sequence $\{R_\nu\}$ converging to $+\infty$ such that $\{u_{R_\nu}\}$ converges to u^* in the C^1_{loc} topology of $K_S(0,\infty) \cup B_S(0,\infty)$. Moreover $u^* > 0$ in $K_S(0,\infty)$, $u^* = 0$ in $B_S(0,\infty)$, u^* is p-harmonic and ∇u^* never vanishes in $\overline{K_S(0,\infty)}$ from (4.7) (which implies that $u \in C^\infty$).

The main point is to prove that there exists $\beta > 0$ such that

$$u^*(r,\sigma) = r^{-\beta} u^*(1,\sigma). \tag{4.13}$$

We now take $R > 1$ and define

$$\Sigma_R = \sup\{C > 0 : Cu^*(x) \le u^*(Rx), \quad \forall x \in \overline{K_S(0,\infty)}\}. \tag{4.14}$$

which exists from Tolksdorf's equivalence principle. If we now assume that

$$\Sigma_R u^*(x) = u^*(Rx) \tag{4.15}$$

does not hold in $\overline{K_S(0,\infty)}$, then

$$\Sigma_R u^*(x) < u^*(Rx) \tag{4.16}$$

in $K_S(0,\infty) \cup B_S(0,\infty)$ from Hopf's comparison principle and the strong comparison principle, and the function θ defined by

$$\theta(\rho) = \min_{|x|=\rho} u^*(Rx)/u^*(x) \tag{4.17}$$

is strictly monotone as in [**31**]. If $\lim_{\rho\to+\infty} \theta(\rho) = \Sigma_R$ then for any ρ there exists $\sigma_\rho \in S$ such that

$$\theta(\rho) = u^*(R\rho\sigma_\rho)/u^*(\rho\sigma_\rho).$$

We can extract a sequence $\{R_{\nu_k}\}$ such that $\lim_{\nu_k\to+\infty} R_{\nu_k}/R_{\nu_{k+1}} = 0$ and we take $\rho_k = R_{\nu_{k+1}}/R_{\nu_k}$. Then

$$\lim_{h\to\infty} \theta(\rho_k) = \lim_{\nu_k\to+\infty} \frac{u(R_{\nu_{k+1}} R\sigma_{\rho_k})}{C(R_{\nu_{k+1}} R)}.$$

As we can assume that $\sigma_{\rho_k} \to \sigma_0 \in \bar{S}$, we get

$$\Sigma_R = u^*(R\sigma_0)/u^*(\sigma_0), \tag{4.18}$$

which contradicts (4.16). If $\lim_{\rho\to 0} \theta(\rho) = \Sigma_R$ the proof is the same.

The last point is to prove that

$$\Sigma_R = R^{-\beta} \tag{4.19}$$

for some $\beta > 0$. It is clear that the function $R \mapsto \Sigma_R$ is decreasing and is C^1 on $(1, +\infty)$ (the case $R \in (0,1)$ being the same). For any $n \in \mathbb{N}^*$ we first have

$$\Sigma_{R^n} u^*(x) = (\Sigma_R)^n u^*(x) = u^*(R^n x),$$

hence $\Sigma_{R^n} = (\Sigma_R)^n$ and, in the same way, $\Sigma_{R^{n/q}} = (\Sigma_R)^{n/q}$ for any $n/q \in Q^{+*}$ which implies that

$$\Sigma_{R^\alpha} = (\Sigma_R)^\alpha \tag{4.20}$$

for any $\alpha \in \mathbb{R}$, $\alpha > 0$. An elementary exercise shows that (4.19) should be true for some $\beta > 0$. We then set

$$\omega(\sigma) = u^*(1, \sigma), \tag{4.21}$$

and ω satisfies (4.2). Moreover $\omega \in C^\infty(S)$. □

Step 2: Uniqueness. If (β', ω') is another solution, then Tolksdorf's equivalence principle implies $\beta = \beta'$. An application of the strict comparison principle and Hopf's comparison principle as in [**31**] yields $\omega = \omega'$.

Write $O(N)$ for the group of isometries of $\mathbb{R}^N$ and let G be any finite subgroup of $O(N)$ generated by reflections through hyperplanes containing O. The group G contains a finite number of reflections through hyperplanes $(H_k)_{k\in K}$ containing O and those hyperplanes divide the space $\mathbb{R}^N$ into a finite number of angular polyhedra $(\mathcal{P}_i)_{i\in I}$, each of them being limited by at most N faces [**6**,4]. Moreover, those angular polyhedra are all equal and G acts transitively on them. The intersections of those angular polyhedra with S^{N-1} are spherical simplexes $(S_i)_{i\in I}$ on which G also acts transitively so we can consider only one model simplex G as a *fundamental domain* for G and we have

THEOREM 4.1. *There exists only one (β, ω) with $\beta > 0$ and $\omega \in C^1(S^{N-1})$, $\omega > 0$ on S with maximum value 1, $\omega = 0$ on ∂S and, if $S_i = g_i(S)$ for some $g_i \in G$, then*

$$\omega|_{S_i} = \det(g_i)\omega|_S \circ g_i^{-1}, \tag{4.22}$$

such that the function u defined by (4.1) is p-harmonic in $\mathbb{R}^N \backslash \{0\}$.

REMARK 4.1: Using Tolksdorf's results [**31**] it is possible to generate p-spherical harmonic function ω with fundamental domain S with $\beta < 0$ (u

is regular) such that (4.22) holds. In the case $p = 2$ it is not true that any spherical harmonic function admits a fundamental domain S with respect to a group of reflections like G. However we don't know whether this is particular to the case $p = 2$ (see [**39**] for details).

PROOF OF THEOREM 4.1: From Proposition 4.1 there exist $\beta > 0$ and $\tilde{\omega} \in C^1(S)$, $\tilde{\omega} > 0$ in S with maximum value 1, $\tilde{\omega} = 0$ on ∂S satisfying (4.2). (The fact that ∂S is just piecewise C^∞ does not affect the existence of $\tilde{\omega}$ as the equivalence principle is satisfied (with some modifications along the edges [**39**]). Moreover $\tilde{\omega} \in C^\infty(\bar{S}\backslash\{A\})$ where A is the set of $(N-3)$-dimensional vertices of S. If $S_i = g_i(S)$ for some $i \in I$ and $g_i \in G$ we then define $\tilde{\omega}_i$ in $C^1(S_i)$ by

$$\tilde{\omega}_i = \det(g_i)\tilde{\omega} \circ g_i^{-1}, \tag{4.23}$$

and $\tilde{\omega}_i$ satisfies (4.2) on S_i as the equation is invariant under $O(N)$. Moreover $\tilde{\omega}_i$ depends only on the class G_i defined by

$$G_i = \{g \in G : S_i = g(S)\}. \tag{4.24}$$

This is due to the fact that all the elements of G_i have the same determinant and to the uniqueness of the positive solution of (4.2) in S_i, vanishing on ∂S_i, with maximum value 1. We can now define the function ω by

$$\omega\,|_{S_i} = \omega_i. \tag{4.25}$$

If S_{i_o} is any of the simplexes $\{S_i\}$ with $a(N-2)$-dimensional edge s_{i_o} in common with S then G_{i_0} contains a reflection g_{i_0} through a hyperplane H_{i_0} containing s_i and $\omega_{i_0} = -\tilde{\omega} \circ g_{i_0}^{-1} = \tilde{\omega} \circ g_{i_0}$. As

$$\tilde{\omega}\,|_{s_{i_0}} = 0 = \omega_{i_0}\,|_{s_{i_0}}$$

the restriction of ω to $S \cup S_{i_0}$ is a solution of (4.2) in $S \cup S_{i_0}$. We can iterate this process and prove that ω satisfies (4.2) in $\mathcal{D}'(S^{N-1}\backslash\{A\})$. As $\operatorname{codim}(A)/S^{N-1} = 2$ and ω is bounded we deduce from Serrin's results [**26**] that there exists no singularity on A and so ω is a regular solution of (4.2) in S^{N-1}. As we shall see from Corollary 4.1, the function ω is not C^∞ on S^{N-1}.

REMARK 4.2: The complete description of the finite subgroups of $O(N)$ generated by reflections through hyperplanes passing through 0 can be found in [**4**]. In the case $N = 3$ there exist five types of subgroups: Type I is generated by the reflections through two planes with angle $\pi/n (n \in \mathbb{N}^*)$; Type II is generated by two reflections through two planes with angle π/n and a reflection through a plane orthogonal to the first ones; Types III, IV and V have, respectively, 24, 48 and 120 elements and are associated to Plato polyhedra [**8**].

In order to go a little further in the case $N = 3$ we introduce the spherical coordinates (r, φ, θ) with $r > 0$, $\varphi \in [0, 2\pi]$, $\theta \in [0, \pi]$ and

$$\begin{cases} x = r\cos\varphi\sin\theta, \\ y = r\sin\varphi\sin\theta, \\ z = r\cos\theta \end{cases} \tag{4.26}$$

If $\omega(\varphi, \theta)$ satisfies (4.2) we then have

$$\int_0^{2\pi}\int_0^{\pi}\left(\beta^2\omega^2 + \omega_\theta^2 + \frac{\omega_\varphi^2}{\sin^2\theta}\right)^{(p-2)/2}\left\{\omega_\theta\varsigma_\theta + \frac{1}{\sin^2\theta}\omega_\varphi\varsigma_\varphi - \beta(\beta(p-1) + p - 3)\omega\varsigma\right\}\sin\theta\, d\theta\, d\varphi = 0 \tag{4.27}$$

for any $\varsigma \in C^1(S^2)$. As θ is the geodesic distance on S^2 between the north pole $(0,0,1)$ and the point (x, y, z) it is natural to try to find solutions of (4.27) under the form

$$\omega(\varphi, \theta) = (\sin\theta)^\delta v(\varphi). \tag{4.28}$$

A straightforward computation shows that $\delta = \beta > 0$ (as we just want to deal with regular ω) and that v is a 2π-periodic solution of

$$\left((\delta^2 v + v_\varphi^2)^{(p-2)/2} v_\varphi\right)_\varphi + \delta\left((\delta - 1)(p-1) - 1\right)\left(\delta^2 v^2 + v_\varphi^2\right)^{(p-2)/2} v = 0. \tag{4.29}$$

It must be noticed that if we take $N = 2$ and $\beta = -\delta > 0$ in equation (4.2) we obtain the same equation as above. This equation has been completely integrated by Kroll and Mazja [**19**].

THEOREM 4.2. *Assume $p > 1$. Then for any integer $k > 1$ there exist $\delta_k > 0$ and $v_k : \mathbb{R} \mapsto \mathbb{R}$ with at least period $2\pi/k$ of class C^∞ satisfying (4.29); δ_k is the root ≥ 1 of*

$$\left(1 - \tfrac{1}{k}\right)^2\left(\delta^2 + \tfrac{2-p}{p-1}\delta\right) = (\delta - 1)^2. \tag{4.30}$$

(δ_k, v_k) *is unique up to translation and homothety over* v_k.

Thanks to Theorem 4.2 we obtain all the regular p-harmonic functions in $\mathbb{R}^2$ under the form

$$u(r,\varphi) = r^{\delta}v(\varphi) \tag{4.31}$$

and in $\mathbb{R}^3$ under the form

$$u(r,\varphi,\theta) = (r\sin\theta)^{\delta}v(\varphi). \tag{4.32}$$

In the particular case $p = N$ the N-harmonic equation is invariant under the conformal transformation $x \mapsto x/|x|^2$ so Theorem 4.1 provides us with regular and singular solutions under the form

$$u(x) = |x|^{\pm\beta}\omega(x/|x|). \tag{4.33}$$

We then deduce from Theorem 4.2 and (4.33)

COROLLARY 4.1. *Assume* $N = p = 3$. *Then for any integer* $k > 1$ *there exists a regular p-harmonic function* u_k *and a singular p-harmonic function* $\tilde{u}_k$ *under the following forms:*

$$u_k(r,\varphi,\theta) = r^{\delta_k}(\sin\theta)^{\delta_k}v_k(\varphi) \tag{4.34}$$

$$\tilde{u}_k(r,\varphi,\theta) = r^{-\delta_k}(\sin\theta)^{\delta_k}v_k(\varphi) \tag{4.35}$$

where δ_k *is the root* ≥ 1 *of (4.30) and* $v_k : \mathbb{R} \mapsto \mathbb{R}$ *is a solution with at least period* $2\pi/k$ *of (4.29);* u_k *and* $\tilde{u}_k$ *are unique up to homothety over* v_k *and* $SO(2)$ *action over* (φ,θ).

If we look for p-harmonic singular functions in the plane under the form (4.1) then $\omega(x/|x|) = \omega(\varphi)$ satisfies

$$\left((\beta^2\omega^2 + \omega_\varphi^2)^{(p-2)/2}\omega_\varphi\right)_\varphi + \left((\beta+1)(p-1)-1\right)\beta(\beta^2\omega^2 + \omega_\varphi^2)^{(p-2)/2}\omega = 0. \tag{4.36}$$

on S^1. This equation has been completely integrated by Kichenassamy and Veron [**17**] and they proved the following:

THEOREM 4.3. *Assume $N = 2$ and $p > 1$. Then for each positive integer k there exist a $\beta_k > 0$ and $\omega_k : \mathbb{R} \mapsto \mathbb{R}$ with at least period $2\pi/k$ of class C^∞ such that*

$$u_k(x) = |x|^{-\beta k}\omega_k(x/|x|) \tag{4.37}$$

is p-harmonic in $\mathbb{R}^2\backslash\{0\}$; β_k is the positive root of

$$(\beta+1)^2 = \left(1+\tfrac{1}{k}\right)^2\left(\beta^2 - \beta\tfrac{2-p}{p-1}\right). \tag{4.38}$$

(β_k, ω_k) is unique up to translation and homothety over ω_k.

If we look now for solutions of (0.7) under the form (4.1) then it is clear that $\beta = p/(q+1-p)$ for homogeneity reasons (we always assume $q > p-1$) and ω must satisfy

$$\begin{aligned} -\operatorname{div}_{S^{N-1}}\left(\left(\beta^2\omega^2 + |\nabla_\tau\omega|^2\right)^{(p-2)/2}\nabla_\tau\omega\right) + \omega|\omega|^{q-1} \\ = \left(\frac{p}{q+1-p}\right)\left(\frac{pq}{q+1-p} - N\right)\left(\beta^2\omega^2 + |\nabla_\tau\omega|^2\right)^{(p-2)/2}\omega, \end{aligned} \tag{4.39}$$

on S^{N-1}. Using the shooting method of Proposition 4.1 we have the following: if S is a connected open subset of S^{N-1} with a regular boundary ∂S let (β_s, ω_s) be the couple obtained in Proposition 4.1.

PROPOSITION 4.2. *Assume $0 < p-1 < q$, $p/(q+1-p) > \beta_s$. Then there exists at least one C^1 function ω positive in S, vanishing on ∂S satisfying (4.39).*

SKETCH OF THE PROOF: For $\epsilon > 0$ small enough we construct the unique solution u of

$$\begin{cases} -\operatorname{div}(|Du|^{p-2}Du) + u|u|^{q-1} = 0 & \text{in } K_S(1,\infty), \\ u = \epsilon g & \text{on } \bar{S} \cup B_S(1,\infty), \\ \limsup\limits_{|x|\to+\infty} |x|^{p/(q+1-p)}u(x) < +\infty, & \end{cases} \tag{4.40}$$

where g is defined by (4.3). Then $0 < u < \gamma_{N,p,q}|x|^{-p/(q+1-p)}$ in $K_S(1,\infty)$. An important point is to prove that u does not decrease too quickly and this is obtained in showing that for $\rho > 0$ small enough the function

$$x \mapsto v(x) = \rho|x|^{-p/(q+1-p)}\omega_s(x/|x|) \tag{4.41}$$

satisfies

$$\left\{\begin{aligned} -\operatorname{div}(|Du|^{p-2}Du)+v^q &\leq 0 \quad \text{in } K_S(1,\infty), \\ v &\leq u \quad \text{on } \bar{S}\cup B_S(1,\infty). \end{aligned}\right. \tag{4.42}$$

Hence $v \leq u$ and

$$\rho\omega_w(x/|x|) \leq |x|^{p/(q+1-p)}u(x) \leq \gamma_{N,p,q}. \tag{4.43}$$

We then set $u_R(x) = R^{p/(q+1-p)}u(Rx)$ and there exists a function u^* satisfying (0.7) in $K_S(0,\infty)$ and a sequence $R_\nu \to +\infty$ such that $\{u_{R_\nu}\}$ converges to u^* in the C^1_{loc} topology of $K_S(0,\infty) \cup B_S(0,\infty)$. Moreover

$$u^*(Rx) = \Gamma_R u^*(x) \tag{4.44}$$

which implies $\Gamma_R = R^{-p/(q+1-p)}$ and

$$u^*(x) = u^*(r,\sigma) = r^{-p/(q+1-p)}\omega(\sigma). \tag{4.45}$$

As (4.39) is invariant under the action of $O(N)$ we have the following:

THEOREM 4.4. *Assume G is any finite subgroup of $O(N)$ generated by reflections through hyperplanes containing O and let S be its fundamental simplicial domain on S^{N-1}; assume also $0 < p-1 < q$ and $\beta_s < p/(q+1-p)$. Then there exists a $\omega \in C^1(S^{N-1})$ satisfying (4.35) such that $\omega > 0$ on S, $\omega = 0$ on ∂S and if $S_i = g_i(S)$ for some $g_i \in G$ then*

$$\omega\,|_{S_i} = \det(g_i)\omega\,|_S \circ g_i^{-1} \tag{4.46}$$

In the particular case where $N = 2$ we get

COROLLARY 4.2. *Assume $N = 2$, $p > 1$, k is a positive integer and β_k is the positive root of (4.34). If $\beta_k < p/(q+1-p) \leq \beta_{k+1}$ there exist at least k connected components S_j of nontrivial solutions of (4.39) of class C^1 with period $2\pi/j$ ($j = 1\ldots k$) generated in the following way: $S_j = \{\omega_j(\cdot+\alpha), \alpha \in \mathbb{R}\}$.*

REMARK 4.3: There are still many unsolved problems associated with anisotropic singularities of solutions of (0.1) or (0.7). Two of the most exciting are probably the following:

I. Assume u is p-harmonic in $B_1(0)\backslash\{0\}$ such that

$$|x|^\alpha |u(x)| \leq C \tag{4.47}$$

near 0 for some $\alpha > 0$ and C. Does there exist a $\beta > 0$ and a solution ω of (4.2) on S^{N-1} such that

$$\lim_{|x|\to 0} |x|^{\beta} u(x) = \omega(x/|x|)? \tag{4.48}$$

II. Assume u satisfies (0.7) in $B_1(0)\backslash\{0\}$. Does there exist a ω satisfying (4.39) such that

$$\lim_{|x|\to 0} |x|^{p/(q+1-p)} u(x) = \omega(x/|x|)? \tag{4.49}$$

When $p = 2$, (4.48) holds; as for (4.49), we just have partial answers except when $N = 2$.

Department de Mathématiques, Faculté de Sciences
Parc de Grandmont, F 37200 Tours, FRANCE

This work was partially supported by NSF Grant DMS 8600710.

REFERENCES

1. P. Aviles, *Local behavior of solutions of some elliptic equations*, Comm. Math. Phys. **108** (1987), 177–192.
2. Ph. Benilan and H. Brezis, *Nonlinear problems related to the Thomas-Fermi equation*, unpublished work; see also [**5a**].
3. B. Bojarski and T. Iwaniec, *p-harmonic equations and quasilinear mappings*, "Sonderforschungsbereich 72," Univ. Bonn, 1983.
4. N. Bourbaki, "Groups et algèbres de Lie," Chap. 4-6, Hermann, 1968.
5. H. Brezis and E.T. Lieb, *Long-range atomic potentials in Thomas-Fermi theory*, Comm. Math. Phys. **65** (1980), 231–246.

5a. H. Brezis, *Some variational problems of the Thomas-Fermi type*, in "Variational Inequalities and Complementary Conditions," R.W. Cottle, F. Gianessi and J.L. Lions, eds., Wiley Interscience, 1980.

6. E. Cartan, *Complément au mémoire "Sur la géométrie des groupes simples"*, Ann. di Mat. **5** (1928), 253–260.
7. P.T. Chruschiel, *Conformally minimal foliations of three-dimensional Riemannian manifolds and the energy of the gravitational field*, Ph.D. thesis, Polish Acad. Sci., Inst. Th. Phys., (1981).
8. H.S.M. Coxeter, "Regular polytopes," Dover Publ., Inc., 1973.
9. M. Dobrowolsi, *Nichtlineare eckenprobleme und finite elemente methode*, Z.A.M.M. **64** (1984), 270–271.
10. L.C. Evans, *A new proof of local $C^{1,\alpha}$ regularity for solutions of certain degenerate elliptic partial differential equations*, J. Diff. Equ. **50** (1980), 315–338.
11. R.H. Fowler, *Further studies on Emden's and similar differential equations*, Quant. J. Math. **2** (1931), 259–288.
12. A. Friedman and L. Veron, *Solutions singulières d'équations quasilinéaires elliptiques*, Cr. Acad. Sci. Paris t. **302 I** (1986), 145–148.
13. A. Friedman and L. Veron, *Singular solutions of some quasilinear elliptic equations*, Arch. Rat. Mech. Anal. **96** (1986), 259–287.
14. B. Gidas and J. Spruck, *Global and local behavior of positive solutions of nonlinear elliptic equations*, Comm. Pure Appl. Math. **34** (1980), 525–598.
15. M. Guedda, *Propriétés locales et globales des solutions d'équations quasilinéaires elliptiques*, Ph.D. thesis, Univ. Tours, (1987).
16. M. Guedda and L. Veron, *Local and global properties of solutions of quasilinear elliptic equations*, to appear.
17. S. Kichenassamy and L. Veron, *Singular solutions of the p-Laplace equation*, Math. Ann. **275** (1986), 599–615.
18. I.N. Kroll, *On the behavior of the solutions of a quasilinear equation near the null salient point of the boundary*, Proc. Steklov Inst. Math. **125** (1973), 130–136.
19. I.N. Kroll and V.G. Mazja, *The lack of continuity and Hölder continuity of the solutions of a certain quasilinear equation*, Sem. Math. V.A. Steklov Math. Inst. **14** (1969), 44–45.
20. J.L. Lewis, *Regularity of the derivatives of solutions to certain degenerate elliptic equations*, Ind. Univ. Math. J. **32** (1983), 849–858.
21. P.L. Lions, *Isolated singularities in semilinear problems*, J. Diff. Equ. **38** (1980), 441–450.
22. K. McLeod and J. Serrin, to appear.
23. W.M. Ni and J. Serrin, *Existence and nonexistence theorems for ground states of quasilinear partial differential equations: the anomalous case*, Acad. Naz. dei Lincei **77** (1986), 231–257.
24. W.M. Ni and J. Serrin, *Nonexistence theorems for singular solutions of quasilinear partial differential equations*, Comm. Pure Appl. Math. **38** (1986), 379–399.

25. V.G. Reshetniak, *Mappings with bounded deformations as extremals of Dirichlet-type integrals*, Sibirsk. Math. Zh. **9** (1966), 652–666.
26. J. Serrin, *Local behavior of solutions of quasilinear equations*, Acta Math. **111** (1964), 247–302.
27. J. Serrin, *Isolated singularities of solutions of quasilinear equations*, Acta Math. **113** (1965), 219–240.
28. J. Serrin and L. Veron, unpublished work (see [**16**, Appendix]).
29. A. Sommerfeld, *Asymptotische integrazion der differenzialgleichnug der Thomas-Fermischen atoms*, Z. für Phys. **78** (1932), 283–308.
30. P. Tolksdorf, *Regularity for a more general class of quasilinear elliptic equations*, J. Diff. Equ. **51** (1984), 126–150.
31. P. Tolksdorf, *On the Dirichlet problem for quasilinear equations in domain with conical boundary points*, Comm. Part. Diff. Equ. **8** (1983), 773–817.
32. N.S. Trudinger, *On Harnack-type inequalities and their applications to quasilinear elliptic equations*, Comm. Pure Appl Math. **20** (1967), 721–747.
33. K. Uhlenbeck, *Regularity for a class of nonlinear elliptic systems*, Acta Math. **138** (1977), 219–240.
34. N.N. Ural'ceva, *Degenerate elliptic systems*, Sem. Math. V.A. Stecklov Math. Inst. **7** (1968), 83–99.
35. J.L. Vazquez, *An a priori interior estimate for the solutions of a nonlinear problem representing weak diffusion*, Nonlinear Anal. **5** (1981), 95–103.
36. J.L. Vazquez and L. Veron, *Removable singularities of some strongly nonlinear elliptic equations*, Manuscripta Math. **33** (1980), 129–144.
37. J.L. Vazquez and L. Veron, *Isolated singularities of some semilinear elliptic equations*, J. Diff. Equ. **60** (1985), 301–321.
38. L. Veron, *Singular solutions of some nonlinear elliptic equations*, Nonlinear Anal. **5** (1981), 225–242.
39. L. Veron, *Anisotropic singularities of some degnerate elliptic equations*, in preparation.
40. G.N. Watson and E.T. Whiteaker, "A course of modern analysis," Cambridge Univ. Press, 1927.

MIX
Papier aus verantwortungsvollen Quellen
Paper from responsible sources
FSC® C105338

If you have any concerns about our products,
you can contact us on
ProductSafety@springernature.com

In case Publisher is established outside the EU,
the EU authorized representative is:
Springer Nature Customer Service Center GmbH
Europaplatz 3, 69115 Heidelberg, Germany

Printed by Libri Plureos GmbH
in Hamburg, Germany